T0387896

AC to AC Converters
Modelling, Simulation, and Real-Time Implementation Using SIMULINK®

AC to AC Converters
Modelling, Simulation, and Real-Time Implementation Using SIMULINK®

Narayanaswamy P. R. Iyer
Electronics Consultant
Sydney, NSW, Australia

CRC Press
Taylor & Francis Group
Boca Raton London New York

CRC Press is an imprint of the
Taylor & Francis Group, an **informa** business

CRC Press
Taylor & Francis Group
6000 Broken Sound Parkway NW, Suite 300
Boca Raton, FL 33487-2742

© 2019 by Taylor & Francis Group, LLC
CRC Press is an imprint of Taylor & Francis Group, an Informa business

No claim to original U.S. Government works

Printed on acid-free paper

International Standard Book Number-13: 978-0-367-19750-6 (Hardback)

Printed and bound in Great Britain by
TJ International Ltd, Padstow, Cornwall

To my family members: Mythili Iyer, Ramnath,

Dr. Sudha, Rekha Karthik, Nikita and Nidhi.

Contents

Preface

Power electronic converters can be broadly classified as AC to DC, DC to AC, DC to DC and AC to AC converters. AC to AC converters can be further classified as AC controllers or AC regulators, cycloconverters and matrix converters. AC controllers and cycloconverters are fabricated using silicon controlled rectifiers (SCR), whereas matrix converters are built using semiconductor bidirectional switches. The modelling of the first three of the power electronic converters mentioned above and that of AC controllers were dealt with in my text book titled *Power Electronic Converters: Interactive Modelling Using Simulink*, CRC Press, USA, 2018.

This text book provides a summary of AC to AC converter modelling excluding AC controllers. The software Simulink® by MathWorks, Inc., USA is used to develop the models of AC to AC converters presented in this text book. The term *model* in this text book refers to SIMULINK model. This text book is mostly suitable for researchers and practising professional engineers in the industry working in the area of AC to AC converters.

Chapter 1 provides various circuit configurations for semiconductor bidirectional switches, three-phase AC to three-phase AC (3 × 3) matrix converter using common emitter bidirectional switches. The table showing various switching combinations and its significance, objective and novelty, research methodology and book outline are also presented in this chapter.

Models with simulation results are presented in the following order.

In Chapter 2, models for 3 × 3 matrix converter are presented using Venturini, optimum Venturini and advanced modulation algorithms. A case study for the speed control and brake by plugging of three-phase induction motor fed by matrix converter is presented along with a block diagram schematic for real-time implementation.

Model for a 3 × 3 multilevel matrix converter with three flying capacitors per output phase is presented in Chapter 3.

Models for direct and indirect space vector modulation of a 3 × 3 matrix converter are presented in Chapters 4 and 5, respectively. Direct space vector modulation includes models for direct asymmetrical and symmetrical space vector modulation.

Models for single and dual programmable AC to DC rectifier using three-phase matrix converter topology are presented in Chapter 6. In this chapter, additional models for speed control and brake by plugging of separately excited DC (SEDC) motor, and single and dual pure sine-wave AC power supply using single and dual programmable AC to DC rectifier topology are presented. Model simulation results indicate that this single and dual rectifier topology can be used for hybrid electric vehicle (HEV) and electric traction applications. A block diagram for the real-time implementation of dual programmable rectifier topology driving an SEDC motor is presented. The applications presented in this chapter are totally new findings.

Model for delta-sigma-modulated 3×3 matrix converter is presented in Chapter 7. In this model, an R–L load is used. A case study of the delta-sigma-modulated 3×3 matrix converter driving a three-phase induction motor load is also presented.

In Chapter 8, model for the direct single-phase AC to three-phase AC matrix converter driving an induction motor load is presented. Here an L–C filter is connected to a single-phase AC power supply. A compensation capacitor is connected in series with the capacitor of the L–C filter. The junction formed by the inductor L and capacitor C of filter, filter capacitor and compensation capacitor and the tail end of the compensation capacitor forms the three-phase AC voltage nodes that are given to nine bidirectional switches forming a 3×3 matrix converter.

In Chapter 9, models for a new method of generating variable-frequency single-phase and three-phase PWM AC voltage from a constant-frequency single-phase and three-phase AC power supply are presented. In this method, the DC link voltage is converted to two square wave AC voltages having $180°$ phase difference with frequency equal to that of the AC mains supply frequency which are then converted to variable-frequency variable magnitude pulse width-modulated square wave AC voltage using semiconductor bidirectional switches. This method is my original contribution.

In Chapter 10, real-time hardware-in-the-loop (HIL) simulation of a three-phase AC to single-phase AC matrix converter using dSPACE DS1104 hardware controller board is presented. Real-time experimental results are compared with model simulation results.

Models using simple boost and maximum boost control strategy for three-phase voltage-fed Z-source direct matrix converter (ZSDMC) and that for three-phase Quasi Z-source indirect matrix converter (QZSIMC) are presented in Chapter 11.

A novel single-phase/three-phase sine-wave fixed-frequency AC to PWM single-phase/three-phase sine-wave variable-frequency AC converter without an intermediate DC link voltage source which can also be used as an AC to DC converter with polarity reversal capability in the DC mode is my original contribution which I have also submitted as a patent. An original derivation for the RMS and average value of a uniform pulse width-modulated sine-wave AC voltage source is also presented in this chapter. Models for this single-phase/three-phase sine-wave AC to AC and AC to DC converter are presented with simulation results in Chapter 12.

In Chapter 13, models for single-phase and three-phase SCR cycloconverters, three-phase indirect matrix converters using three-phase conventional inverter and three-phase diode-clamped three-level inverter configuration and models for solid-state transformers (SSTs) using dual active bridge topology and direct AC to AC converter topology are presented.

In Appendix A, essential derivations for matrix converter relating to Venturini algorithm, indirect space vector modulation algorithm and three-phase voltage-fed ZSDMC algorithm are presented.

Most of the models presented in this text book from Chapters 1 to 10 excluding that in Sections 6.5 to 6.8 are from my doctoral research thesis work. Models in Sections 6.5, 6.6, 11.2 and Chapter 12 were developed by me while working as a research fellow in the Faculty of Engineering, The University of Nottingham, UK. Models in Sections 6.7, 6.8, 11.3, Chapter 13 and the derivations relating to three-phase voltage-fed ZSDMC in Appendix A are developed by me here as an electronics consultant.

I have referred many literature references to develop the models presented in this text book. These literature references are gratefully acknowledged at the end of each chapter.

Model files in this text book are available on the website https://www. crcpress.com/AC-to-AC-Converters-Modelling-Simulation-and-Real-Time-Implementation-Using-Simulink/Iyer/p/book/9780367197506. Instructors/ supervisors and professional engineers working in the power electronics area can download a copy of these model files.

I am grateful to Dr. Vassilios G. Agelidis, formerly Energy Australia Chair in Electrical Power Engineering, School of Electrical and Information Engineering (EIE), the University of Sydney, Redfern, NSW for introducing me to the research area of matrix converter and for granting Australian Postgraduate Award (APA) and Norman Price Fellowship to carry out this research work. As he left the University of Sydney, NSW, I transferred my research work to the Department of Electrical and Computer Engineering (ECE), Curtin University, Perth, WA.

I am grateful to Prof. Chem V. Nayar, formerly Professor, presently Emeritus Professor, Department of ECE, Curtin University, Perth, WA and also Managing Director, Regenpower Pty. Ltd., Perth, WA for extending help, supervision, support and for providing APA and departmental fellowship to continue on the same research area of matrix converter. Many thanks to Dr. Sumedha Rajakaruna, senior lecturer, Department of ECE, Curtin University, Perth, WA for his co-supervision, help and support. I wish to extend my thanks to Dr. Kevin Fynn, Professor and Head, School of Electrical Engineering and Computing, Curtin University, Perth, WA for chairing my research seminar session.

I am also grateful to Prof. Pat Wheeler and Prof. Alberto Castellazzi, both of the School of Electrical and Electronics Engineering, The University of Nottingham, UK for providing facilities and support while I was a research fellow in their Faculty of Engineering.

I would like to thank Ms. Nora Konopka, Editorial Director, Engineering, Taylor & Francis for help and support.

My wife Mythili Iyer, in spite of her busy office and domestic schedule, gave me moral and administrative support while writing this text book. I would like to extend a special, heartfelt thanks to her for all her support and cooperation without which I could not have completed writing this text book.

Narayanaswamy P. R. Iyer
B.Sc. (Engg) (Elec); M.Sc. (Engg);
M.E. (UTS); Ph.D. (Curtin); C.P. Engg. (Australia)
Sydney, NSW
July 2018

MATLAB® is a registered trademark of The MathWorks, Inc. For product information, please contact:

The MathWorks, Inc.
3 Apple Hill Drive
Natick, MA 01760-2098, USA
Tel: 508-647-7000
Fax: 508-647-7001
E-mail: info@mathworks.com
Web: www.mathworks.com

Author

Narayanaswamy P. R. Iyer received his M.E. degree by Research and Ph.D. degree both in the area of power electronics and drives from the University of Technology Sydney, NSW and Curtin University of Technology, Perth, WA, Australia, respectively. He received his B.Sc. degree in electrical engineering and M.Sc. degree in power systems engineering from University of Kerala and University of Madras, respectively. He worked as a part-time faculty in the Department of Electrical Engineering, UNSW, Kensington, NSW and UTS, NSW. Earlier he had worked as a full-time faculty in the Department of Electrical Engineering, GEC, Trichur, Kerala, VEC, Vellore and RREC, Chennai. He was a research fellow in the Faculty of Engineering, PEMC group, the University of Nottingham, England during July 2012 to March 2014. Presently, he is an electronics consultant managing his own consultancy organization. He has more than two decades of experience in the modelling of electrical and electronic circuits, power electronic converters and electric drives using various software packages such as MATLAB/SIMULINK, PSIM, PSCAD, PSPICE, and MICROCAP and has published several papers in this area in leading conferences and journals. He has to his credit several discoveries such as "Three-Phase Clipped Sinusoid PWM Inverter", "Single Programmable and Dual Programmable Rectifier Using Three-Phase Matrix Converter Topology", "A Novel AC to AC Converter Using a DC Link" and a submitted patent on "A Single-Phase AC to PWM Single-Phase AC and DC Converter" with alternative title "Swamy Converter". He is a chartered professional engineer of Australia.

1

Introduction

1.1 Background

The most desirable features of the power electronic frequency converters are (a) simple and compact power circuit, (b) variable output voltage magnitude and frequency, (c) sinusoidal input and output currents, (d) unity power factor operation for any type of load and (e) regeneration capability. Matrix converter (MC) meets all these requirements and is the most sought after converter for industrial drives application. MCs are essentially forced commutated cycloconverters with inherent four-quadrant operation consisting of a matrix of bidirectional switches such that there is a switch for each possible connection between the input and output lines. MC directly converts the AC input voltage at any given frequency to AC output voltage with arbitrary amplitude at any unrestricted frequency without the need for a DC link capacitor storage element at the input side.

The early work, prior to the development of MC using bidirectional semiconductor switches, was the cycloconverters using thyristors with forced commutation circuits which act as bidirectional switches [1]. But the power circuit of these thyristor cycloconverters were too bulky and the performance was poor. The introduction of bidirectional switch using power transistors and Insulated Gate Bipolar Transistors (IGBT) made easy realization of the MC. The real development of the MC starts with the work of Venturini and Alesina who proposed a mathematical analysis and introduced the low-frequency modulation matrix concept to describe the low-frequency behaviour of the MC [2–4]. In this way, the output voltages can be obtained by multiplication of the modulation matrix or transfer matrix with the input voltages. One limitation of the MC is that the maximum output voltage available is limited to 86.6% of the input voltage in the linear modulation range.

As mentioned above, it is the development of the bidirectional switches using semiconductor components that makes MC really attractive. Although a number of topologies are available for bidirectional switches, the common-emitter IGBT topology is used here in this textbook. Some of the IGBT topologies used as bidirectional switches are shown in Figure 1.1.

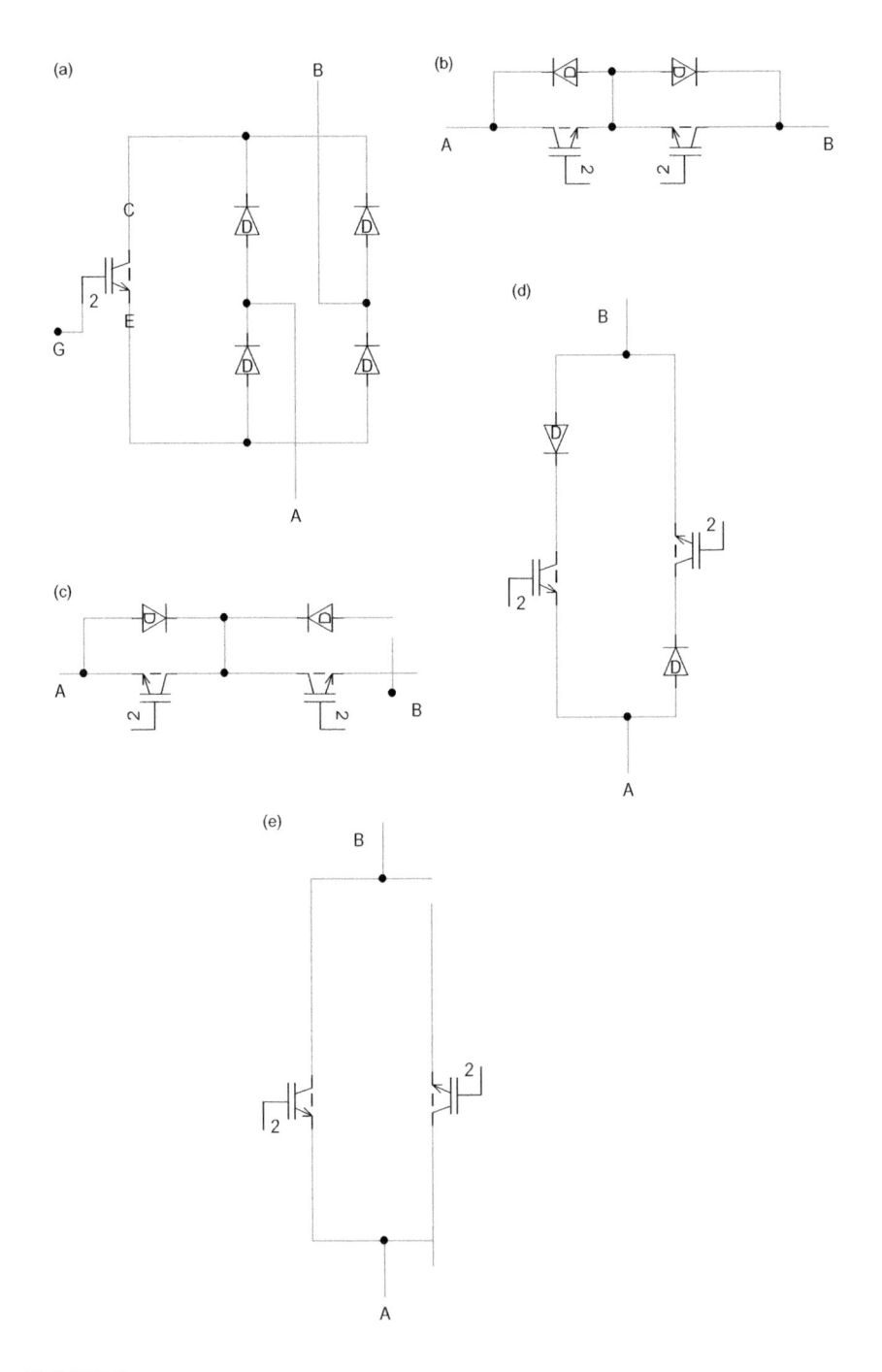

FIGURE 1.1
Bidirectional switch topology: (a) diode embedded, (b) common emitter, (c) common collector, (d) two antiparallel IGBTs in series with diodes and (e) reverse blocking IGBTs or RB-IGBTs.

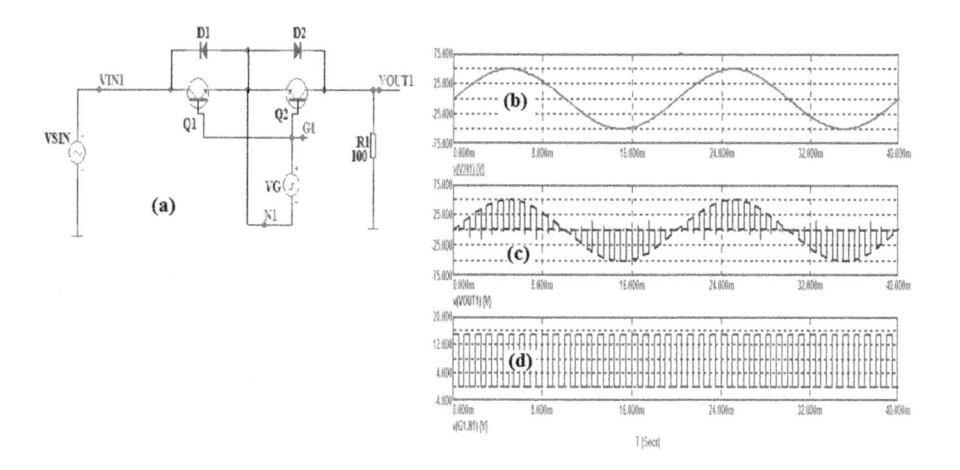

FIGURE 1.2
Correct method of applying gate pulse to a bidirectional switch.

These topologies are (a) diode embedded, (b) common emitter, (c) common collector, (d) antiparallel in series with diodes and (e) reverse blocking or RB-IGBT.

The correct method of applying gate pulse to a common-emitter IGBT bidirectional switch is shown in Figure 1.2a. The sinusoidal voltage input, the pulse-width-modulated output voltage across the load and the gate switching pulses are shown in Figure 1.2b, c and d, respectively. A three-phase AC to three-phase AC MC consists of nine bidirectional switches (18 IGBTs each with a diode in parallel as in Figure 1.1b). The topology of the three-phase AC to three-phase AC MC is shown in Figure 1.3. The three-phase input voltage terminals are marked A, B and C, and the corresponding three-phase AC output voltage terminals are marked a, b and c, respectively. The bidirectional switches are represented as S_{Kj}, where $K \in A, B, C$ and $j \in a, b, c$, respectively. Thus, a three-phase AC to three-phase AC MC can be represented as 3×3 MC. A 3×3 MC has 27 switching combinations as shown in Table 1.1. In this table, corresponding to Group II-c, C C A means that the input phase C is connected to output phase a and b and input phase A is connected to output phase c. Referring to Figure 1.3, output phase to phase voltages v_{ab}, v_{bc} and v_{ca} are, respectively, zero (V_{CC}), V_{CA} and V_{AC} or $-V_{CA}$. Also input currents i_A, i_B and i_C are the load currents $i_c, 0$ and $-i_c,$ respectively. This is illustrated diagrammatically in Figure 1.4 for Group II-a of Table 1.1.

The position of the nine bidirectional switches corresponding to Group II-a in Table 1.1 is shown in Figure 1.4. The switches S_{Aa}, S_{Cb} and S_{Cc} are closed, and all others are open. Applying Kirchhoff's law to Figure 1.4, the following set of equations will result:

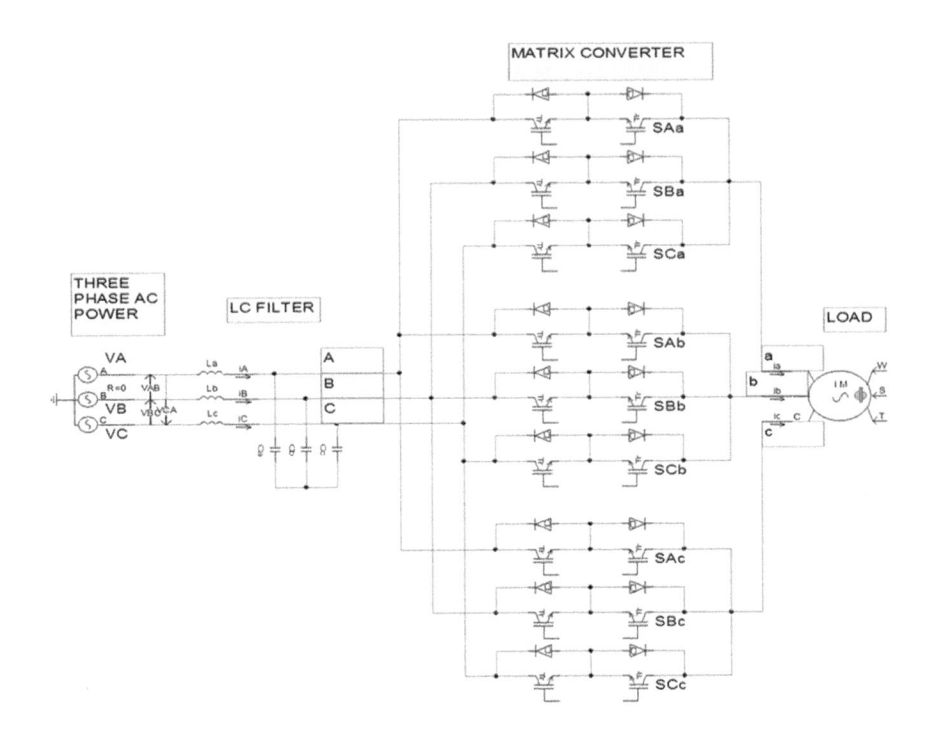

FIGURE 1.3
Three-phase AC to three-phase AC matrix converter.

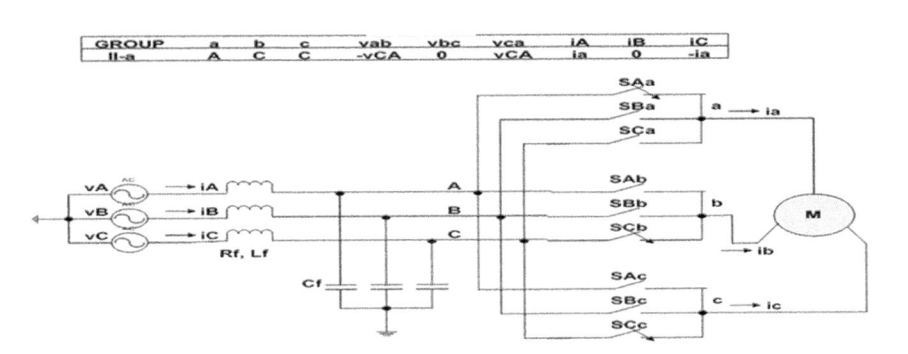

FIGURE 1.4
Three-phase matrix converter driving a three-phase AC motor load.

$$v_{ab} = v_{AC} = -v_{CA}$$
$$v_{bc} = 0 \qquad\qquad (1.1)$$
$$v_{ca} = v_{CA}$$

TABLE 1.1

Three-Phase AC to Three-Phase AC MC Switching Combinations

Group	a	b	c	v_{ab}	v_{bc}	v_{ca}	i_A	i_B	i_C	S_{Aa}	S_{Ba}	S_{Ca}	S_{Ab}	S_{Bb}	S_{Cb}	S_{Ac}	S_{Bc}	S_{Cc}
I	A	B	C	v_{AB}	v_{BC}	v_{CA}	i_a	i_b	i_c	1	0	0	0	1	0	0	0	1
I	A	C	B	$-v_{CA}$	$-v_{BC}$	$-v_{AB}$	i_a	i_c	i_b	1	0	0	0	0	1	0	1	0
I	B	A	C	$-v_{AB}$	$-v_{CA}$	$-v_{BC}$	i_b	i_a	i_c	0	1	0	1	0	0	0	0	1
I	B	C	A	v_{BC}	v_{CA}	v_{AB}	i_b	i_c	i_a	0	1	0	0	0	1	1	0	0
I	C	A	B	v_{CA}	v_{AB}	v_{BC}	i_c	i_a	i_b	0	0	1	1	0	0	0	1	0
I	C	B	A	$-v_{BC}$	$-v_{AB}$	$-v_{CA}$	i_c	i_b	i_a	0	0	1	0	1	0	1	0	0
II-a	A	C	C	$-v_{CA}$	0	v_{CA}	i_a	0	$-i_a$	1	0	0	0	0	1	0	0	1
II-a	B	C	C	v_{BC}	0	$-v_{BC}$	0	i_a	$-i_a$	0	1	0	0	0	1	0	0	1
II-a	B	A	A	$-v_{AB}$	0	v_{AB}	$-i_a$	i_a	0	0	1	0	1	0	0	1	0	0
II-a	C	A	A	v_{CA}	0	$-v_{CA}$	$-i_a$	0	i_a	0	0	1	1	0	0	1	0	0
II-a	C	B	B	$-v_{BC}$	0	v_{BC}	0	$-i_a$	i_a	0	0	1	0	1	0	0	1	0
II-a	A	B	B	v_{AB}	0	$-v_{AB}$	i_a	$-i_a$	0	1	0	0	0	1	0	0	1	0
II-b	C	A	C	v_{CA}	$-v_{CA}$	0	i_b	0	$-i_b$	0	0	1	1	0	0	0	0	1
II-b	C	B	C	$-v_{BC}$	v_{BC}	0	0	i_b	$-i_b$	0	0	1	0	1	0	0	0	1
II-b	A	B	A	v_{AB}	$-v_{AB}$	0	$-i_b$	i_b	0	1	0	0	0	1	0	1	0	0
II-b	A	C	A	$-v_{CA}$	v_{CA}	0	$-i_b$	0	i_b	1	0	0	0	0	1	1	0	0
II-b	B	C	B	v_{BC}	$-v_{BC}$	0	0	$-i_b$	i_b	0	1	0	0	0	1	0	1	0
II-b	B	A	B	$-v_{AB}$	v_{AB}	0	i_b	$-i_b$	0	0	1	0	1	0	0	0	1	0
II-c	C	C	A	0	v_{CA}	$-v_{CA}$	i_c	0	$-i_c$	0	0	1	0	0	1	1	0	0
II-c	C	C	B	0	$-v_{BC}$	v_{BC}	0	i_c	$-i_c$	0	0	1	0	0	1	0	1	0
II-c	A	A	B	0	v_{AB}	$-v_{AB}$	$-i_c$	i_c	0	1	0	0	1	0	0	0	1	0
II-c	A	A	C	0	$-v_{CA}$	v_{CA}	$-i_c$	0	i_c	1	0	0	1	0	0	0	0	1
II-c	B	B	C	0	v_{BC}	$-v_{BC}$	0	$-i_c$	i_c	0	1	0	0	1	0	0	0	1
II-c	B	B	A	0	$-v_{AB}$	v_{AB}	i_c	$-i_c$	0	0	1	0	0	1	0	1	0	0
III	A	A	A	0	0	0	0	0	0	1	0	0	1	0	0	1	0	0
III	B	B	B	0	0	0	0	0	0	0	1	0	0	1	0	0	1	0
III	C	C	C	0	0	0	0	0	0	0	0	1	0	0	1	0	0	1

$$i_A = i_a$$
$$i_B = 0$$
$$i_C = -i_a \tag{1.2}$$

Also in Figure 1.4, assigning closed switch as logic 1 and open switch as logic 0, switches S_{Aa} to S_{Cc} in order can be assigned the position: 1 0 0 0 0 1 0 0 1.

Referring to Figure 1.3, in terms of the switch position the output voltage corresponding to phase a and the input current corresponding to phase A can be expressed as follows:

$$v_a = S_{Aa} * v_A + S_{Ba} * v_B + S_{Ca} * v_C \tag{1.3}$$

$$i_A = S_{Aa} * i_a + S_{Ab} * i_b + S_{Ac} * i_c \tag{1.4}$$

One of the essential requirements for switching three-phase AC to three-phase AC MC is that two or more bidirectional switches connected to any one output phase should NOT be closed simultaneously, as this will cause dangerously high short-circuit current. Similarly, any one bidirectional switch connected to each output phase should remain closed to provide a current path with inductive load. This necessitates saw-tooth carrier signal to be compared with modulation functions for each of the bidirectional switches in a comparator and the resulting output is given to logic gates in such a manner that the gate pulses to the three bidirectional switches connected to any one output phase occur in sequence or one after the other.

1.2 Objectives and Novelty

The following are the objectives of the work presented in this textbook. This novelty is presented under the 'Explore applications' subsection:

- Develop models for MC
  - To develop models for three-phase AC to three-phase AC conventional MC topology connected to R–L load using the Venturini, optimum Venturini and other advanced modulation algorithms and for speed control and brake by plugging of three-phase induction motor (IM).
  - To develop models for three-phase AC to three-phase AC multi-level MC (MMC) topology with three flying capacitors (FCs) per output phase, using the Venturini algorithm.
  - To develop models for direct and indirect space-vector-modulated three-phase AC to three-phase AC MC topology.

- To develop models for the single and dual programmable AC to DC rectifier using three-Phase MC topology, speed control and brake by plugging of separately excited DC Motor, single and dual variable-voltage variable-frequency AC power supply.
- To develop models for three-phase AC to three-phase AC MC, when switching is carried out by sample time-based modulation technique also known as delta-sigma modulation.
- To develop model for single-phase AC to three-phase AC MC topology
- To develop models for a single-phase/three-phase AC to single-phase/three-phase AC converter using (a) DC link voltage source and (b) directly using three winding transformer without any DC link voltage source.
- To develop model for real-time hardware-in-the-loop (HIL) simulation of a three-phase AC to single-phase AC MC using dSPACE DS1104 hardware controller board.
- To develop models for three-phase voltage-fed Z-source direct matrix converter (ZSDMC) using (a) Simple Boost Control and (b) Maximum Boost Control strategy and that for three-phase quasi Z-source indirect matrix converter (QZSIMC).
- To develop models for selected three-phase cycloconverter, indirect MC (IMC) and solid-state transformer (SST) topologies.
- Hardware implementation
  - To implement in real time, a three-phase AC to single-phase AC MC using dSPACE DS1104 hardware controller board.
- Explore applications
  - Applications of three-phase AC to three-phase AC MC for industrial AC drives [5–10], wind energy conversion [11–13], aircraft control applications [14,15] and railway electric traction power supply [16] are already reported in the literature. In this textbook, additional applications for the speed control and brake by plugging of single and two separately excited DC motors fed by single and dual programmable AC to DC rectifier suitable for hybrid electric vehicle and electric traction are presented [17,18]. The application of single and dual programmable rectifier topology as a single- and dual-mode variable-magnitude variable-frequency pure sine-wave AC power supply is presented. A new sine-wave AC to AC cum AC to DC converter without a DC link is presented [19–23]. In addition, another application of three-phase AC to three-phase AC MC for the speed control and brake by plugging or by reversing the phase sequence of three-phase IM is presented.

1.3 Research Methodology

In this textbook, only the following research methods are presented:

- Software approach:
  - Develop models for MC using SIMULINK [24].
- Hardware implementation:
  - Real-time HIL simulation using dSPACE [25].

1.4 Book Outline

This textbook mainly provides an account of the three-phase AC to three-phase AC MC modelling using the software SIMULINK [24]. A chapter on modelling cycloconverters, IMC and SST is presented. Unless specified otherwise the term *model* in this textbook refers to SIMULINK model. The literatures referred for developing models in this textbook are mentioned at the end of each chapter.

The outline and novelty of each chapter in this textbook are discussed below.

The three-phase AC to three-phase AC MC low-frequency modulation matrix concept was first developed by Alesina and Venturini [2–4,26,27]. Chapter 2 presents an overview of this Venturini and optimum Venturini modulation algorithm and models of three-phase AC to three-phase AC MC based on these algorithms. An advanced modulation algorithm for three-phase AC to three-phase AC MC proposed by Ned Mohan et al. [7,28–30] is presented in Chapter 2 with model simulation results [31–33]. A case study on the speed control and brake by plugging of three-phase IM fed by MC is presented in this chapter along with a block diagram for the real-time HIL simulation scheme using dSPACE DS1104 hardware controller board.

Based on three-phase AC to three-phase AC MMC with two and three FCs per output phase [34,35], a model is presented for the three-phase MMC with three FCs with simulation results in Chapter 3. A model representation for three-phase MMC using Boolean logic is also presented in this chapter [36,37].

Models for direct asymmetrical (ASVM) and direct symmetrical (SSVM) space-vector modulation [38] of three-phase AC to three-phase AC MC are presented along with Simulation results in Chapter 4.

Indirect space-vector modulation (ISVM) [39] of three-phase AC to three-phase AC MC considers that the above MC is equivalent to a three-phase AC to DC rectifier and a DC to three-phase AC inverter. Based on this technique, a model has been developed for three-phase AC to three-phase AC ISVM MC and is presented with simulation results in Chapter 5.

Based on the single and dual fixed AC to DC rectifier using three-phase AC to three-phase AC MC topology [40–42] reported in the literature, models and simulation results for single and dual programmable AC to DC rectifier using the above MC topology are presented in Chapter 6. Derivations for the single and dual DC output voltage magnitude have been presented using phasor diagrams which are original contributions [17,18]. A case study for the speed control and brake by plugging of separately excited DC motor using single and dual programmable AC to DC rectifier is presented with simulation results. Another case study for single and dual variable-magnitude variable-frequency pure sine-wave AC power supply is presented in this chapter.

A model developed for three-phase MC using sample time-based modulation technique, which is also known as delta-sigma modulation technique [43–45], is presented in Chapter 7. This model is developed based on Z-transform approach. The model is presented with simulation results for three-phase AC to three-phase AC MC with delta-sigma modulation using the Venturini algorithm for both R–L load and three-phase IM load [46].

Based on the single-phase AC to three-phase AC MC topology, without the need for rectification and inversion, reported in the literature [47], a model has been developed for single-phase AC to three-phase AC MC and presented in Chapter 8 [48]. Simulation results are presented for both R–L load and three-phase IM load.

A novel AC to AC converter using a DC link is a new method developed by the author of this textbook [19]. Here, models and simulation results for both single-phase/three-phase AC voltages at supply frequency to single-phase/three-phase AC output voltages with a variable frequency and magnitude are presented [19]. The model essentially consists of N-P MOSFET pairs driven by operational amplifier zero crossing comparators to produce single-phase/three-phase square-wave AC output voltage at supply frequency which are given as input to IGBT bidirectional switches driven by gate pulse having frequency equal to that of the desired output frequency and 50% duty cycle. This gives a variable-frequency square-wave AC output voltage. Pulse-width modulation (PWM) is achieved by conventional technique using square pulse carrier. The model and simulation results are presented in Chapter 9. A section is provided with simulation results to explain the operation of supply frequency N-P MOSFET inverter and that of the variable-output-frequency converter using IGBT bidirectional switches.

Real-time HIL simulation of a three-phase AC to single-phase AC MC using dSPACE DS1104 hardware controller board is presented in Chapter 10 [20]. Here, a three-phase square-wave AC voltage is used as the input, and the output is a single-phase PWM square-wave AC voltage. Real-time simulation results are compared with model simulation results.

With conventional MC, the voltage gain is limited to 0.866. To increase the voltage gain above unity, Z-source MC is used. Here, a Z-source L–C network is placed in-between the AC supply voltage and the MC bidirectional switches [21]. With a conventional MC, the switches connected to the same output phase cannot be closed simultaneously as this causes dangerous short-circuit current to flow. With Z-source MC, switches connected to the same output phase can be closed simultaneously providing a shoot through (ST) state [21]. During ST state, AC mains supply is switched off. During the non-shoot through (NST) state, AC mains supply is switched on and the bidirectional switches are switched as for the conventional MC [21]. In this way, it is possible to boost the output voltage much higher than the input voltage [21]. In the same way, an LC network with a diode switch is placed between front-end three-phase rectifier and back-end three-phase inverter in a three-phase QZSIMC, and an ST state is created by closing the switches in the upper and lower legs of back-end inverter simultaneously with AC mains supply switched off [22]. During NST state, AC mains supply is switched on and the inverter switches are turned on as in a conventional three-phase inverter. In this way, it is possible to achieve voltage gain greater than unity in a three-phase QZSIMC [22]. Models for three-phase voltage-fed ZSDMC and three-phase QZSIMC are presented with simulation results in Chapter 11.

A combined PWM sine-wave AC to AC and AC to DC converter is my original idea for which a patent has been submitted [19,23].

In Chapter 12, a combined pulse-width-modulated single-phase/three-phase sine-wave AC to single-phase/three-phase variable-voltage variable-frequency sine-wave AC converter which can also be used as a sine-wave AC to DC converter with provision for reversing the polarity of the DC output voltage is presented. Simulation results indicate that both variable-frequency variable-voltage PWM AC output voltage in the AC mode and a PWM variable DC output voltage with polarity reversal capability in the DC mode can be obtained from the same converter by properly transferring the gate drive using a selector switch.

Modelling of cycloconverters, IMC and SST are presented in Chapter 13. Cycloconverters are AC to AC converters which operate as frequency changers giving an AC output voltage at a different frequency from that of the input AC voltage source [1,49]. IMCs convert the input AC voltage with a known frequency to DC and then convert the DC to AC voltage using semiconductor switches, giving the same or different output frequency [49]. SST is basically a high-frequency transformer (HFT) isolated AC to AC conversion technique [50,51]. In this chapter, models for single-phase and three-phase cycloconverters, three-phase conventional and multilevel IMC and SST models using dual active bridge (DAB) [49–51], and direct AC to AC converter topology are presented [23].

The Appendix A provides the selected MC derivations. These are the Venturini algorithm, ISVM algorithm and three-phase voltage-fed ZSDMC algorithm.

References

1. B.R. Pelly: *Thyristor Phase-Controlled Converters and Cycloconverters: Operation, Control and Performance,* John Wiley & Sons, New York, 1971.
2. A. Alesina and M.G.B. Venturini: Solid state power conversion: a Fourier analysis approach to generalized transformer synthesis, *IEEE Transactions on Circuits and Systems,* Vol. CAS-28, No. 4, 1981, pp. 319–330.
3. A. Alesina and M. Venturini: Intrinsic amplitude limits and optimum design of 9-switches direct PWM AC-AC converters, IEEE-PESC, Kyoto, Japan, Vol. 2, 1988, pp. 1284–1291.
4. A. Alesina and M. Venturini: Analysis and design of optimum amplitude nine-switch direct AC-AC converters, *IEEE Transactions on Power Electronics,* Vol. 4, 1989, pp. 101–112.
5. S. Sunter and J.C. Clare: Development of a matrix converter induction motor drive, *IEEE Melecon,* Antalya, 1994, pp. 833–836.
6. S. Sunter and J.C. Clare: A true four quadrant matrix converter induction motor drive with servo performance, *Proceedings of IEEE PESC,* Italy, 1996, pp. 146–151.
7. K.K. Mohapatra, T. Satish and N. Mohan: A speed-sensorless direct torque control scheme for matrix converter driven induction motor, *IEEE-IECON,* Paris, France, 2006, pp. 1435–1440.
8. C. Klumpner, P. Nielsen, I. Boldea and F. Blaaberg: A new matrix converter motor (MCM) for industry applications, *IEEE Transactions on Industrial Electronics,* Vol. 49, No. 2, 2002, pp. 325–335.
9. H.J. Cha and P. Enjetti: Matrix converter–fed ASDs, *IEEE Industry Applications Magazine,* Vol. 10, 2004, pp. 33–39.
10. T. Podlesak, D.C. Katsis, P.W. Wheeler, J.C. Clare, L. Empringham and M. Bland: A 150-kVA vector-controlled matrix converter induction motor drive, *IEEE Transactions on Industry Applications,* Vol. 41, No. 3, 2005, pp. 841–847.
11. R. Cardenas, R. Pena, J. Clare and P. Wheeler: Control of the reactive power supplied by a matrix converter, *IEEE Transactions on Energy Conversion,* Vol. 24, No. 1, 2009, pp. 301–303.
12. R.K. Gupta, G.F. Castelino, K.K. Mohapatra and N. Mohan: A novel integrated three-phase, switched multi-winding power electronic transformer converter for wind power generation system, *IEEE-IECON,* Porto, Portugal, 2009, pp. 4481–4486.
13. D.V. Nicolae, C.G. Richards and P. Ehlers: Small power three to one phase matrix converters for wind generators, *International Symposium on Power Electronics, Electrical Drives, Automation and Motion – SPEEDAM,* Pisa, Italy, 2010, pp. 1339–1343.

14. C.R. Whitley, G.K. Towers, P. Wheeler, J. Clare, K. Bradley, M. Apap and L. Empringham: A matrix converter based electro-hydrostatic actuator, *IEE Seminar on Matrix Converter*, London, UK, 2003, pp. 1–6.
15. P.W. Wheeler, J.C. Clare, M. Apap, L. Empringham, K.J. Bradley, C. Whitley and G. Towers: A matrix converter based permanent magnet motor drive for an electro-hydrostatic aircraft actuator, *IEEE-IECON*, Blacksburg, VA, 2003, pp. 2072–2077.
16. D. Pavel, P. Martin and C. Marek: High voltage matrix converter topology for multi-system locomotives, *IEEE-ECCE*, Atlanta, GA, 2010, pp. 1640–1645.
17. N.P.R. Iyer: Modelling, simulation and real time implementation of a three phase AC to AC matrix converter, Ph.D. thesis, Ch. 10 and Ch. 14, Department of ECE, Curtin University, Perth, WA, Australia, February 2012.
18. N.P.R. Iyer: A dual programmable AC to DC rectifier using three-phase matrix converter topology—analysis aspects, *Electrical Engineering*, Vol. 100, No. 2, 2018, pp. 1183–1194. DOI: 10.1007/s00202-017-0572-9.
19. N.P.R. Iyer: Modelling, simulation and real time implementation of a three phase AC to AC matrix converter, Ph.D. thesis, Ch. 13, Department of ECE, Curtin University, Perth, WA, Australia, February 2012.
20. N.P.R. Iyer: Modelling, simulation and real time implementation of a three phase AC to AC matrix converter, Ph.D. thesis, Ch. 6, Department of ECE, Curtin University, Perth, WA, Australia, February 2012.
21. B. Ge, Q. Lei, W. Qian and F.Z. Peng: A family of Z-source matrix converters, *IEEE Transactions on Industrial Electronics*, Vol. 59, No. 1, 2012, pp. 35–46.
22. S. Liu, B. Ge, H.A. Rub, X. Jiang and F.Z. Peng: A novel indirect quasi-Z-source matrix converter applied to induction motor drives, *IEEE Energy Conversion Congress and Exposition (ECCE)*, Denver, CO, September 2013, pp. 2440–2444.
23. N.P.R. Iyer: A single phase AC to PWM single phase AC and DC converter (alternative title: "Swamy converter"), Patent submitted to Technology Transfer Office, The University of Nottingham, UK, 2013.
24. The Mathworks Inc.: www.mathworks.com, MATLAB/SIMULINK user manual, MATLAB R2017b, USA, 2017.
25. dSPACE GmbH: www.dspace.com, dSPACE users' manual, 2010.
26. P. Wheeler, J. Rodriguez, J.C. Clare, L. Empringham and A. Weinstein: Matrix converters – a technology review, *IEEE Transactions on Industrial Electronics*, Vol. 49, No. 2, 2002, pp. 276–288.
27. P. Wheeler, J. Clare, L. Empringham, M. Apap and M. Bland: Matrix converters, *Power Engineering Journal*, Vol. 16, 2002, pp. 273–282.
28. K.K. Mohapatra, P. Jose, A. Drolia, G. Aggarwal, S. Thuta and N. Mohan: A novel carrier-based PWM scheme for matrix converters that is easy to implement, *IEEE-PESC*, Recife, Brazil, 2005, pp. 2410–2414.
29. T. Sathish, K.K. Mohapatra and N. Mohan: Steady state over-modulation of matrix converter using simplified carrier based control, *IEEE-IECON*, Taipei, Taiwan, 2007, pp. 1817–1822.
30. S. Thuta, K.K. Mohapatra and N. Mohan: Matrix converter over-modulation using carrier-based control: maximizing the voltage transfer ratio, *IEEE-PESC*, Rhodes, Greece, 2008, pp. 1727–1733.
31. N.P.R. Iyer: Modelling, simulation and real time implementation of a three phase AC to AC matrix converter, Ph.D. thesis, Ch. 3, Ch. 4 and Ch. 14, Department of ECE, Curtin University, Perth, WA, Australia, February 2012.

32. N.P.R. Iyer: Carrier based modulation technique for three phase matrix converters – state of the art progress, *IEEE_SIBIRCON-2010*, Irkutsk, Listvyanka, Russia, July 11–15, 2010, pp. 659–664.

33. N.P.R. Iyer and C.V. Nayar: Performance comparison of a three phase AC to three phase AC matrix converter using different carrier based switching algorithms, *International Journal of Engineering Research & Technology (IJERT)*, Vol. 6, No. 5, 2017, pp. 827–832.

34. Y. Shi, X. Yang, Q. He and Z. Wang: Research on a novel capacitor clamped multilevel matrix converter, *IEEE Transactions on Power Electronics*, Vol. 20, No. 5, 2005, pp. 1055–1065.

35. J. Rzasa: Capacitor clamped multilevel matrix converter controlled with Venturini method, *13th International Power Electronics and Motion Control Conference (EPE-PEMC)*, Poznan, Poland, 2008, pp. 357–364.

36. N.P.R. Iyer: Modelling, simulation and real time implementation of a three phase AC to AC matrix converter, Ph.D. thesis, Ch. 7 and 14, Department of ECE, Curtin University, Perth, WA, Australia, February 2012.

37. N.P.R. Iyer: Performance comparison of a three-phase multilevel matrix converter with three flying capacitors per output phase with a three-phase conventional matrix converter, *Electrical Engineering*, Vol. 99, No. 2, 2017, pp. 775–789. DOI: 10.1007/s00202-016-0500-4.

38. D. Casadei, G. Serra A. Tani and L. Zarri: Matrix converter modulation strategies: a new general approach based on space-vector representation of the switching state, *IEEE Transactions on Industrial Electronics*, Vol. 49, No. 2, 2002, pp. 370–381.

39. L. Huber and D. Borojevic: Space vector modulated three-phase to three-phase matrix converter with input power factor correction, *IEEE Transactions on Industry Applications*, Vol. 31, No. 6, 1995, pp. 1234–1245.

40. D.G. Holmes and T.A. Lipo: Implementation of a controlled rectifier using AC–AC matrix converter theory, *IEEE-PESC*, Milwaukee, WI, 1989, pp. 353–359.

41. D.G. Holmes and T.A. Lipo: Implementation of a controlled rectifier using AC–AC matrix converter theory, *IEEE Trans Power Electronics*, Vol. 7, No. 1, 1992, pp. 240–250.

42. S. Huseinbegovic and O. Tanovic: Matrix converter based AC/DC rectifier, *IEEE-SIBIRCON II*, Irkutsk, Russia, 2010, pp. 653–658.

43. A. Hirota and M. Nakaoka: A low noise three phase matrix converter introducing delta-sigma modulation scheme, *37th IEEE-PESC*, Jeju, South Korea, 2006, pp. 1–6.

44. A. Hirota, S. Nagai, B. Saha and M. Nakaoka: Fundamental study of a simple control AC-AC converter introducing delta-sigma modulation approach, *IEEE-ICIT*, Chengdu, China, 2008, pp. 1–5.

45. A. Hirota, S. Nagai and M. Nakaoka: Suppressing noise peak single phase to three phase AC-AC direct converter introducing delta-sigma modulation technique, *IEEE-PESC*, Rhodes, Greece, 2008, pp. 3320–3323.

46. N.P.R. Iyer: Modelling, simulation and real time implementation of a three phase AC to AC matrix converter, Ph.D. thesis, Ch. 11 and Ch. 14, Department of ECE, Curtin University, Perth, WA, Australia, February 2012.

47. K. Iino, K. Kondo and Y. Sato: An experimental study on induction motor drive with a single phase-three phase matrix converter, *13th European Conference on Power Electronics and Applications*, EPE, Barcelona, Spain, 2009, pp. 1–9.

48. N.P.R. Iyer: Modelling, simulation and real time implementation of a three phase AC to AC matrix converter, Ph.D. thesis, Ch. 12 and Ch. 14, Department of ECE, Curtin University, Perth, WA, Australia, February 2012.
49. A.K. Chattopadhyay: AC-AC converters. In: *Power Electronics Handbook*, ed. M.H. Rashid, Chapter 18, pp. 497–506, Elsevier, 2011.
50. Y. Du, S. Baek, S. Bhattacharya and A.Q. Huang: High-voltage high-frequency transformer design for a 7.2kV to 120V/240V 20kVA solid state transformer, *IEEE-IECON*, Glendale, AZ, November 2010, pp. 487–492.
51. H. Qin and J.W. Kimball: AC-AC dual active bridge converter for solid state transformer, *IEEE Energy Conversion Congress and Exposition (ECCE)*, San Jose, CA, November 2009, pp. 3039–3044.

2

Carrier-Based Modulation Algorithms for Matrix Converters

2.1 Introduction

Recently, considerable interest is shown in the development of AC to AC converters, which is also known as 'matrix converters' (MC) for adjustable speed drive applications [1–8]. MCs are essentially forced commutated cycloconverters consisting of a matrix of bidirectional switches such that there is a switch for each possible connection between the input and output lines. For a three-phase AC to three-phase AC MC, there are nine bidirectional semiconductor switches (18 semiconductor switches: each with a diode in parallel). Although several topologies for bidirectional switches have been proposed, the common-emitter bidirectional switch topology is used here. The switching of the bidirectional switches in an MC is really complicated. Several algorithms have been proposed for switching the MC bidirectional switches. This chapter presents the model of the three-phase AC to three-phase AC MC using the Venturini, optimum Venturini and advanced modulation algorithms. The model is developed using the software SIMULINK [17]. The simulation results are presented.

2.2 Model of Three-Phase AC to Three-Phase AC Matrix Converter

A three-phase AC to three-phase AC MC is shown in Figure 2.1. While operating the MC, two essential points must be remembered. The three or any two combinations of the bidirectional switches in any one output phase should NOT be closed at the same instant of time [4–5]. Referring to Figure 2.1, if any two or all of switches S_{Aa}, S_{Ba} and S_{Ca} are closed simultaneously, the input lines are short circuited causing dangerous short-circuit currents through the bidirectional switches. Similarly with inductive loads all the three bidirectional switches connected to an output phase should NOT be open simultaneously [4–5]. Any one switch in each output phase must remain closed. MC directly converts the

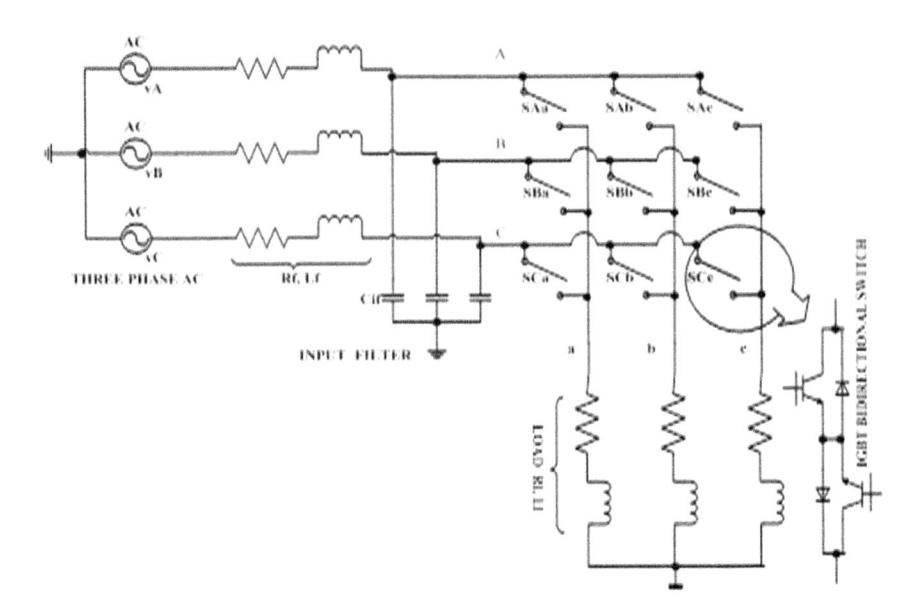

FIGURE 2.1
Three-phase matrix converter.

AC input voltage at any given frequency to AC output voltage with arbitrary amplitude at any unrestricted frequency without the need for a DC link capacitor storage element at the input side. Sinusoidal input and output currents can be obtained with unity power factor for any load. It has regeneration capability [1–5]. One limitation of the MC is that the maximum output voltage available is limited to 86.6% of the input voltage in the linear modulation range.

The introduction of bidirectional switch using power transistors made easy realization of the MC. The real development of the MC starts with the work of Venturini and Alesina who proposed a mathematical analysis and introduced the low-frequency modulation matrix concept to describe the low-frequency behaviour of the MC [1–3]. In this, the output voltages are obtained by multiplication of the modulation matrix or transfer matrix with the input voltages. Models for three-phase AC to three-phase AC MCs using the Venturini modulation algorithm is reported [9,10]. These models are developed in SIMULINK [17].

The 3 × 3 MC shown in Figure 2.1 connects the three-phase AC source to the three-phase load. The switching function for a 3 × 3 MC can be defined as follows [4–5]:

$$S_{Kj} = \begin{cases} 1 & \text{when } S_{Kj} \text{ is closed} \\ 0 & \text{when } S_{Kj} \text{ is open} \end{cases} \tag{2.1}$$

$$K \in \{A, B, C\} \text{ and } j \in \{a, b, c\}$$

The above constraint can be expressed in the following form:

$$S_{Aj} + S_{Bj} + S_{Cj} = 1$$
$$j \in \{a,b,c\}$$

(2.2)

With the above restrictions, a 3 × 3 MC has 27 possible switching states, which are given in Table 1.1 of Chapter 1 [11–16].

The mathematical expression that represents the operation of a three-phase AC to three-phase AC MC can be expressed as follows [4–5]:

$$
\begin{bmatrix} v_a(t) \\ v_b(t) \\ v_c(t) \end{bmatrix} = \begin{bmatrix} S_{Aa}(t) & S_{Ba}(t) & S_{Ca}(t) \\ S_{Ab}(t) & S_{Bb}(t) & S_{Cb}(t) \\ S_{Ac}(t) & S_{Bc}(t) & S_{Cc}(t) \end{bmatrix} * \begin{bmatrix} v_A(t) \\ v_B(t) \\ v_C(t) \end{bmatrix}
$$

(2.3)

$$
\begin{bmatrix} i_A(t) \\ i_B(t) \\ i_C(t) \end{bmatrix} = \begin{bmatrix} S_{Aa}(t) & S_{Ba}(t) & S_{Ca}(t) \\ S_{Ab}(t) & S_{Bb}(t) & S_{Cb}(t) \\ S_{Ac}(t) & S_{Bc}(t) & S_{Cc}(t) \end{bmatrix}^T * \begin{bmatrix} i_a(t) \\ i_b(t) \\ i_c(t) \end{bmatrix}
$$

(2.4)

where v_a, v_b and v_c and i_A, i_B and i_C are the output voltages and input currents, respectively. To determine the behaviour of the MC at output frequencies well below the switching frequency, a modulation duty cycle can be defined for each switch. The modulation duty cycle M_{Kj} for the switch S_{Kj} in Figure 2.1 is defined as

$$M_{Kj} = \frac{t_{Kj}}{T_s}$$

(2.5)

$$K \in \{A,B,C\} \text{ and } j \in \{a,b,c\}$$

where t_{Kj} is the on time for the switch S_{Kj} between input phase $K \in \{A,B,C\}$ and $j \in \{a,b,c\}$ and T_s is the period of the pulse-width-modulated (PWM) switching signal or sampling period. In terms of the modulation duty cycle, equations 2.2–2.4 can be rewritten as follows:

$$
\begin{bmatrix} v_a(t) \\ v_b(t) \\ v_c(t) \end{bmatrix} = \begin{bmatrix} M_{Aa}(t) & M_{Ba}(t) & M_{Ca}(t) \\ M_{Ab}(t) & M_{Bb}(t) & M_{Cb}(t) \\ M_{Ac}(t) & M_{Bc}(t) & M_{Cc}(t) \end{bmatrix} * \begin{bmatrix} v_A(t) \\ v_B(t) \\ v_C(t) \end{bmatrix}
$$

(2.6)

$$\begin{bmatrix} i_A(t) \\ i_B(t) \\ i_C(t) \end{bmatrix} = \begin{bmatrix} M_{Aa}(t) & M_{Ba}(t) & M_{Ca}(t) \\ M_{Ab}(t) & M_{Bb}(t) & M_{Cb}(t) \\ M_{Ac}(t) & M_{Bc}(t) & M_{Cc}(t) \end{bmatrix}^T * \begin{bmatrix} i_a(t) \\ i_b(t) \\ i_c(t) \end{bmatrix} \qquad (2.7)$$

$$M_{Aj} + M_{Bj} + M_{Cj} = 1$$

$$j \in \{a,b,c\} \qquad (2.8)$$

2.3 Venturini and Optimum Venturini Modulation Algorithms

The modulation problem normally encountered in MC can be stated as follows: The input voltages and output currents are given as follows [1–5]:

$$v_i = \begin{bmatrix} v_A \\ v_B \\ v_C \end{bmatrix} = V_{im} * \begin{bmatrix} \cos(\omega_i t) \\ \cos\left(\omega_i t - \dfrac{2\pi}{3}\right) \\ \cos\left(\omega_i t - \dfrac{4\pi}{3}\right) \end{bmatrix} \qquad (2.9)$$

$$i_o = \begin{bmatrix} i_a \\ i_b \\ i_c \end{bmatrix} = I_{om} * \begin{bmatrix} \cos(\omega_o t + \varphi_o) \\ \cos\left(\omega_o t + \varphi_o - \dfrac{2\pi}{3}\right) \\ \cos\left(\omega_o t + \varphi_o - \dfrac{4\pi}{3}\right) \end{bmatrix} \qquad (2.10)$$

where ω_i, ω_o and φ_o are the input and output angular frequencies and output phase displacement angle. The problem is to find a modulation matrix $M(t)$ such that equations 2.11 and 2.12 below and the constraint in equation 2.8 are satisfied:

$$v_o = M(t) * v_i = \begin{bmatrix} v_a \\ v_b \\ v_c \end{bmatrix} = q * V_{im} * \begin{bmatrix} \cos(\omega_o t + \varphi_o) \\ \cos\left(\omega_o t + \varphi_o - \dfrac{2\pi}{3}\right) \\ \cos\left(\omega_o t + \varphi_o - \dfrac{4\pi}{3}\right) \end{bmatrix} \qquad (2.11)$$

where $q * V_{im} = V_{om}$

$$i_i = \left[M(t)\right]^T * i_o = \begin{bmatrix} i_A \\ i_B \\ i_C \end{bmatrix} = \frac{q * \cos(\varphi_o)}{q * \cos(\varphi_i)} * I_{om} * \begin{bmatrix} \cos(\omega_i t + \varphi_i) \\ \cos\left(\omega_i t + \varphi_i - \dfrac{2\pi}{3}\right) \\ \cos\left(\omega_i t + \varphi_i - \dfrac{4\pi}{3}\right) \end{bmatrix} \quad (2.12)$$

where q is the voltage transfer ratio and φ_i is the input phase displacement angle. The solution of equations 2.9–2.12 can be expressed as in equations 2.13 and 2.14 below for $\varphi_i = \varphi_o$ and for $\varphi_i = -\varphi_o$, respectively:

$$M_1 = \frac{1}{3} \begin{bmatrix} 1 + 2q * \cos(\omega_m t) & 1 + 2q * \cos\left(\omega_m t - \dfrac{2\pi}{3}\right) \\ 1 + 2q * \cos\left(\omega_m t - \dfrac{4\pi}{3}\right) & 1 + 2q * \cos(\omega_m t) \\ 1 + 2q * \cos\left(\omega_m t - \dfrac{2\pi}{3}\right) & 1 + 2q * \cos\left(\omega_m t - \dfrac{4\pi}{3}\right) \end{bmatrix}$$

$$\begin{bmatrix} 1 + 2q * \cos\left(\omega_m t - \dfrac{4\pi}{3}\right) \\ 1 + 2q * \cos\left(\omega_m t - \dfrac{2\pi}{3}\right) \\ 1 + 2q * \cos\left(\omega_m t - \dfrac{4\pi}{3}\right) \end{bmatrix} \quad (2.13)$$

where $\omega_m = (\omega_o - \omega_i)$

$$M_2 = \frac{1}{3} \begin{bmatrix} 1 + 2q * \cos(\omega_m t) & 1 + 2q * \cos\left(\omega_m t - \dfrac{2\pi}{3}\right) \\ 1 + 2q * \cos\left(\omega_m t - \dfrac{2\pi}{3}\right) & 1 + 2q * \cos\left(\omega_m t - \dfrac{4\pi}{3}\right) \\ 1 + 2q * \cos\left(\omega_m t - \dfrac{4\pi}{3}\right) & 1 + 2q * \cos(\omega_m t) \end{bmatrix}$$

$$\begin{bmatrix} 1 + 2q * \cos\left(\omega_m t - \dfrac{4\pi}{3}\right) \\ 1 + 2q * \cos(\omega_m t) \\ 1 + 2q * \cos\left(\omega_m t - \dfrac{2\pi}{3}\right) \end{bmatrix} \quad (2.14)$$

where $\omega_m = -(\omega_o - \omega_i)$.

This basic solution gives the direct transfer function approach. During each switching sequence T_s, the average output voltage is equal to the demand or target output voltage. The target output voltage must fit into the input voltage envelope for any output frequency. Also combining the two solutions in equations 2.13 and 2.14 gives a method for input displacement factor control. This can be expressed as follows:

$$M(t) = \alpha_1 * M_1(t) + \alpha_2 * M_2(t) \tag{2.15}$$

$$(\alpha_1 + \alpha_2) = 1 \tag{2.16}$$

For unity input phase displacement factor, $\alpha_1 = \alpha_2 = 0.5$. For unity input displacement factor, the modulation function can be expressed as follows:

$$M_{Kj} = \frac{t_{Kj}}{T_s} = \left[\frac{1}{3} + \frac{2v_K * v_j}{3V_{im}^2} \right] \tag{2.17}$$

for $K = A, B, C$ and $j = a, b, c$

This method has little practical significance because of the 50% voltage ratio limitation. Venturini's optimum method employs a common-mode addition technique defined in the following equation:

$$v_o = q * V_{im} * \begin{bmatrix} \cos(\omega_o t) - \dfrac{1}{6}\cos(3\omega_o t) + \dfrac{1}{2\sqrt{3}}\cos(3\omega_i t) \\[2ex] \cos\left(\omega_o t + \dfrac{4\pi}{3}\right) - \dfrac{1}{6}\cos(3\omega_o t) + \dfrac{1}{2\sqrt{3}}\cos(3\omega_i t) \\[2ex] \cos\left(\omega_o t + \dfrac{2\pi}{3}\right) - \dfrac{1}{6}\cos(3\omega_o t) + \dfrac{1}{2\sqrt{3}}\cos(3\omega_i t) \end{bmatrix} \tag{2.18}$$

The target output voltage using equation 2.18 is 86.6% of the input voltage. In this case, equation 2.17 can be modified to

$$M_{Kj} = \frac{t_{Kj}}{T_s} = \left[\frac{1}{3} + \frac{2v_K * v_j}{3V_{im}^2} + \frac{4q}{9\sqrt{3}} * \sin(\omega_i t + \beta_K)\sin(3\omega_i t) \right]$$

for $K = A, B, C$ and $j = a, b, c$ and $\beta_K = 0, \dfrac{2\pi}{3}, \dfrac{4\pi}{3}$ for $K = A, B, C$, respectively.

$$\tag{2.19}$$

In this equation, v_j includes the common-mode addition defined in equation 2.18.

2.4 Model Development

To study the behaviour of the three-phase AC to three-phase AC MC using the Venturini and optimum Venturini modulation algorithms, a model of this MC is developed in SIMULINK [17]. The data shown in Table 2.1 are used to develop the model for all the above algorithms. The MC switching is developed based on equations 2.13 and 2.17–2.19 [22–24].

2.4.1 Model of a Matrix Converter Using Venturini First Method

The model of the three-phase MC is developed using equation 2.13 for the data shown in Table 2.1. The input and output voltages are defined in equations 2.9 and 2.11, respectively. The input voltage in equation 2.9 is made to lag by $\pi/2\,\mathrm{rad}$, so that this becomes a three-phase sine-wave input voltage. The nine modulation functions defined in equation 2.13 are developed using Embedded MATLAB Function in SIMULINK. This is shown in Program segment 2.1. These modulation functions are compared with a 2 kHz saw-tooth carrier generator, and the gate pulses for the nine bidirectional switches are developed satisfying the constraint in equation 2.8 using another Embedded MATLAB Function, as shown in Program segment 2.2 [22–23].

Program Segment 2.1

```
function [MAa,MBa,MCa,MAb,MBb,MCb,MAc,MBc,MCc] =
fcn(q,fi,fo,fsw,t)
%% Dr. Narayanaswamy.P.R.Iyer
%% Venturini Algorithm for Three Phase Matrix Converter -
First method.
MAa = (1 + 2*q*cos((2*pi*fo*t - 2*pi*fi*t)))*1/(3);
MBa = (1 + 2*q*cos((2*pi*fo*t - 2*pi*fi*t - 2*pi/(3))))*1/(3);
MCa = (1 + 2*q*cos((2*pi*fo*t - 2*pi*fi*t - 4*pi/(3))))*1/(3);
MAb = (1 + 2*q*cos((2*pi*fo*t - 2*pi*fi*t - 4*pi/(3))))*1/(3);
MBb = (1 + 2*q*cos((2*pi*fo*t - 2*pi*fi*t)))*1/(3);
MCb = (1 + 2*q*cos((2*pi*fo*t - 2*pi*fi*t - 2*pi/(3))))*1/(3);
MAc = (1 + 2*q*cos((2*pi*fo*t - 2*pi*fi*t - 2*pi/(3))))*1/(3);
MBc = (1 + 2*q*cos((2*pi*fo*t - 2*pi*fi*t - 4*pi/(3))))*1/(3);
MCc = (1 + 2*q*cos((2*pi*fo*t - 2*pi*fi*t)))*1/(3);
```

TABLE 2.1

Matrix Converter Parameters

Sl. No.	Modulation Index q	RMS Line-to-Neutral Input Voltage (V)	Input Frequency (Hz)	Output Frequency (Hz)	Carrier Frequency (kHz)
1	0.4	220	50	50	2

Program Segment 2.2

```
function [tAa,tBa,tCa,tAb,tBb,tCb,tAc,tBc,tCc] = fcn(MAa,MBa,
MCa,MAb,MBb,MCb,MAc,MBc,MCc,vsaw)
%% Dr.Narayanaswamy.P.R.Iyer
%%Venturini Algorithm for Three Phase Matrix Converter - First
method.
if(vsaw < MAa )
    tAa = 1;
else
    tAa = 0;
end
if(vsaw < (MAa+MBa))
    tABab = 1;
else
    tABab = 0;
end
tBa = (~tAa & tABab);
tCa = ~(tABab);
if (vsaw < MAb)
    tAb = 1;
else
    tAb = 0;
end
if (vsaw < (MAb + MBb))
    tBAba = 1;
else
    tBAba = 0;
end
tBb = (tBAba & ~tAb);
tCb = ~(tBAba);
if (vsaw < MAc)
    tAc = 1;
else
    tAc = 0;
end
if (vsaw < (MAc + MBc))
    tBAca = 1;
else
    tBAca = 0;
end
tBc = (tBAca & ~tAc);
tCc = ~(tBAca);
```

In the source code above in Program segment 2.2, the less than (<) and the plus (+) sign can be generated using op. amp. comparator and summer and the logic operation AND (&) and NOT (~) can be developed using integrated circuits AND and NOT gates, respectively.

The nine bidirectional switches are the insulated gate bipolar transistor (IGBT)/diode model in the power electronics library and the three-phase programmable voltage source is from electrical sources library, respectively, in the Specialized Technology block set in Simscape Power Systems.

The three-phase cosine-wave input voltage as defined in equation 2.9 phase delayed by 90° to get three-phase sine wave with data as in Table 2.1 can be generated by entering $[381.04 \quad 0 \quad 50]$ in the box named 'Positive-sequence:' in the three-phase programmable voltage source model.

The model of the three-phase MC using the Venturini first method developed in SIMULINK is shown in Figure 2.2. A balanced three-phase R–L load of 50 Ω and 0.5 H is used. A series RLC filter to resonate at a carrier switching frequency of 2 kHz is used. This R–L load and RLC filter are from the Elements library in the Specialized Technology block set. The three-phase V–I measurement block is from the Measurements library in the Specialized Technology block set.

2.4.2 Simulation Results

The simulation of the MC using the Venturini first method using the data in Table 2.1 is carried out using SIMULINK. The fixed-step ode3 (Bogacki–Shampine) solver is used. Harmonic spectrum of the line-to-neutral output voltage, phase A input current and line-to-line output voltage are shown in Figures 2.3–2.5, respectively. The simulation results for the three-phase line-to-neutral output voltage, phase A input current, three-phase line-to-line output voltage and three-phase load current, saw-tooth carrier, and gate pulse are shown in Figures 2.6–2.11, respectively. The simulation results from the harmonic spectrum are tabulated in Table 2.2 [22–24].

2.4.3 Model of a Matrix Converter Using Venturini Second Method

The model of the three-phase MC is developed using equation 2.17 for the data shown in Table 2.1. Unity input phase displacement factor is assumed. The input voltage defined in equation 2.9 is made to phase lag by 90° to get three-phase sine-wave input voltage. The nine modulation functions defined in equation 2.17 are developed using Embedded MATLAB Function in SIMULINK, as shown in Program segment 2.3. These modulation functions are compared with a 2-kHz saw-tooth carrier generator and the gate pulses for the nine bidirectional switches are developed satisfying the constraint in equation 2.8 using another Embedded MATLAB Function as shown in Program segment 2.2 [23]. The model of the MC using Venturini second method is shown in Figure 2.12. This is the same as explained in Section 2.4.1. Validity of equation 2.17 is proved in the Appendix A.

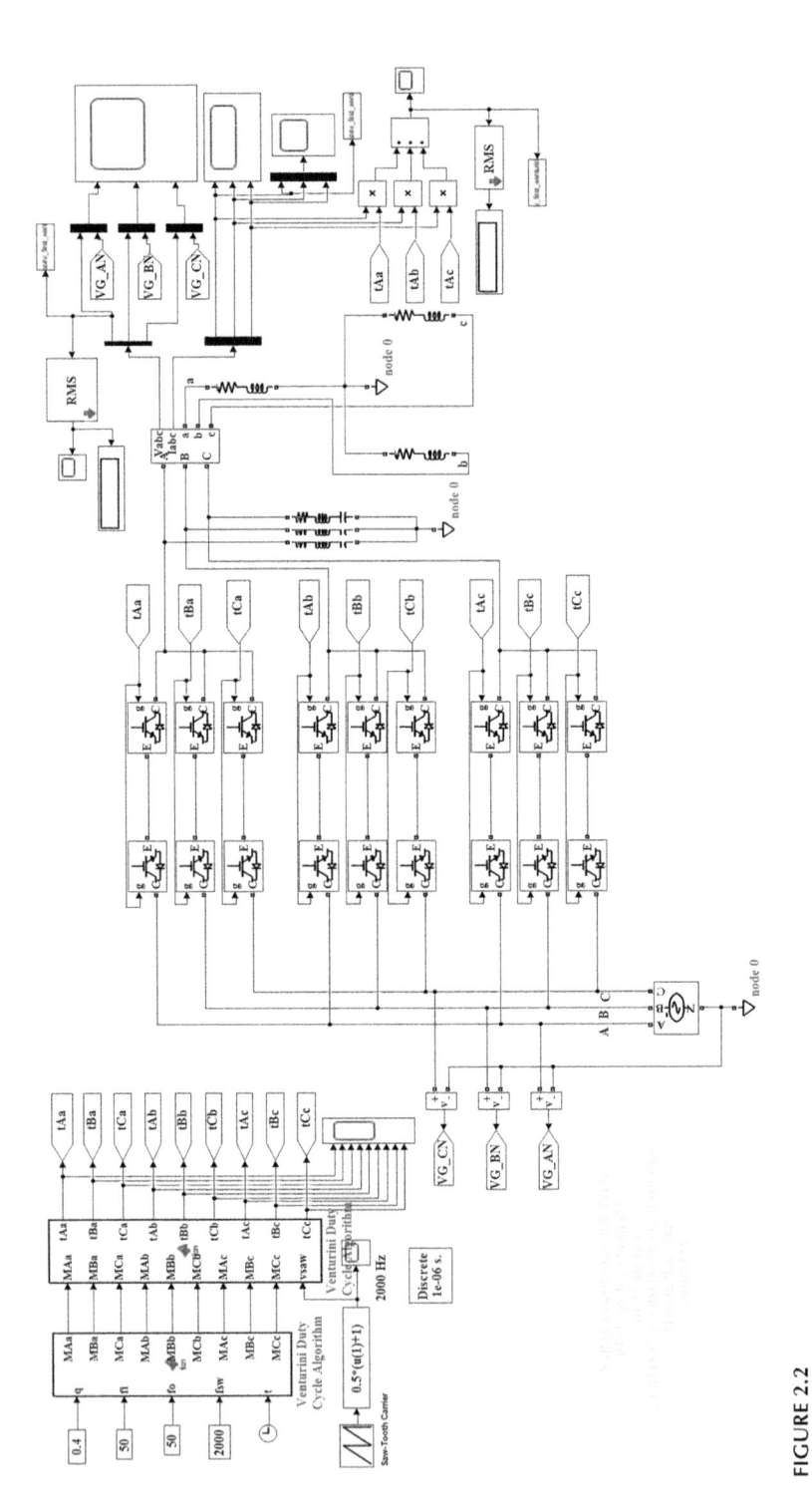

FIGURE 2.2
Model of three-phase AC to three-phase AC matrix converter using the Venturini algorithm first method.

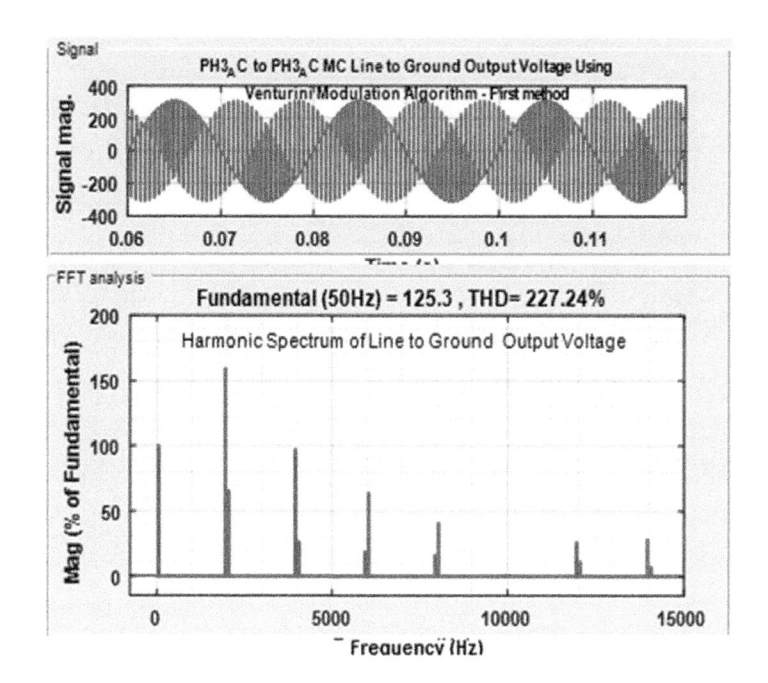

FIGURE 2.3
Line-to-neutral output voltage and harmonic spectrum – Venturini first method.

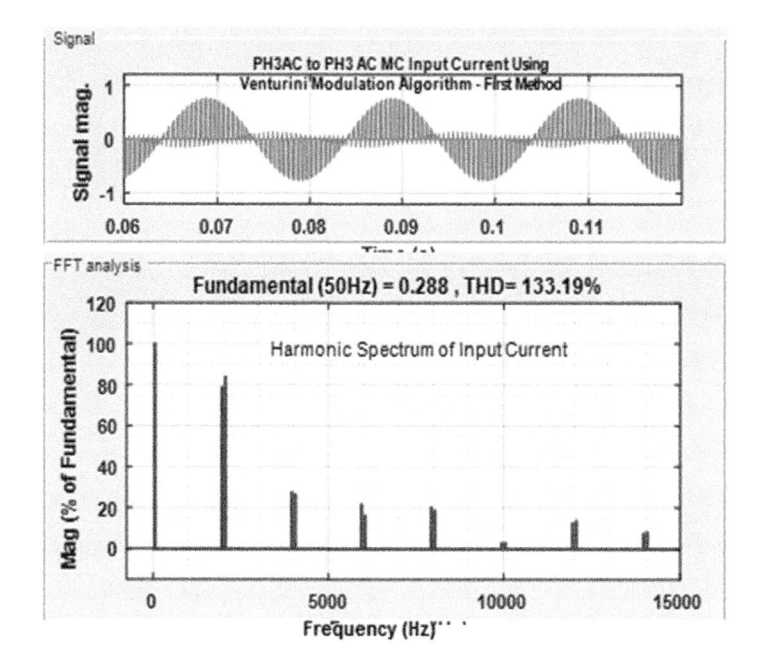

FIGURE 2.4
Phase A input current and harmonic spectrum – Venturini first method.

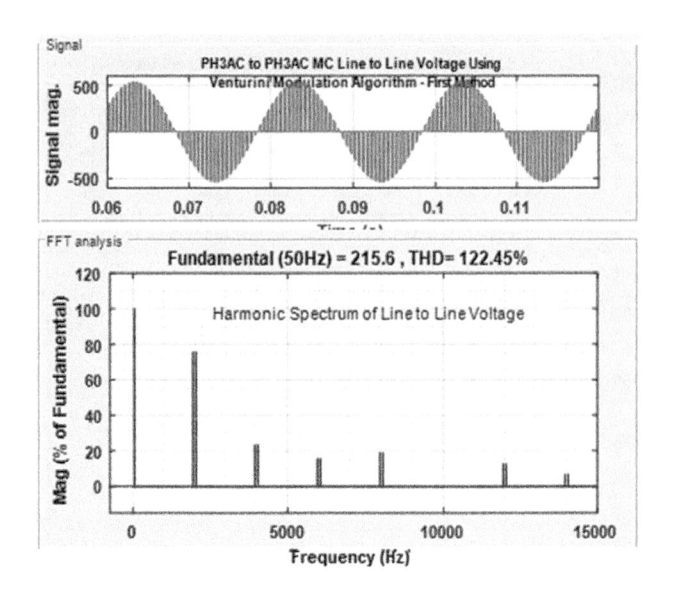

FIGURE 2.5
Line-to-line output voltage and harmonic spectrum – Venturini first method.

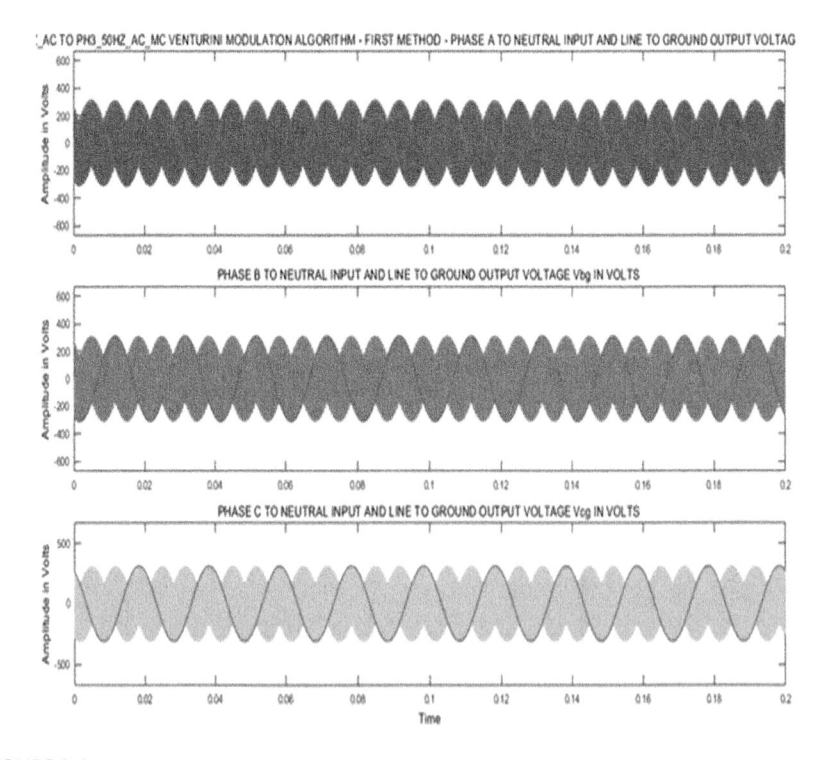

FIGURE 2.6
Three-phase line-to-neutral output voltage of matrix converter – Venturini first method.

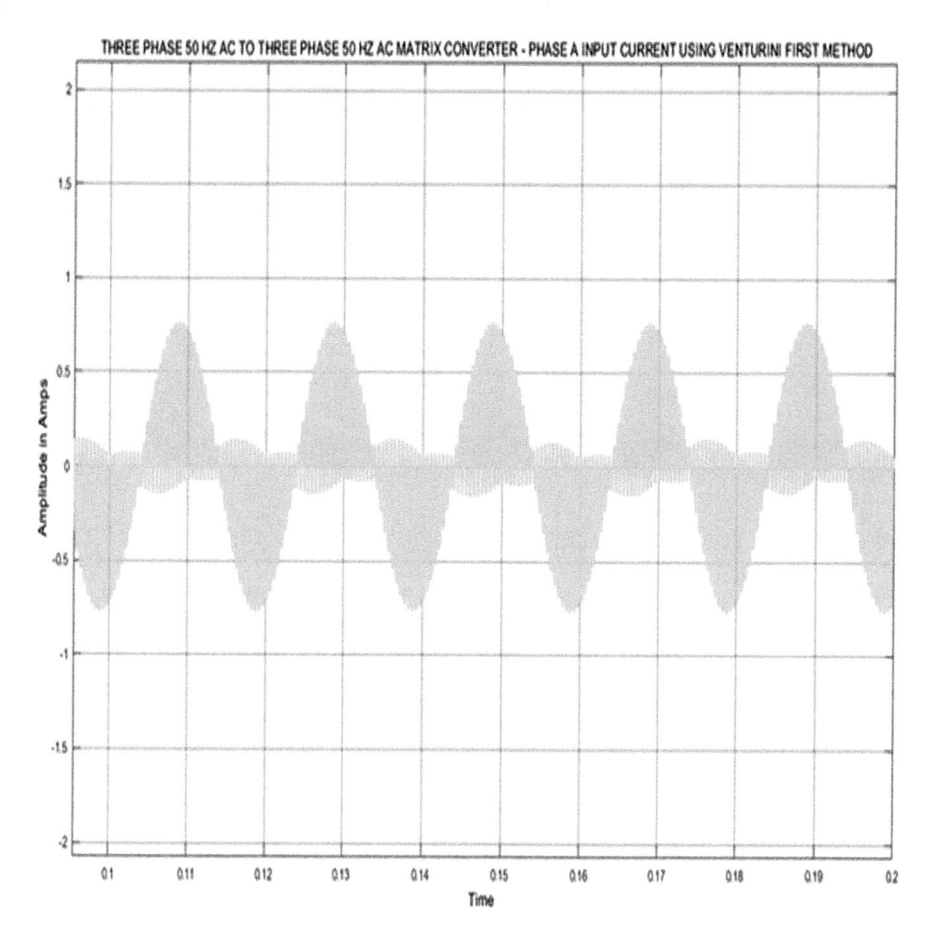

FIGURE 2.7
Phase A input current of matrix converter – Venturini first method.

2.4.4 Simulation Results

The simulation of the MC using the Venturini second method is carried out using SIMULINK [17]. The data shown in Table 2.1 with an output frequency value of 50 Hz replaced by 20 Hz are used. The variable-step ode15s (Stiff/NDF) solver is used. Harmonic spectrum of the line-to-neutral output voltage, phase A input current and line-to-line output voltage are shown in Figures 2.13–2.15, respectively. The simulation results for the three-phase line-to-neutral output voltage, phase A input current, three-phase line-to-line output voltage and three-phase load current are shown in Figures 2.16–2.19, respectively. The simulation results from the harmonic spectrum are tabulated in Table 2.3 [22–24].

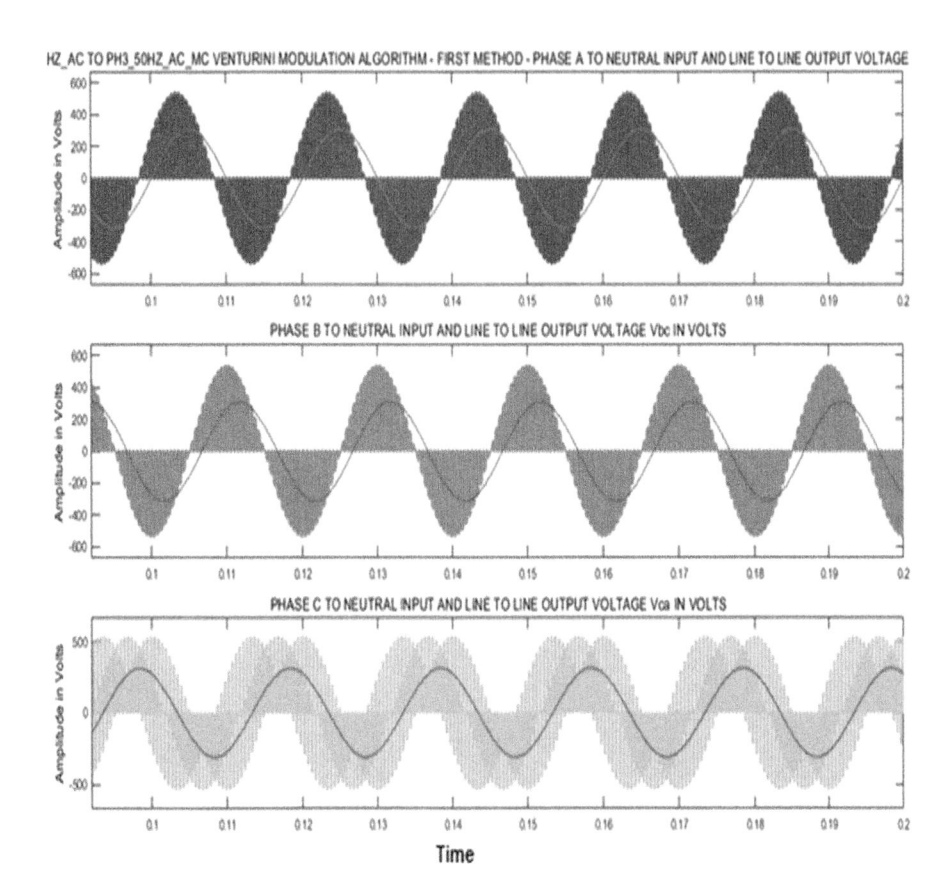

FIGURE 2.8
Three-phase line-to-line output voltage of matrix converter – Venturini first method.

Program Segment 2.3

```
function [MAa,MBa,MCa,MAb,MBb,MCb,MAc,MBc,MCc] =
fcn(q,Vim,fi,fo,t)
%% Dr.Narayanaswamy.P.R.Iyer.
%%Three Phase AC to Three Phase AC MC - Venturini algorithm
second method.
ViA =  Vim*sin(2*pi*fi*t);
ViB =  Vim*sin(2*pi*fi*t - 2*pi/(3));
ViC =  Vim*sin(2*pi*fi*t - 4*pi/(3));
voa = q*Vim*(sin(2*pi*fo*t));
vob = q*Vim*(sin(2*pi*fo*t - 2*pi/(3)));
voc = q*Vim*(sin(2*pi*fo*t - 4*pi/(3)));
%%switch modulation
MAa = (1/(3) + (2*ViA*voa)/(3*Vim*Vim));
MBa = (1/(3) + (2*ViB*voa)/(3*Vim*Vim));
MCa = (1/(3) + (2*ViC*voa)/(3*Vim*Vim));
MAb = (1/(3) + (2*ViA*vob)/(3*Vim*Vim));
```

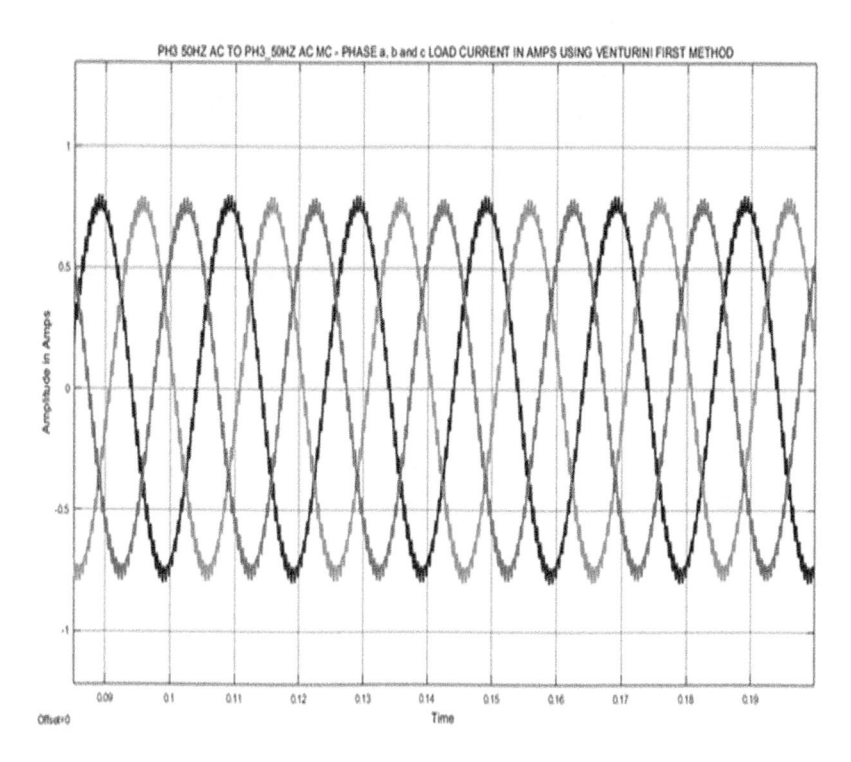

FIGURE 2.9
Three-phase load current of matrix converter – Venturini first method.

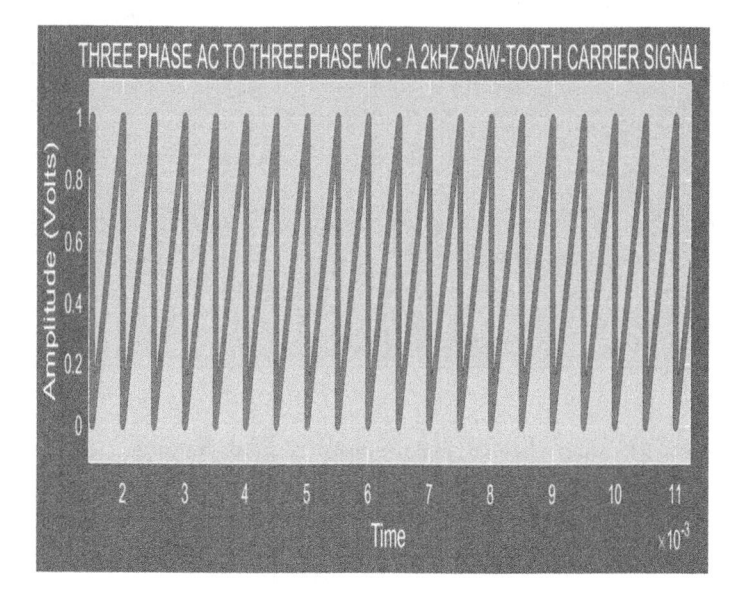

FIGURE 2.10
A 2 kHz saw-tooth carrier signal.

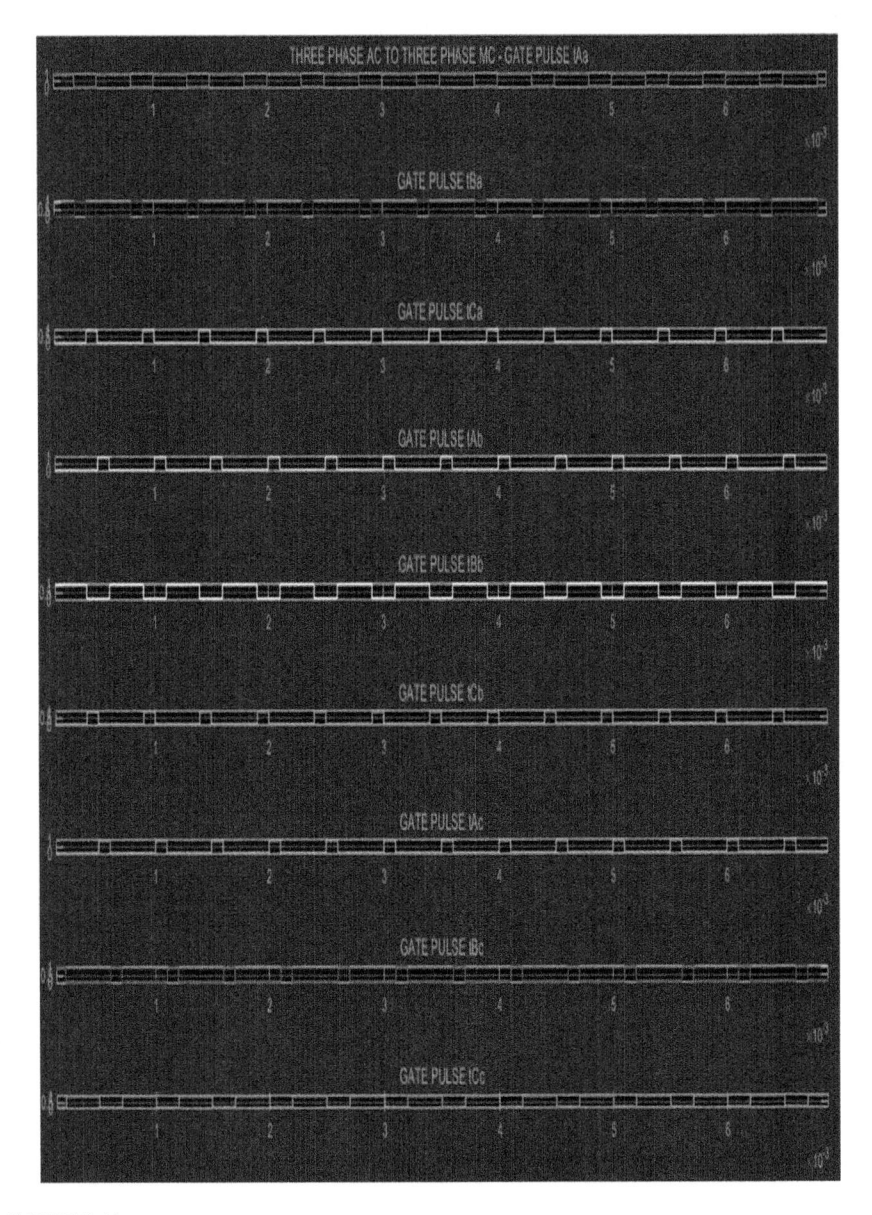

FIGURE 2.11

Gate pulse pattern for three-phase AC to three-phase AC matrix converter.

```
MBb = (1/(3) + (2*ViB*vob)/(3*Vim*Vim));
MCb = (1/(3) + (2*ViC*vob)/(3*Vim*Vim));
MAc = (1/(3) + (2*ViA*voc)/(3*Vim*Vim));
MBc = (1/(3) + (2*ViB*voc)/(3*Vim*Vim));
MCc = (1/(3) + (2*ViC*voc)/(3*Vim*Vim));
```

TABLE 2.2

Simulation Results for MC – Venturini First Method

Sl. No.	Algorithm	THD of Line-to-Neutral Output Voltage (p.u.)	THD Line-to-Line Output Voltage (p.u.)	THD of Input Current (p.u.)
1	Venturini first method	2.27	1.22	1.33

2.4.5 Model of a Matrix Converter Using the Optimum Venturini Modulation Algorithm

The model of the three-phase MC is developed using equation 2.19 with the input and output voltages as defined in equations 2.9 and 2.18, respectively, for the data given in Table 2.1. The nine modulation functions defined in equation 2.19 are developed using Embedded MATLAB Function in SIMULINK, as shown in Program segment 2.4. These modulation functions are compared with a 2 kHz saw-tooth carrier generator and the gate pulses for the nine bidirectional switches are developed satisfying the constraint in equation 2.8 using another Embedded MATLAB Function, as shown in Program segment 2.2 [22–23]. The SIMULINK model is shown in Figure 2.20.

2.4.6 Simulation Results

The simulation of the MC using the optimum Venturini method using the data in Table 2.1 is carried out using SIMULINK [17]. The variable-step ode15s (Stiff/NDF) solver is used. Harmonic spectrum of the line-to-neutral output voltage, phase A input current and line-to-line output voltage are shown in Figures 2.21–2.23, respectively. The simulation results for the three-phase line-to-neutral output voltage, three-phase line-to-line output voltage, phase A input current and three-phase load current are shown in Figures 2.24–2.27, respectively. The simulation results from the harmonic spectrum are tabulated in Table 2.4 [22–24].

2.5 Advanced Modulation Algorithm

A novel carrier-based modulation scheme is proposed by Ned Mohan et al. which requires no sector information and look-up table to calculate duty ratios, with an output voltage amplitude 0.866 times that of the input voltage and the input power factor controllable [18–21]. This algorithm is explained below:

Let the three-phase input voltages, $v_i = \begin{bmatrix} v_A & v_B & v_C \end{bmatrix}^T$ be defined as in equation 2.20 and the corresponding output phase voltages $v_o = \begin{bmatrix} v_a & v_b & v_c \end{bmatrix}^T$ be defined as in equation 2.21 given below.

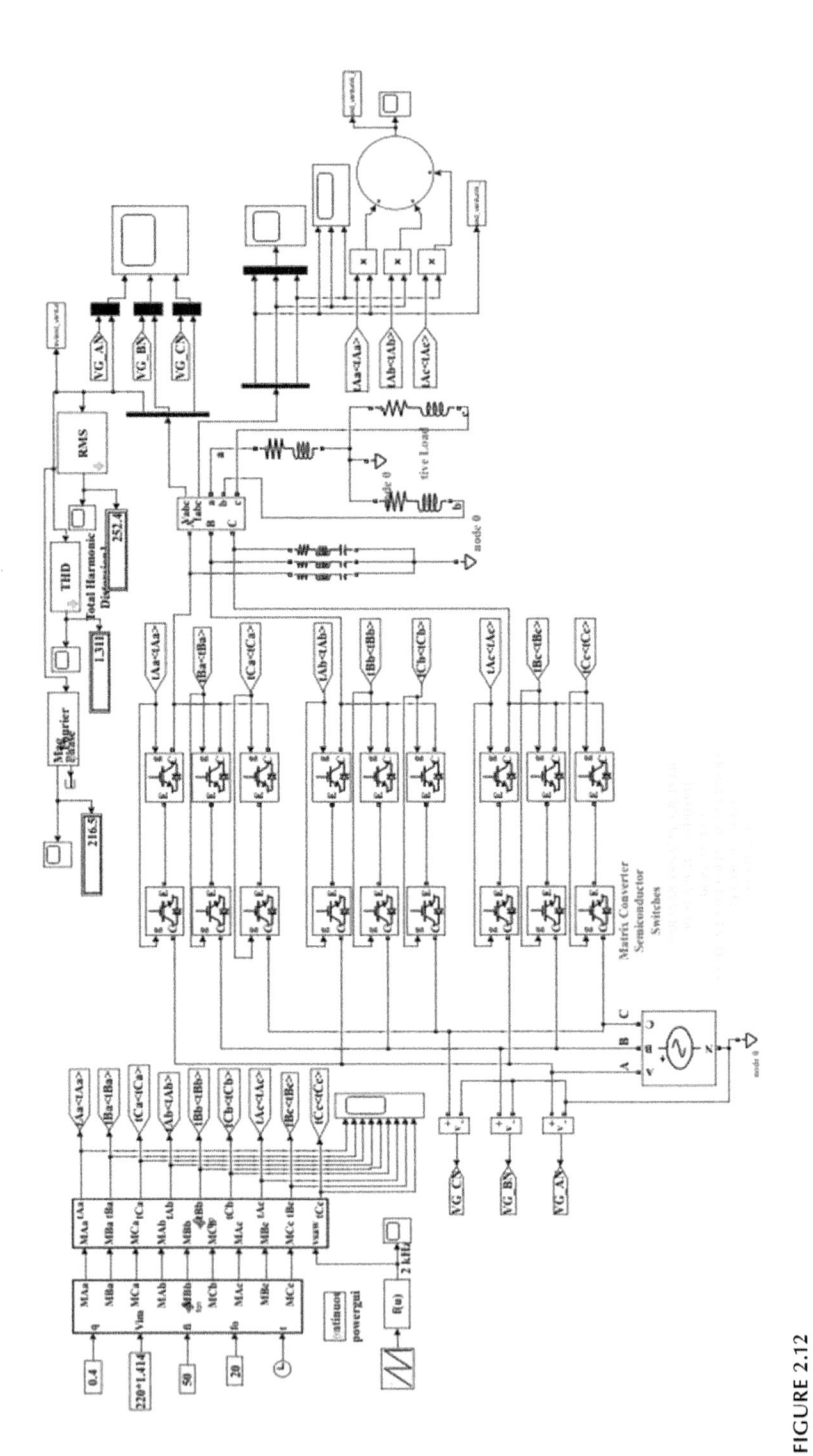

FIGURE 2.12
Model for three-phase AC to three-phase AC matrix converter using Venturini second method.

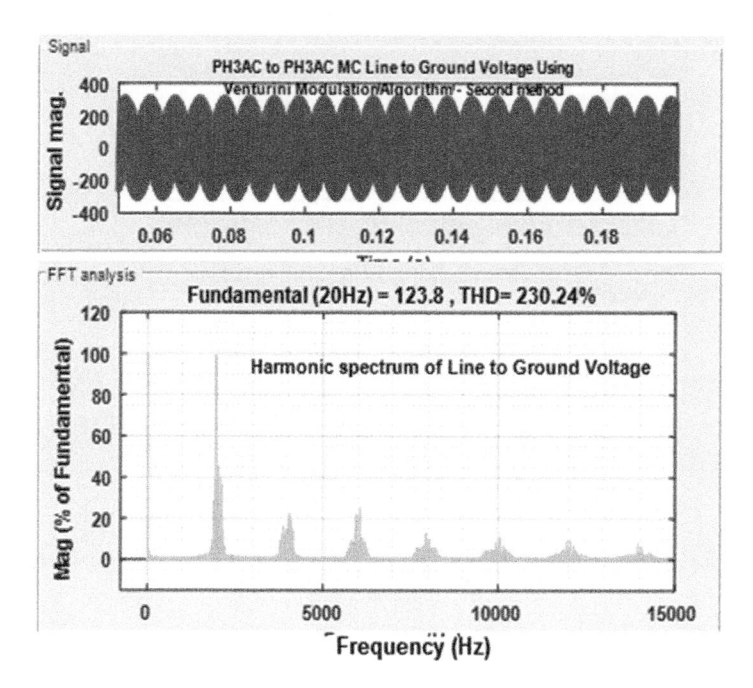

FIGURE 2.13
Line-to-neutral output voltage and harmonic spectrum – Venturini second method.

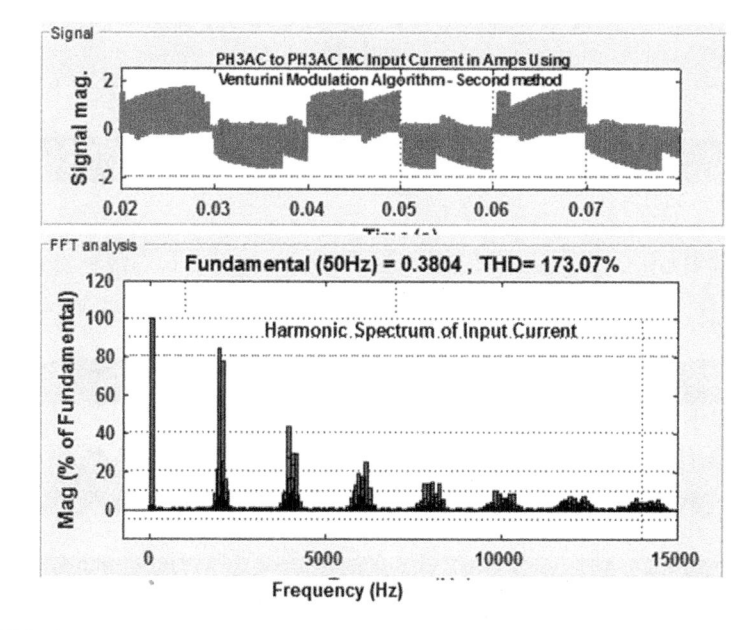

FIGURE 2.14
Phase A input current and harmonic spectrum – Venturini second method.

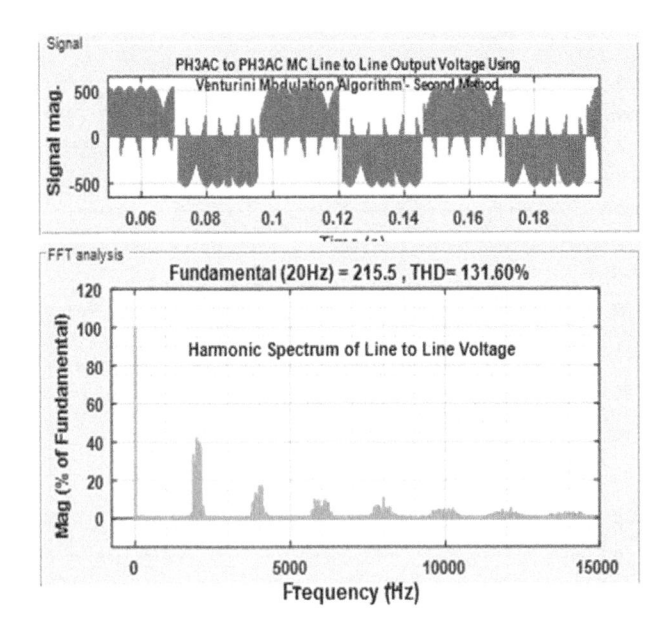

FIGURE 2.15
Line-to-line output voltage and harmonic spectrum – Venturini second method.

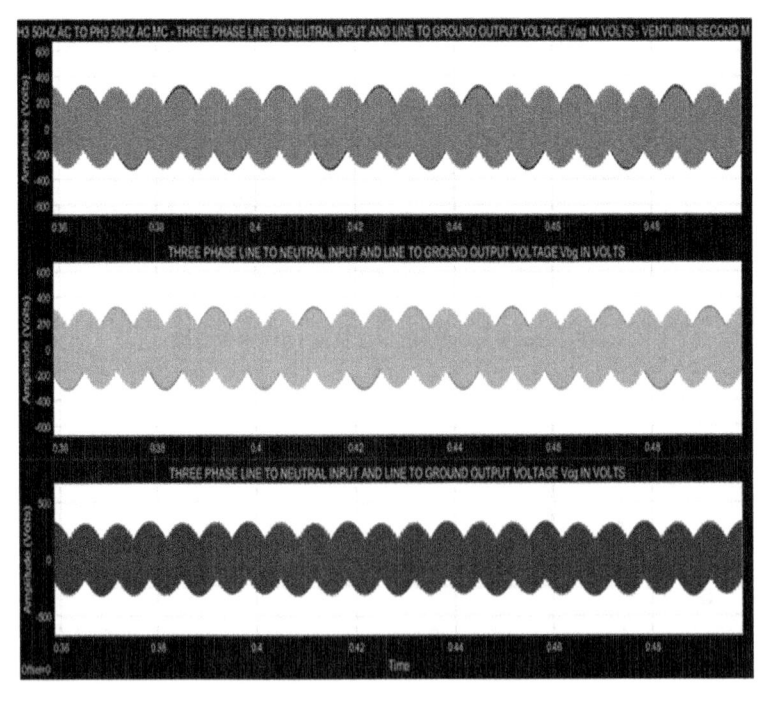

FIGURE 2.16
Three-phase line-to-neutral output voltage of matrix converter – Venturini second method.

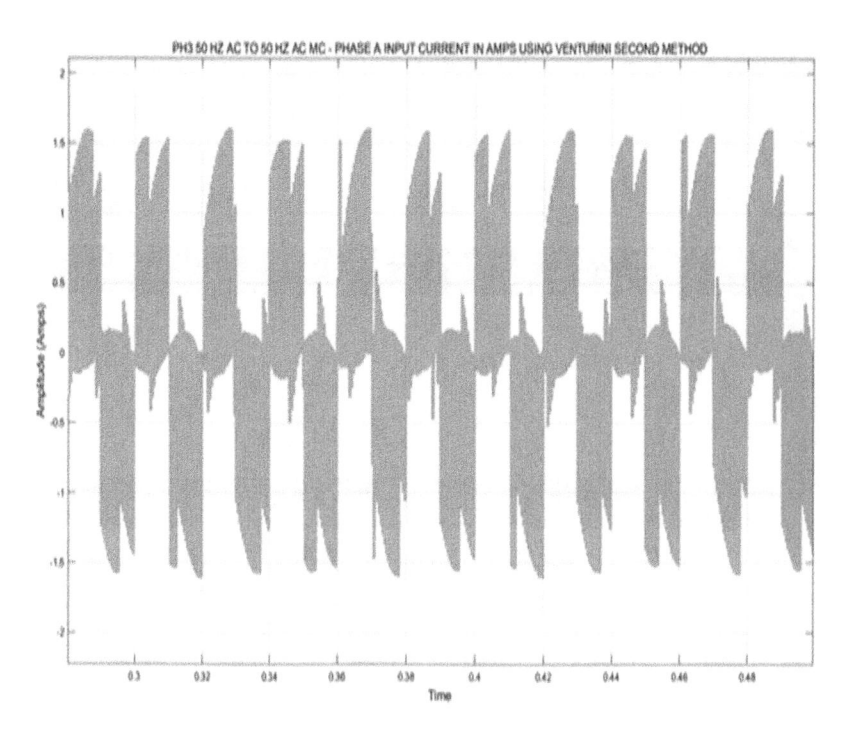

FIGURE 2.17
Phase A input current of matrix converter – Venturini second method.

Program Segment 2.4

```
function [MAa,MBa,MCa,MAb,MBb,MCb,MAc,MBc,MCc] =
fcn(q,Vim,fi,fo,t)
%% Dr.Narayanaswamy.P.R.Iyer.
%%%Three Phase AC to Three Phase AC MC Using Optimum Venturini
Modulation %%Algorithm.
ViA = Vim*cos(2*pi*fi*t);
ViB =  Vim*cos(2*pi*fi*t + 4*pi/(3));
ViC =  Vim*cos(2*pi*fi*t + 2*pi/(3));
voa = q*Vim*(cos(2*pi*fo*t) - (cos(3*2*pi*fo*t)/(6)) +
(cos(3*2*pi*fi*t)/(2*sqrt(3))));
vob = q*Vim*(cos(2*pi*fo*t + 4*pi/(3)) - (cos(3*2*pi*fo*t)/
(6)) + (cos(3*2*pi*fi*t)/(2*sqrt(3))));
voc = q*Vim*(cos(2*pi*fo*t + 2*pi/(3)) - (cos(3*2*pi*fo*t)/
(6)) + (cos(3*2*pi*fi*t)/(2*sqrt(3))));
%%switch modulation
MAa = (1/(3) + (2*ViA*voa)/(3*Vim*Vim) + (4*q*(sin(2*pi*fi*t))
*sin(3*2*pi*fi*t))/(9*sqrt(3)));
MBa = (1/(3) + (2*ViB*voa)/(3*Vim*Vim) + (4*q*(sin(2*pi*fi*t +
2*pi/(3)))*sin(3*2*pi*fi*t))/(9*sqrt(3)));
MCa = (1/(3) + (2*ViC*voa)/(3*Vim*Vim) + (4*q*(sin(2*pi*fi*t +
4*pi/(3)))*sin(3*2*pi*fi*t))/(9*sqrt(3)));
```

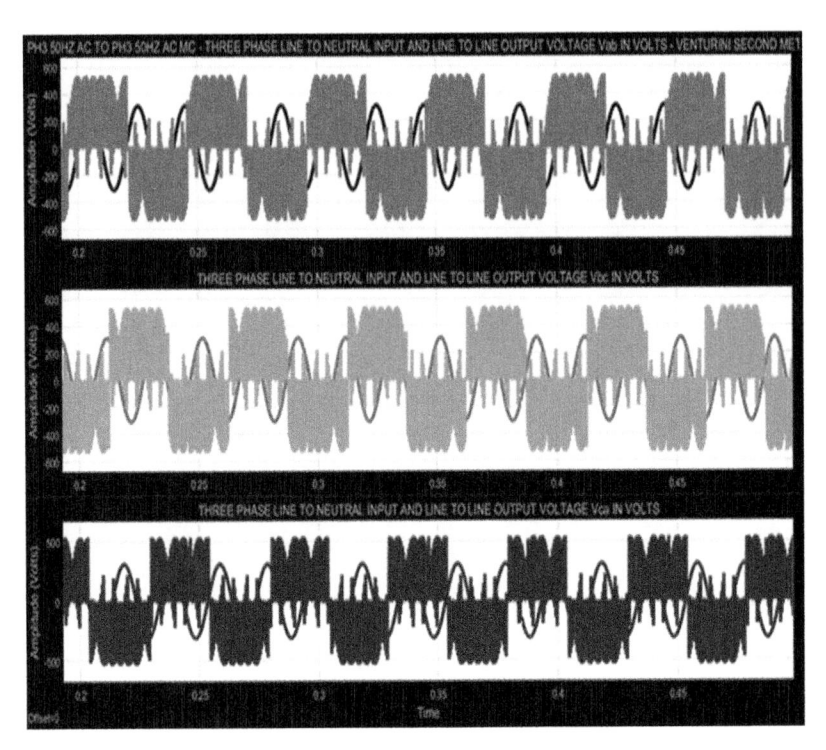

FIGURE 2.18
Three-phase line-to-line voltage of matrix converter – Venturini second method.

```
MAb = (1/(3) + (2*ViA*vob)/(3*Vim*Vim) + (4*q*(sin(2*pi*fi*t))
*sin(3*2*pi*fi*t))/(9*sqrt(3)));
MBb = (1/(3) + (2*ViB*vob)/(3*Vim*Vim) + (4*q*(sin(2*pi*fi*t +
2*pi/(3)))*sin(3*2*pi*fi*t))/(9*sqrt(3)));
MCb = (1/(3) + (2*ViC*vob)/(3*Vim*Vim) + (4*q*(sin(2*pi*fi*t +
4*pi/(3)))*sin(3*2*pi*fi*t))/(9*sqrt(3)));
MAc = (1/(3) + (2*ViA*voc)/(3*Vim*Vim) + (4*q*(sin(2*pi*fi*t))
*sin(3*2*pi*fi*t))/(9*sqrt(3)));
MBc = (1/(3) + (2*ViB*voc)/(3*Vim*Vim) + (4*q*(sin(2*pi*fi*t +
2*pi/(3)))*sin(3*2*pi*fi*t))/(9*sqrt(3)));
MCc = (1/(3) + (2*ViC*voc)/(3*Vim*Vim) + (4*q*(sin(2*pi*fi*t +
4*pi/(3)))*sin(3*2*pi*fi*t))/(9*sqrt(3)));
```

$$v_i = V_{im} * \begin{bmatrix} \sin(\omega_i t) \\ \sin\left(\omega_i t - \dfrac{2\pi}{3}\right) \\ \sin\left(\omega_i t - \dfrac{4\pi}{3}\right) \end{bmatrix} \tag{2.20}$$

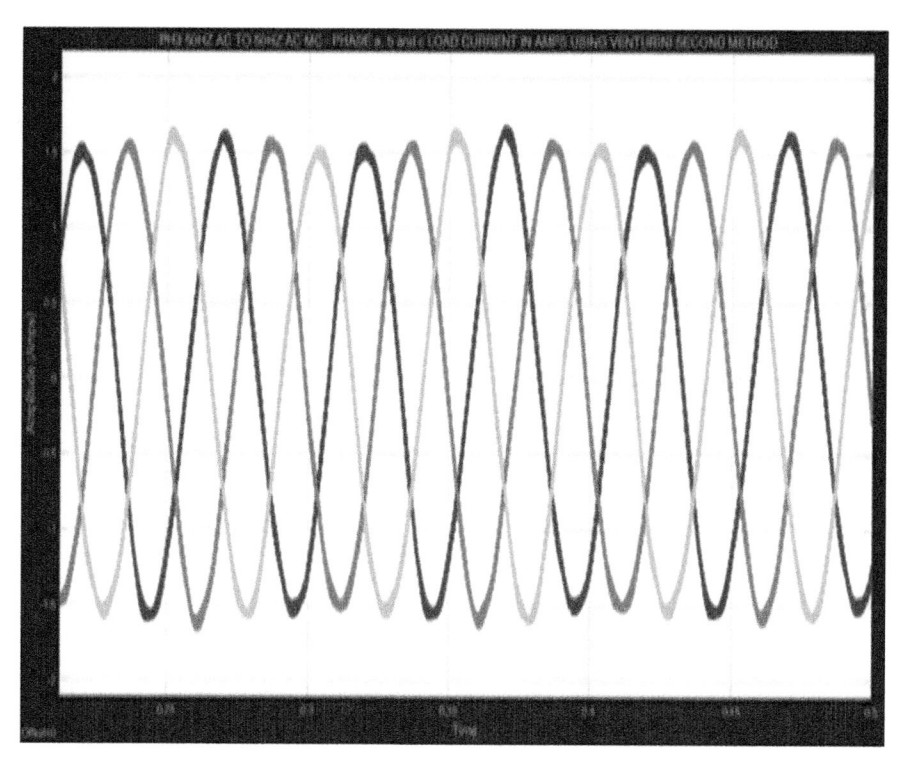

FIGURE 2.19
Three-phase load current of matrix converter – Venturini second method.

TABLE 2.3

Simulation Results for MC – Venturini Second Method

Sl. No.	Algorithm	THD of Line-to-Neutral Output Voltage (p.u.)	THD Line-to-Line Output Voltage (p.u.)	THD of Input Current (p.u.)
1	Venturini second method	2.30	1.316	1.73

$$
v_o = q * V_{im} * \begin{bmatrix} \sin\left(\omega_o t + \varphi_o\right) \\ \sin\left(\omega_o t + \varphi_o - \dfrac{2\pi}{3}\right) \\ \sin\left(\omega_o t + \varphi_o - \dfrac{4\pi}{3}\right) \end{bmatrix} \qquad (2.21)
$$

The duty ratios are chosen such that the output voltages are independent of the input frequency. This is possible by considering the input voltages in stationary reference frame and the output voltages in synchronous

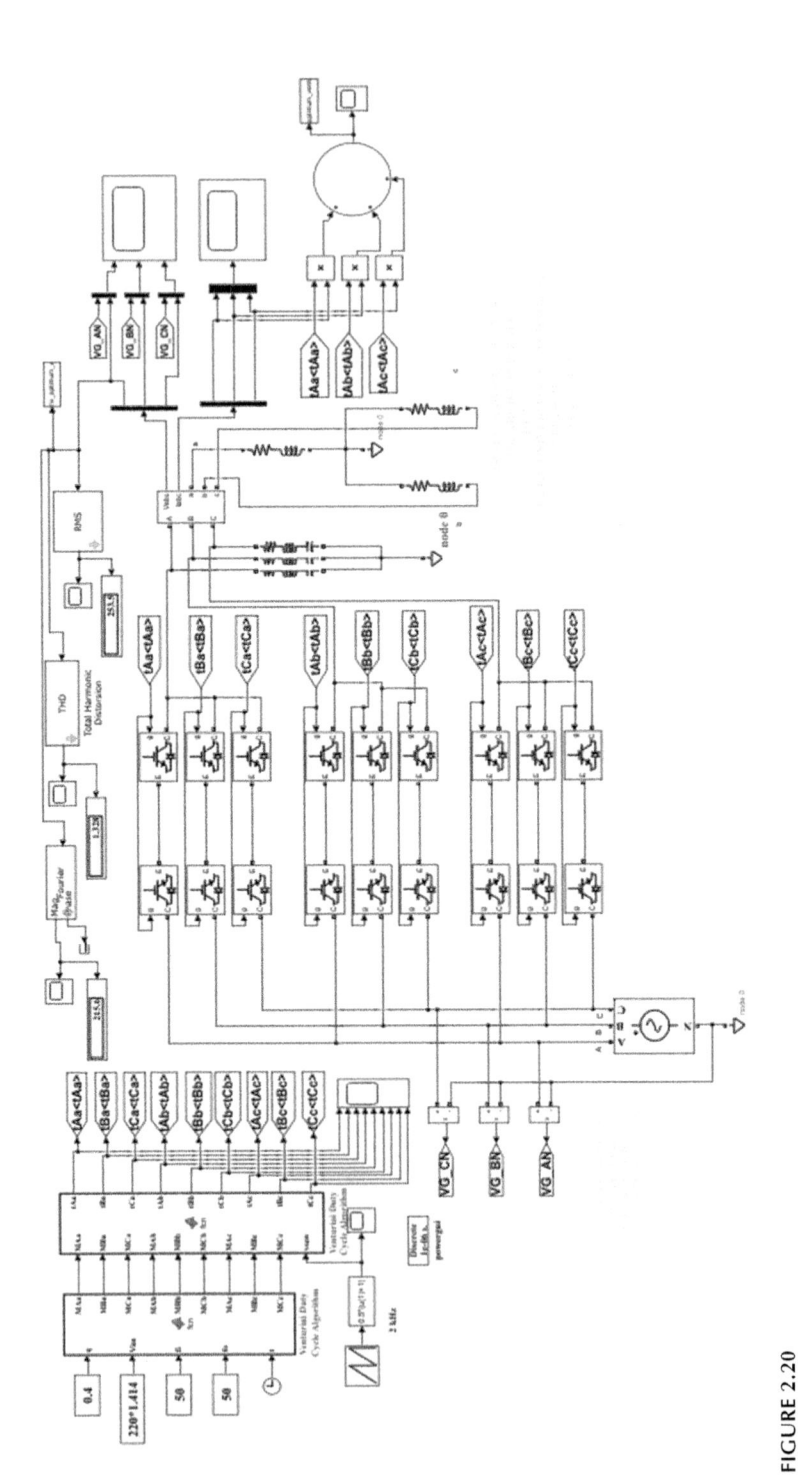

FIGURE 2.20
Model of optimum Venturini modulation algorithm for three-phase AC to three-phase AC matrix converter.

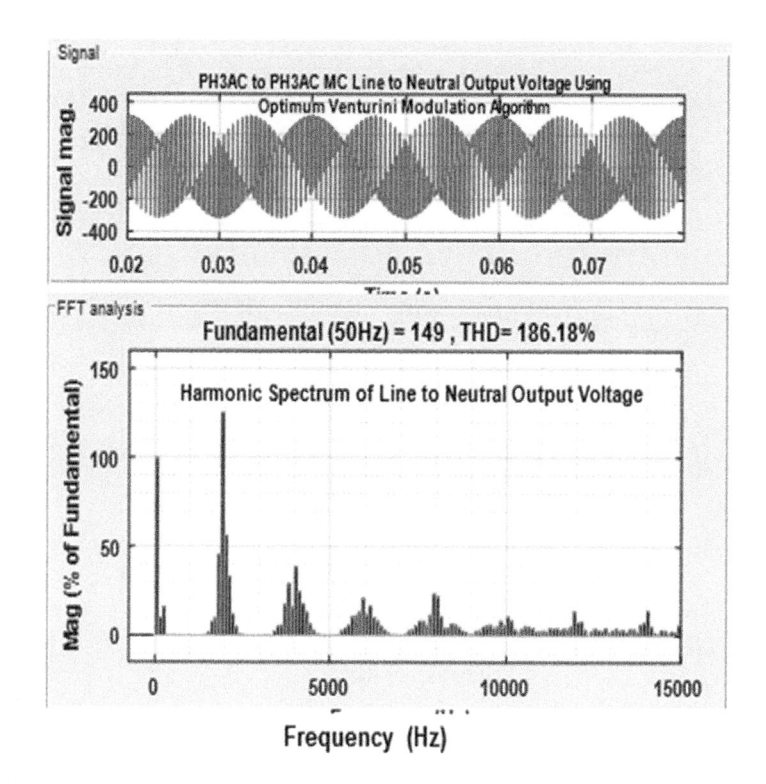

FIGURE 2.21
Line-to-neutral output voltage and harmonic spectrum – optimum Venturini modulation algorithm.

reference frame. Hence, M_{Aa}, M_{Ba} and M_{Ca} are chosen as given in the following equation:

$$\begin{bmatrix} M_{Aa} \\ M_{Ba} \\ M_{Ca} \end{bmatrix} = \begin{bmatrix} k_a * \sin\left(\omega_i t - \varphi_i\right) \\ k_a * \sin\left(\omega_i t - \varphi_i - \dfrac{2\pi}{3}\right) \\ k_a * \sin\left(\omega_i t - \varphi_i - \dfrac{4\pi}{3}\right) \end{bmatrix} \qquad (2.22)$$

Simplifying equations 2.20–2.22, the output voltage equation v_a for phase a reduces to:

$$v_a = \frac{3}{2} k_a V_{im} * \cos\left(\varphi_i\right) \qquad (2.23)$$

Equation 2.23 shows that the output phase voltage v_a is independent of the input frequency but is dependent only on the amplitude of the input voltage.

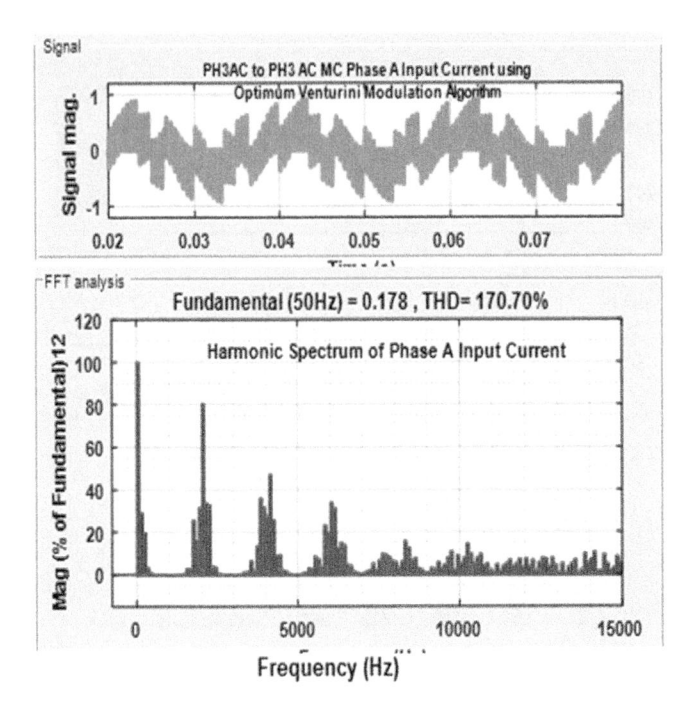

FIGURE 2.22
Phase A input current and harmonic spectrum – optimum Venturini modulation algorithm.

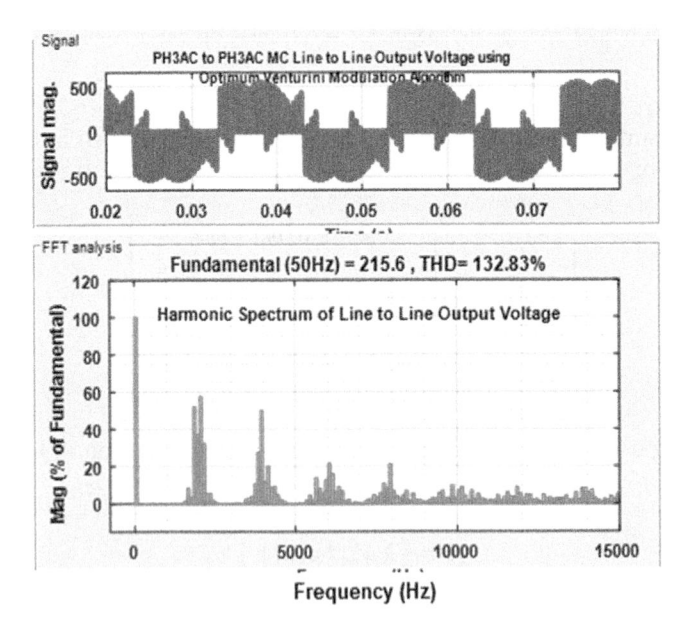

FIGURE 2.23
Line-to-line output voltage and harmonic spectrum – optimum Venturini modulation algorithm.

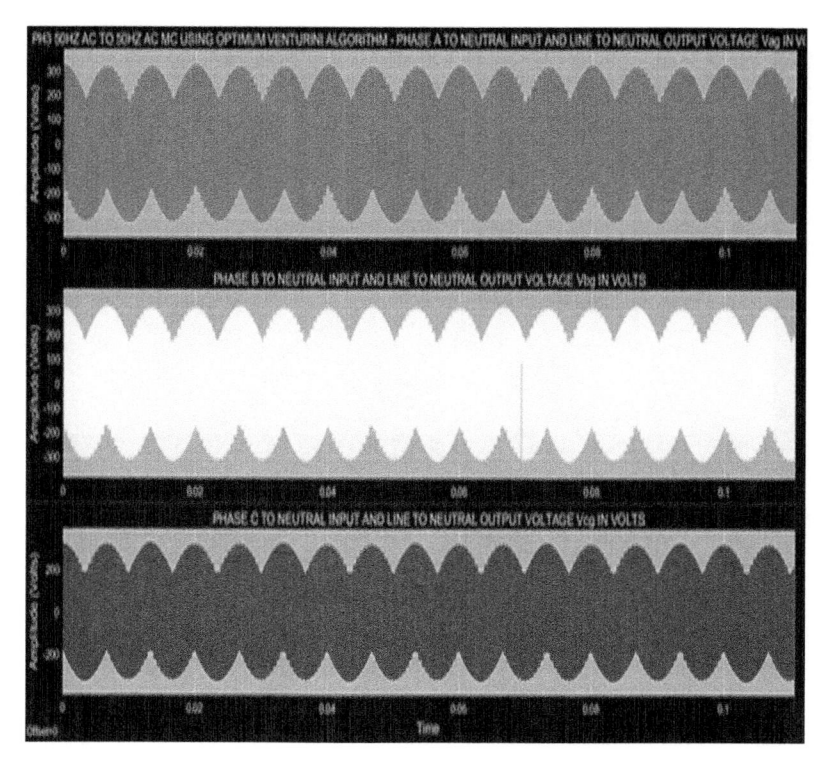

FIGURE 2.24
Three-phase line-to-neutral output voltage of matrix converter – optimum Venturini modulation algorithm.

The modulation index k_a is a function of the output frequency ω_o as defined below:

$$
\begin{bmatrix} k_a \\ k_b \\ k_c \end{bmatrix} = \begin{bmatrix} k * \sin(\omega_o t) \\ k * \sin\left(\omega_o t - \dfrac{2\pi}{3}\right) \\ k * \sin\left(\omega_o t - \dfrac{4\pi}{3}\right) \end{bmatrix}
$$
(2.24)

where k_a, k_b and k_c are the modulation indices for phase a, b and c, respectively.

Using equations 2.23 and 2.24, the output phase voltage v_a simplifies to:

$$
v_a = \left[\frac{3}{2} k V_{im} * \cos(\varphi_i) \sin(\omega_o t) \right]
$$
(2.25)

From equations 2.22 and 2.24, it is clear that the duty ratios of the switches take negative values. But the requirement is that the duty ratios of the

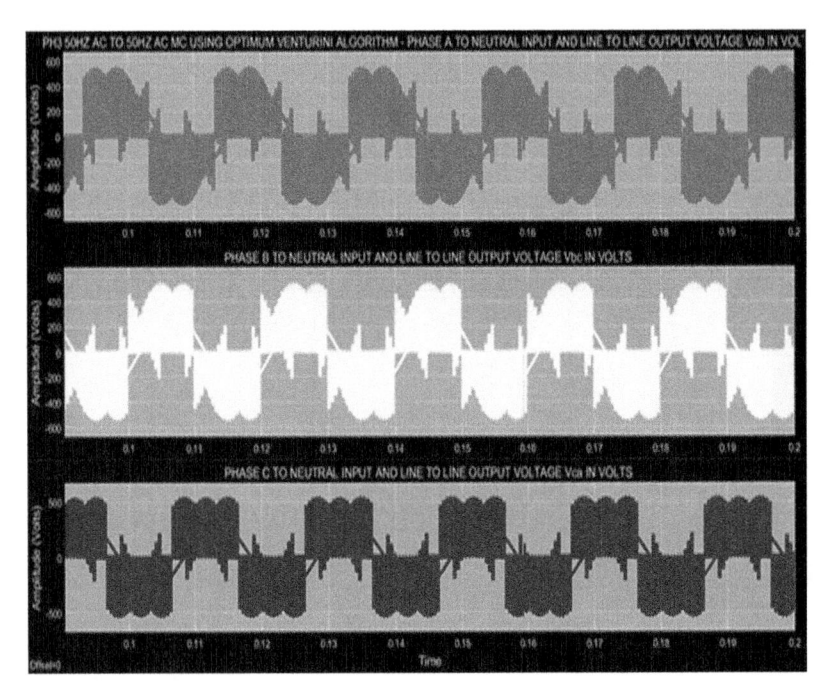

FIGURE 2.25
Three-phase line-to-line output voltage of matrix converter – optimum Venturini modulation method.

switches must lie in the range of 0–1. This is made possible by adding offset duty ratios to the existing duty ratios. Thus, absolute values of duty ratios are added. The offset duty ratio is defined by

$$
\begin{bmatrix}
D_A(t) \\
D_B(t) \\
D_C(t)
\end{bmatrix}
=
\begin{bmatrix}
\left| k_a * \sin(\omega_i t - \varphi_i) \right| \\
\left| k_a * \sin\left(\omega_i t - \varphi_i - \dfrac{2\pi}{3}\right) \right| \\
\left| k_a * \sin\left(\omega_i t - \varphi_i - \dfrac{4\pi}{3}\right) \right|
\end{bmatrix}
\tag{2.26}
$$

Thus, the new duty ratios are defined as follows:

$$
\begin{bmatrix}
M_{Aa} \\
M_{Ba} \\
M_{Ca}
\end{bmatrix}
=
\begin{bmatrix}
D_A(t) + k_a * \sin(\omega_i t - \varphi_i) \\
D_B(t) + k_a * \sin\left(\omega_i t - \varphi_i - \dfrac{2\pi}{3}\right) \\
D_C(t) + k_a * \sin\left(\omega_i t - \varphi_i - \dfrac{4\pi}{3}\right)
\end{bmatrix}
\tag{2.27}
$$

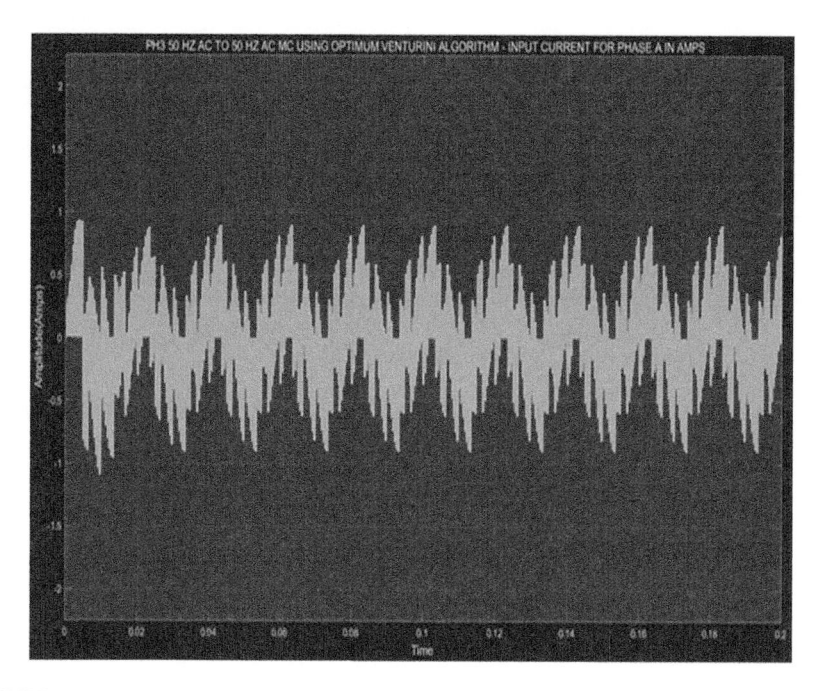

FIGURE 2.26
Phase A input current of matrix converter – optimum Venturini method.

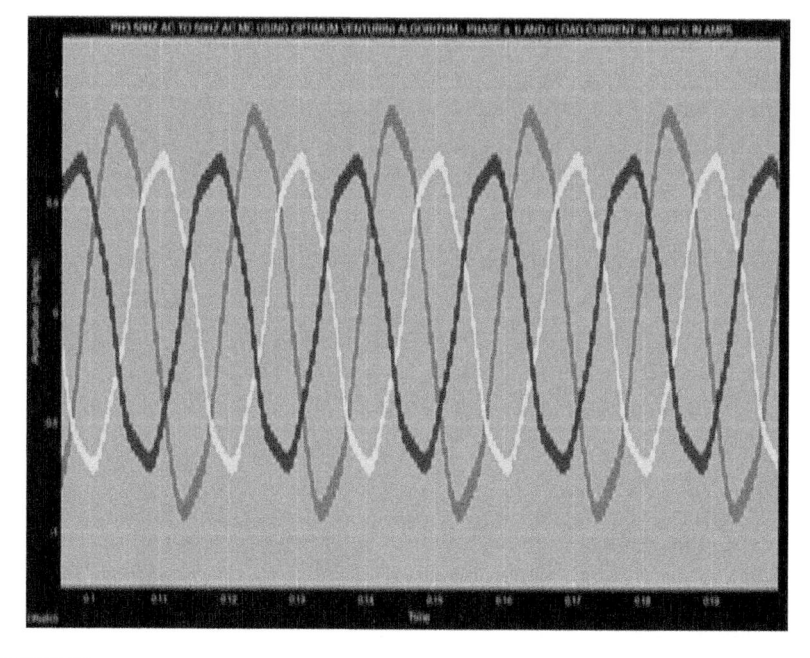

FIGURE 2.27
Three-phase load current of matrix converter – optimum Venturini method.

TABLE 2.4

Simulation Results for MC – Optimum Venturini Method

Sl. No.	Algorithm	THD of Line-to-Neutral Output Voltage (p.u.)	THD Line-to-Line Output Voltage (p.u.)	THD of Input Current (p.u.)
1	Optimum Venturini	1.86	1.328	1.707

Using equation 2.26 in 2.27, in order that the new duty ratio in equation 2.27 lies in the range of 0–1, the following inequality should be satisfied:

$$0 < 2|k_a| = 2k < 1 \tag{2.28}$$

Thus, the maximum value of k_a and k can be 0.5. Using this value, the offset duty ratios are chosen as given below:

$$\begin{bmatrix} D_A(t) \\ D_B(t) \\ D_C(t) \end{bmatrix} = \begin{bmatrix} |0.5 * \sin(\omega_i t - \varphi_i)| \\ \left|0.5 * \sin\left(\omega_i t - \varphi_i - \dfrac{2\pi}{3}\right)\right| \\ \left|0.5 * \sin\left(\omega_i t - \varphi_i - \dfrac{4\pi}{3}\right)\right| \end{bmatrix} \tag{2.29}$$

To utilize the input voltage capability to the full extend, additional common-mode voltage term is added which gives the new modulation index as given below:

$$\begin{bmatrix} M_{Aa} \\ M_{Ba} \\ M_{Ca} \end{bmatrix}$$

$$= \begin{bmatrix} D_A(t) + \left[k_a - \{\max(k_a,k_b,k_c) + \min(k_a,k_b,k_c)\}/2\right] * \sin(\omega_i t - \varphi_i) \\ D_B(t) + \left[k_a - \{\max(k_a,k_b,k_c) + \min(k_a,k_b,k_c)\}/2\right] * \sin\left(\omega_i t - \varphi_i - \dfrac{2\pi}{3}\right) \\ D_C(t) + \left[k_a - \{\max(k_a,k_b,k_c) + \min(k_a,k_b,k_c)\}/2\right] * \sin\left(\omega_i t - \varphi_i - \dfrac{4\pi}{3}\right) \end{bmatrix} \tag{2.30}$$

Similar calculations apply to the other two output phase b and c.

To calculate input power factor, the input current is represented as a function of duty ratios and output currents, as defined in equation 2.7. Hence, the input current in phase A can be expressed as follows:

$$i_A = (k_a i_a + k_b i_b + k_c i_c) * \sin(\omega_i t - \varphi_i) \tag{2.31}$$

In equation 2.31, the modulation index and output currents are at output frequency. Thus, equation 2.31 simplifies to:

$$i_A = \left(\frac{3}{2} k I_O * \cos(\varphi_O) \right) * \sin(\omega_i t - \varphi_i) \qquad (2.32)$$

where I_O is the amplitude of the output current and φ_O is the output power factor angle.

Comparing equation 2.32 with the input phase voltage v_A, it is seen that the input current lags the input phase voltage by an angle of φ_i. Thus, φ_i is chosen to be zero for unity power factor operation. Also from equation 2.25, we have

$$q = \frac{3}{2} k * \cos(\varphi_i) \qquad (2.33)$$

The switching signals corresponding to output phase a are obtained by comparing M_{Aa} and $(M_{Aa} + M_{Ba})$ with a triangular carrier signal whose peak value is one and minimum value is zero. The resulting PWM signals are given to logic gates to obtain the switching pulses for the MC.

2.5.1 Model Development

A model of the three-phase MC using the above advanced modulation algorithm was developed in SIMULINK [17]. The input power factor is unity. The value of modulation index k used is 0.26667 for a q value of 0.4 using equation 2.33. The triangle carrier has a peak value of one and a minimum value of zero. The values for the input voltage, input frequency, output frequency and triangle carrier frequency are shown in Table 2.5.

2.5.2 Model of a Matrix Converter Using the Advanced Modulation Algorithm

The SIMULINK model of the MC using the advanced modulation algorithm for three-phase sine-wave input voltages is shown in Figure 2.28 [22–23]. The modulation index k, input frequency f_i, output frequency f_o, time t and input p.f. angle φ_i are given as inputs to calculate k_a, k_b, k_c and $D_A(t)$, $D_B(t)$ and $D_C(t)$ according to equations 2.24 and 2.29 using Embedded MATLAB Function. This is shown in Program segment 2.5. The minimum and maximum values

TABLE 2.5

Parameters for Matrix Converter

Sl. No.	Modulation Index k	RMS Line-to-Neutral Input Voltage (V)	Input Frequency (Hz)	Output Frequency (Hz)	Carrier Frequency (kHz)
1	0.26667	220	50	20	2

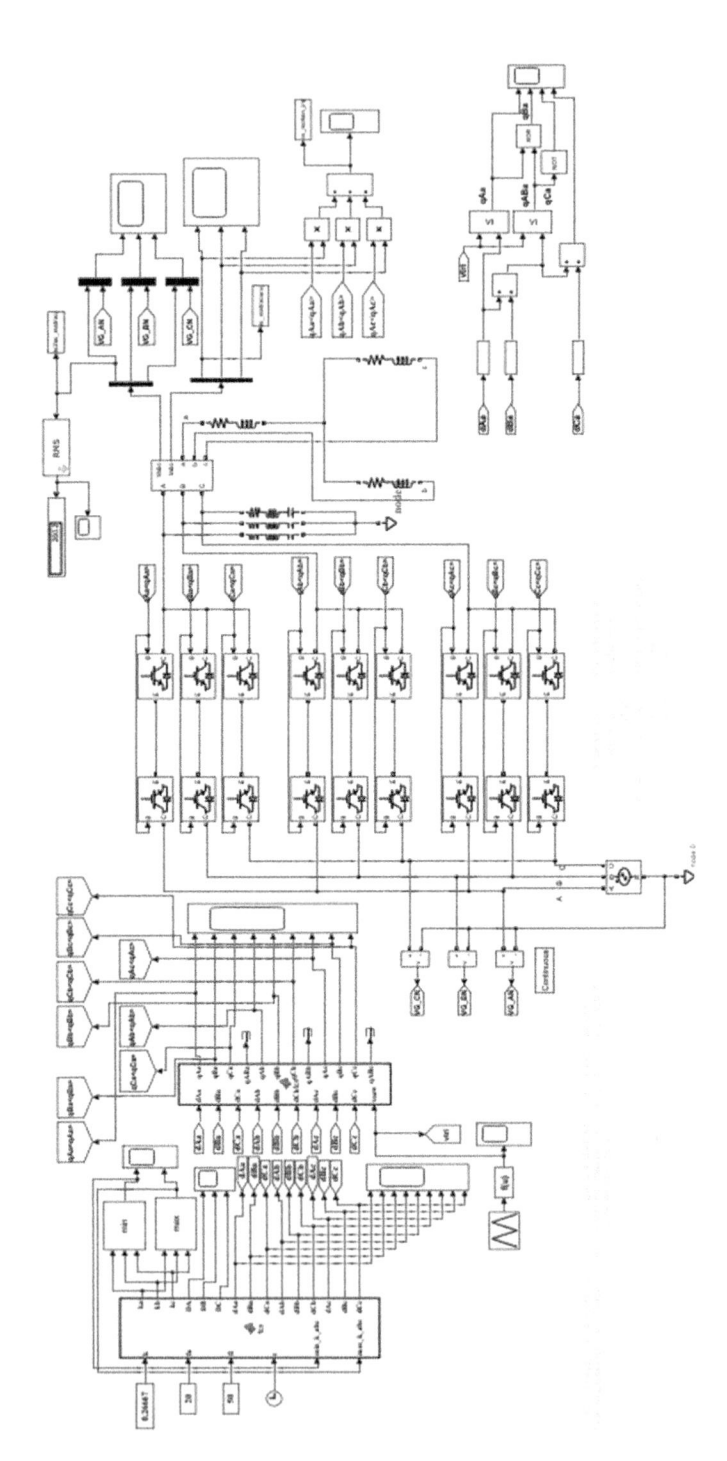

FIGURE 2.28
Model of three-phase AC to three-phase AC matrix converter with three-phase sine-wave voltage input using advanced modulation algorithm.

of the outputs k_a, k_b and k_c are calculated using two MinMax blocks in the SIMULINK block set. The two resulting outputs min_k_abc and max_k_abc together with the above calculated values for $D_A(t)$, $D_B(t)$ and $D_C(t)$ and the input p.f. angle φ_i are used to calculate the duty ratios of the switches according to equation 2.30 using the same Embedded MATLAB Function.

A 2 kHz triangle carrier with a peak value and minimum value of one and zero volt is generated using a Triangle Generator block and an Fcn block. The nine calculated values of the duty ratios for the switches and the 2 kHz triangle carrier are given as inputs to another Embedded MATLAB Function to generate the gate switching pulses for the nine bidirectional switches. The source code for generating the gate switching pulses is given in Program segment 2.6. A method of generating the gate switching pulses using adders, comparators and logic gates is also shown in Figure 2.28. The three-phase sine-wave generator, the nine IGBT bidirectional switch matrix, output series RLC filter, R–L load, etc. are the same as explained in Section 2.4.1.

2.5.3 Simulation Results

The simulation of the model shown in Figure 2.28 was carried out using SIMULINK [17]. The ode15s (stiff/NDF) solver is used. The simulation results for the harmonic spectrum of line-to-neutral output voltage, input current and line-to-line output voltage are shown in Figures 2.29–2.31, respectively. Simulation results for the line-to-neutral output voltage, input current, line-to-line output voltage and load current are shown in Figures 2.32–2.35.

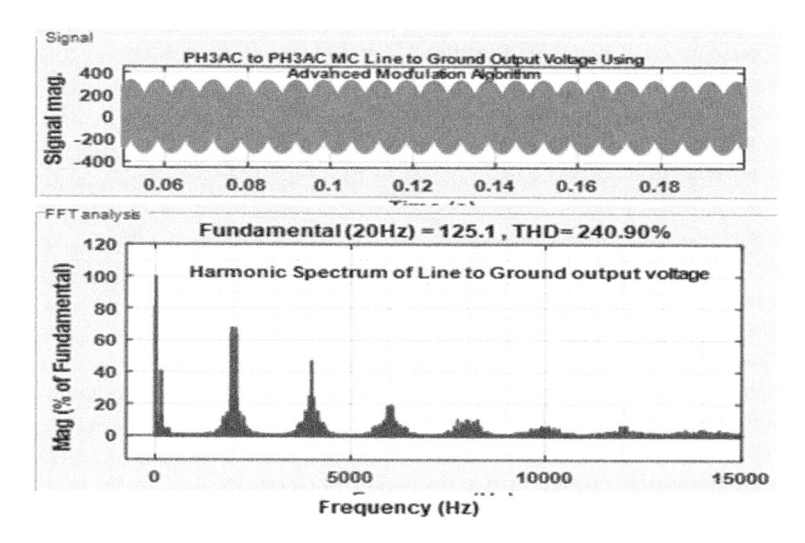

FIGURE 2.29
Line-to-neutral output voltage and harmonic spectrum-advanced modulation algorithm.

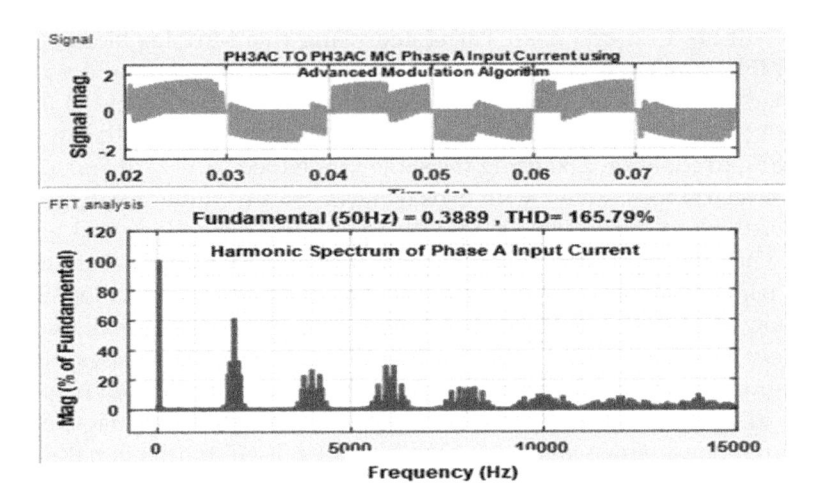

FIGURE 2.30
Phase A input current and harmonic spectrum-advanced modulation algorithm.

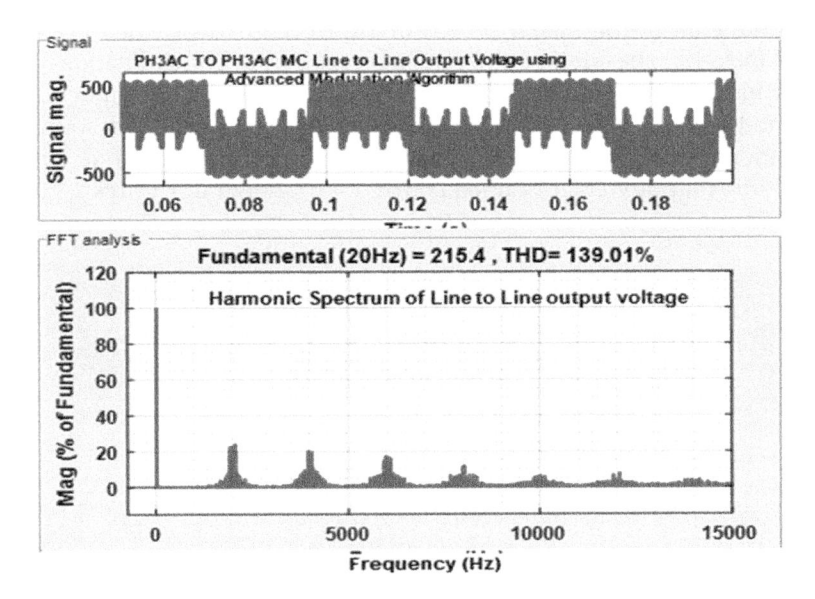

FIGURE 2.31
Line-to-line output voltage and harmonic spectrum-advanced modulation algorithm.

The 2 kHz triangle carrier and gate pulse pattern are the same as shown in Figures 2.36 and 2.37, respectively. Simulation results for a modulation index k of 0.26667 and three-phase sine-wave input voltage are tabulated in Table 2.6 [22–24].

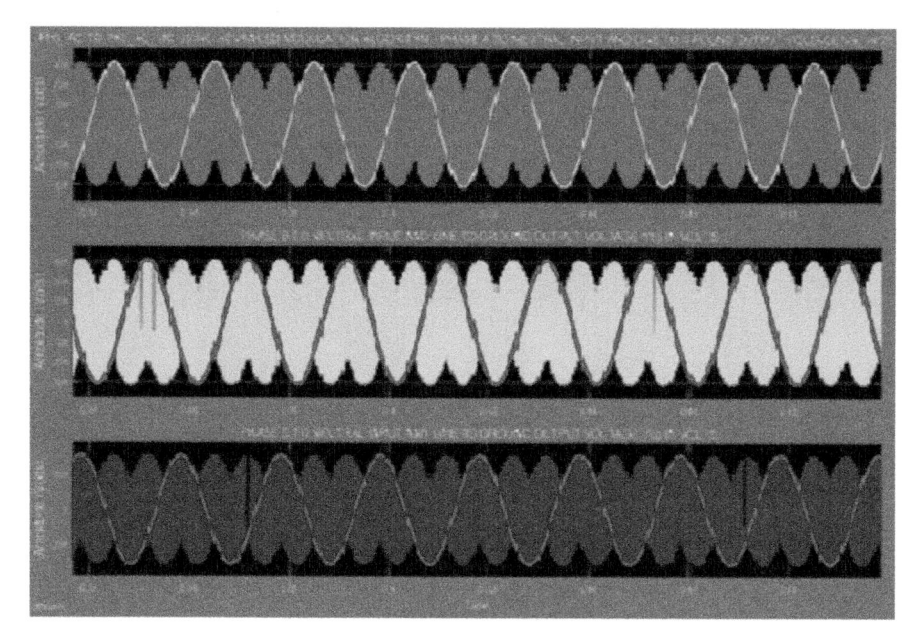

FIGURE 2.32
Three-phase line-to-neutral output voltage of MC – advanced modulation algorithm.

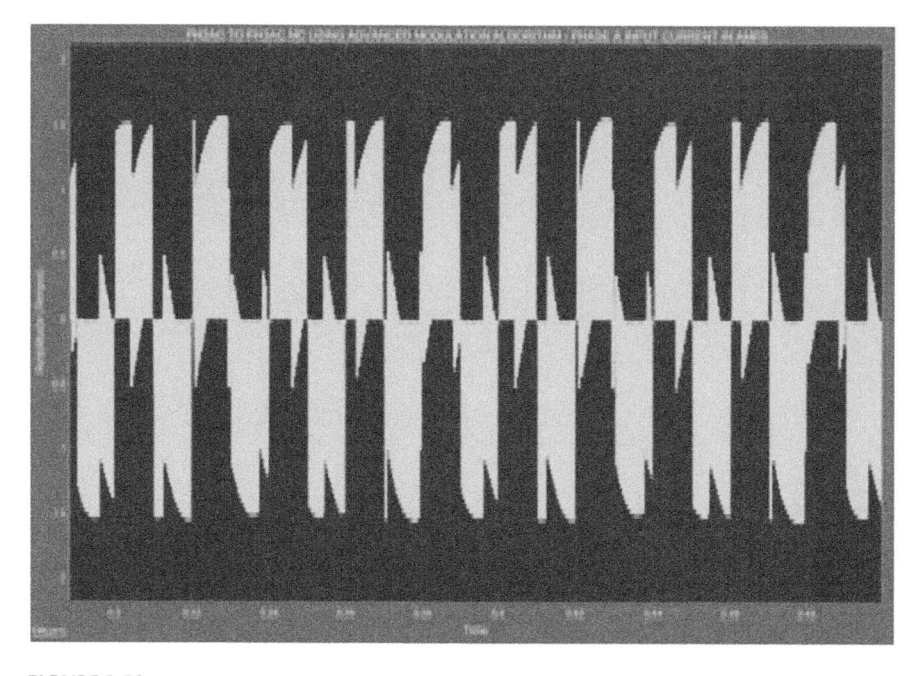

FIGURE 2.33
Phase A input current of MC – advanced modulation algorithm.

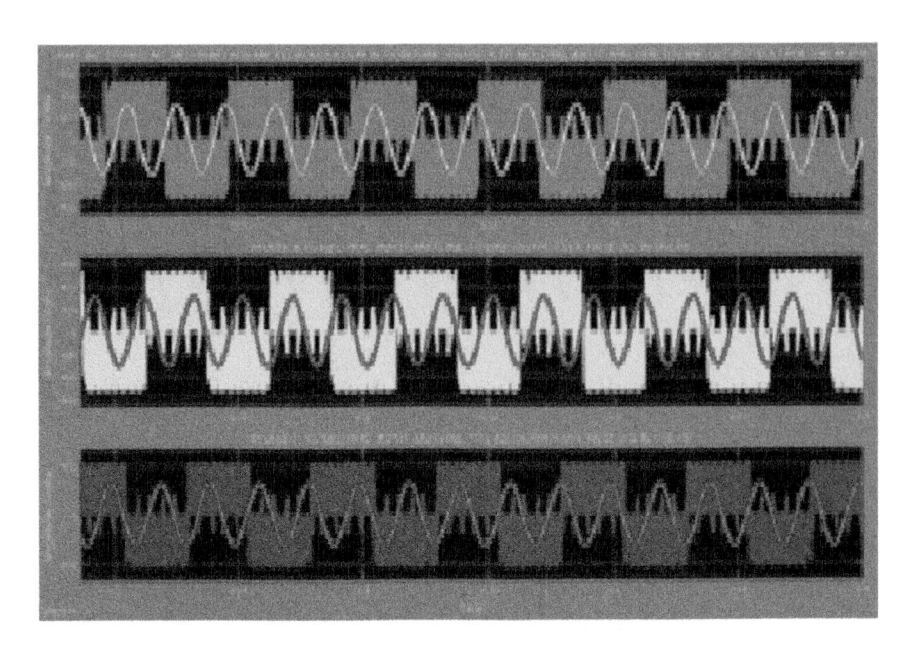

FIGURE 2.34
Line-to-line output voltage of MC – advanced modulation algorithm.

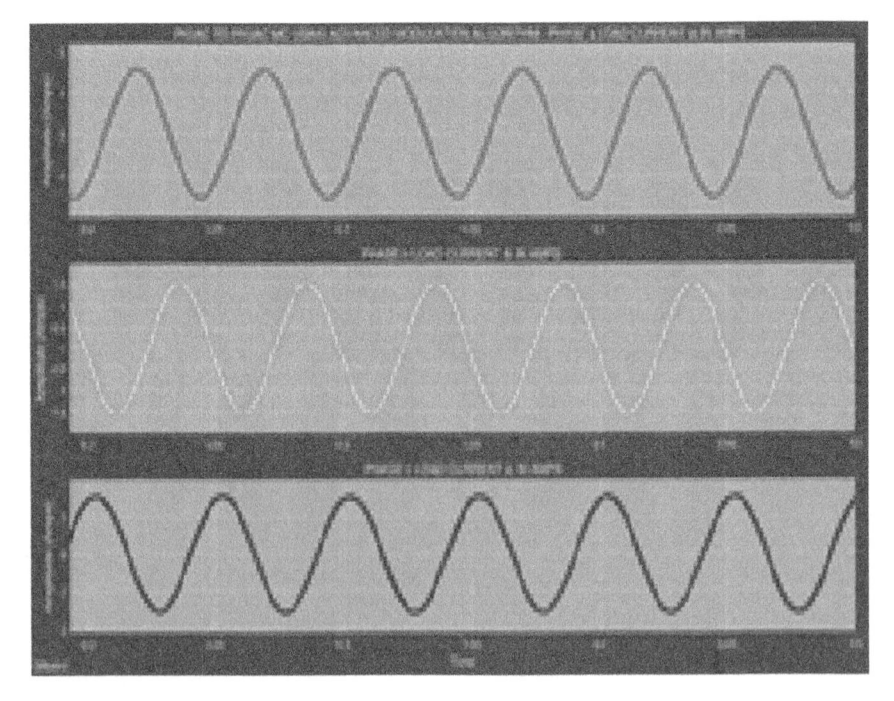

FIGURE 2.35
Three-phase load current of MC – advanced modulation algorithm.

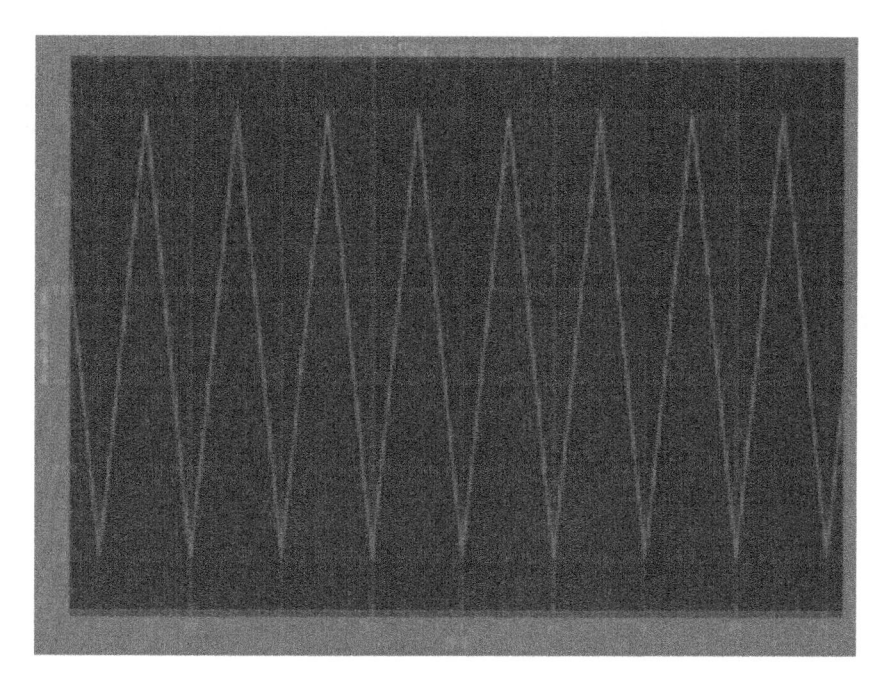

FIGURE 2.36
Triangle carrier for MC – advanced modulation algorithm.

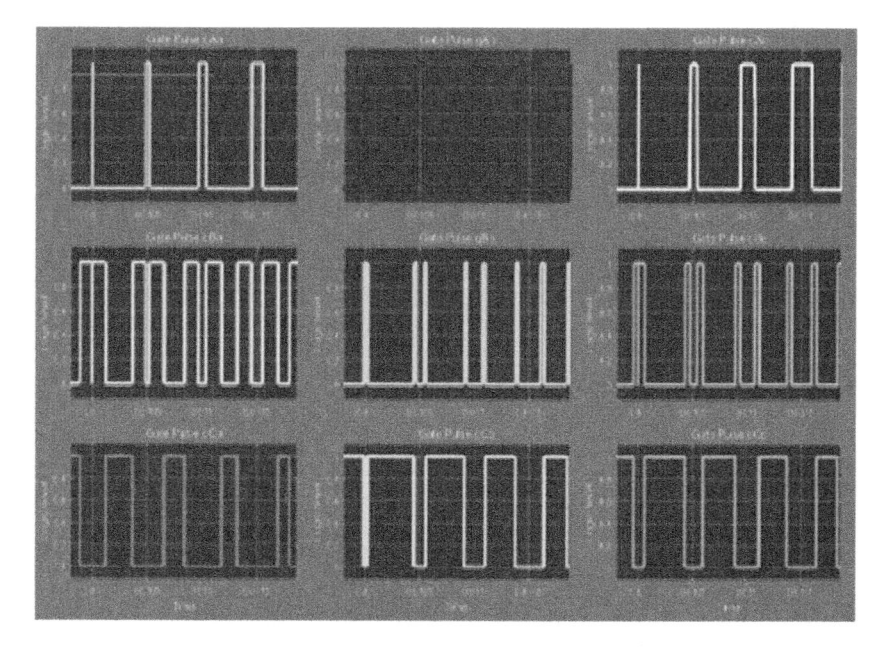

FIGURE 2.37
Gate pulse pattern for MC – advanced modulation algorithm.

TABLE 2.6

Simulation Results for MC – Advanced Modulation

Sl. No.	Algorithm	THD of Line-to-Neutral Output Voltage (p.u.)	THD Line-to-Line Output Voltage (p.u.)	THD of Input Current (p.u.)
1	Advanced modulation	2.409	1.39	1.6579

Program Segment 2.5

```
function [ka,kb,kc,DA,DB,DC,dAa,dBa,dCa,dAb,dBb,dCb,dAc,dBc,
dCc] = fcn(k,fo,fi,t,min_k_abc,max_k_abc)
%%Three Phase AC to Three Phase AC MC - Advanced Modulation
Algorithm
%% Dr.Narayanaswamy. P.R.Iyer.
ka = k*sin(2*pi*fo*t);
kb = k*sin(2*pi*fo*t - 2*pi/(3));
kc = k*sin(2*pi*fo*t - 4*pi/(3));
DA = abs(0.5*sin(2*pi*fi*t));
DB = abs(0.5*sin(2*pi*fi*t - 2*pi/(3)));
DC = abs(0.5*sin(2*pi*fi*t - 4*pi/(3)));
dAa = DA + (ka - 0.5*max_k_abc + 0.5*min_k_abc)*sin(2*pi*fi*t);
dBa = DB + (ka - 0.5*max_k_abc + 0.5*min_k_abc)*sin(2*pi*fi*t -
2*pi/(3));
dCa = DC + (ka - 0.5*max_k_abc + 0.5*min_k_abc)*sin(2*pi*fi*t -
4*pi/(3));
dAb = DA + (kb - 0.5*max_k_abc + 0.5*min_k_abc)*sin(2*pi*fi*t);
dBb = DB + (kb - 0.5*max_k_abc + 0.5*min_k_abc)*sin(2*pi*fi*t -
2*pi/(3));
dCb = DC + (kb - 0.5*max_k_abc + 0.5*min_k_abc)*sin(2*pi*fi*t -
4*pi/(3));
dAc = DA + (kc - 0.5*max_k_abc + 0.5*min_k_abc)*sin(2*pi*fi*t);
dBc = DB + (kc - 0.5*max_k_abc + 0.5*min_k_abc)*sin(2*pi*fi*t -
2*pi/(3));
dCc = DC + (kc - 0.5*max_k_abc + 0.5*min_k_abc)*sin(2*pi*fi*t -
4*pi/(3));
```

Program Segment 2.6

```
function [qAa,qBa,qCa,qABa,qAb,qBb,qCb,qABb,qAc,qBc,qCc,qABc] =
fcn(dAa,dBa,dCa,dAb,dBb,dCb,dAc,dBc,dCc,vsaw)
%%Three Phase AC to Three Phase AC MC - Advanced Modulation
Algorithm.
%% Dr.Narayanaswamy. P.R.Iyer.
if (vsaw <= dAa)
    qAa = 1;
else
    qAa = 0;
end
```

```
if (vsaw <= (dAa+dBa))
    qABa = 1;
else
    qABa = 0;
end
qBa = (qAa & ~(qABa) | qABa & ~(qAa));
qCa = ~(qABa);
if (vsaw <= dAb)
    qAb = 1;
else
    qAb = 0;
end
if (vsaw <= (dAb+dBb))
    qABb = 1;
else
    qABb = 0;
end
qBb = (qAb & ~(qABb) | qABb & ~(qAb));
qCb = ~(qABb);
if (vsaw <= dAc)
    qAc = 1;
else
    qAc = 0;
end
if (vsaw <= (dAc+dBc))
    qABc = 1;
else
    qABc = 0;
end
qBc = (qAc & ~(qABc) | qABc & ~(qAc));
qCc = ~(qABc);
```

2.6 Case Study: Speed Control and Brake by Plugging of Three-Phase Induction Motor Fed by Matrix Converter

Three-phase induction motor (IM) fed by MCs can be used for speed control and brake by plugging or by reversing the phase sequence of the three-phase applied voltage. This can be accomplished in the SIMULINK MC model and can be implemented in dSPACE. Thus, three-phase IM fed by MC can be used in hybrid electric vehicle (HEV) and also in electric traction. This section explores this application.

The SIMULINK model of the three-phase IM fed by MC with provision for speed control and brake by plugging is shown in Figure 2.38. Braking of a three-phase IM by plugging involves reversing the phase sequence of the three-phase applied voltage to the stator terminals of the IM. This is

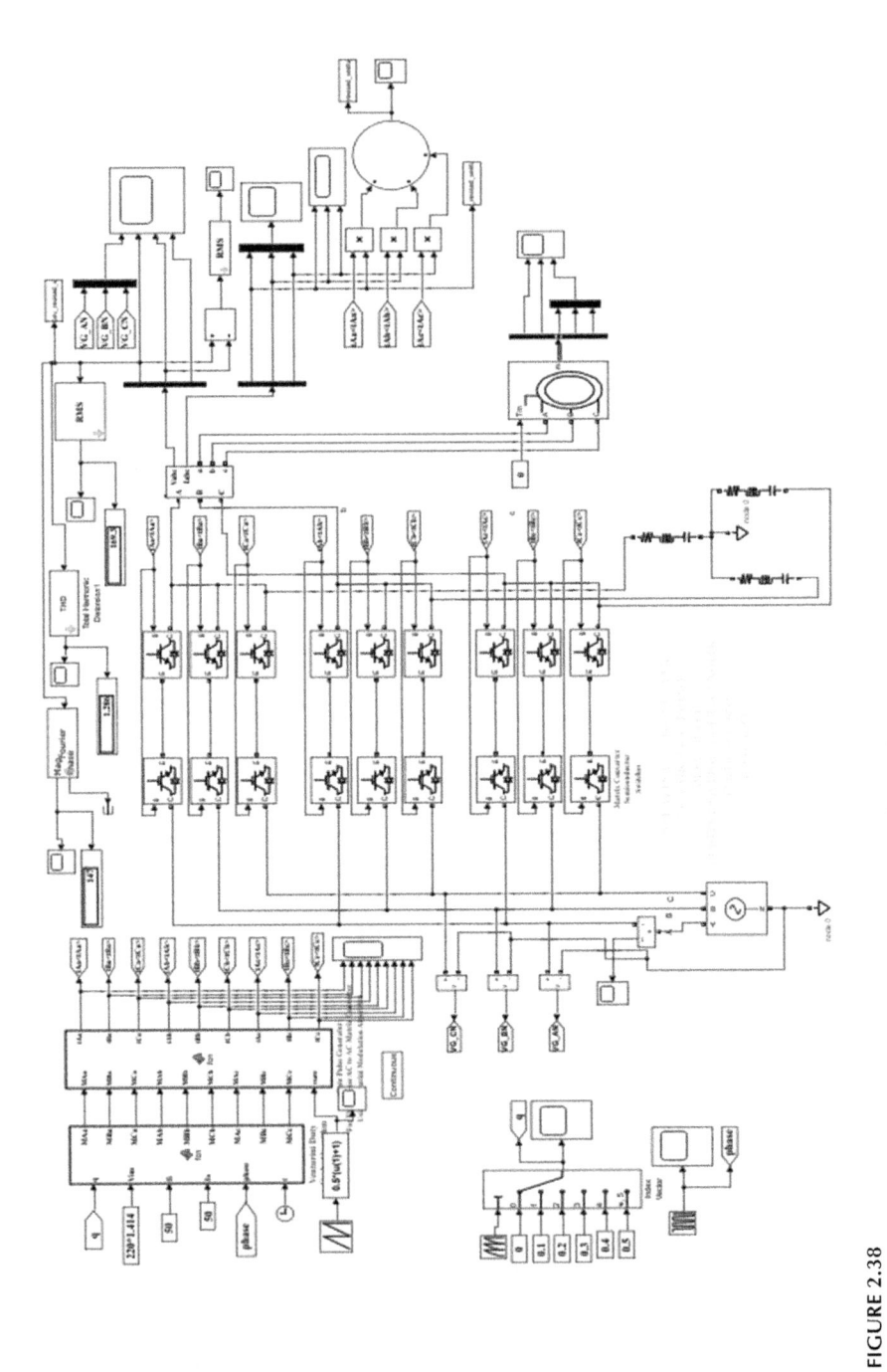

FIGURE 2.38
Speed control and brake by plugging of three-phase IM fed by MC.

done manually using three-pole, double-throw switch. In the method using SIMULINK model, this can be done by varying the model parameter such as the modulation index and the phase of the desired three-phase output voltage. The model shown in Figure 2.38 is developed using the Venturini second method presented in Section 2.4.3. Only relevant changes in the model for speed control and braking are explained below:

Speed control is achieved by varying the magnitude of the three-phase output voltage of the MC which is applied to the terminals of the IM. Magnitude variation is possible by changing the modulation index q in the range from 0 to 0.5. In Figure 2.38, this value of q is varied using a multiport switch. For braking, three-phase output voltage phase sequence has to be reversed. Referring to equation 2.11, the phase sequence of output voltage v_b and v_c can be changed by swapping the phase $2\pi/3$ with proper sign. In Figure 2.38, this is achieved by using a repeating sequence block. The three-phase IM used has the parameter shown in Table 2.7.

2.6.1 Simulation Results

The simulation of the model shown in Figure 2.38 is carried out using SIMULINK [17]. The ode23tb (stiff/TR-BDF2) solver is used. The simulation results are shown in Figure 2.39a–h [23].

The variation of modulation index q achieved using multiport switch and a repeating sequence block is shown in Figure 2.40a and the swapping of output voltage phase angle from $+2\pi/3$ to $-2\pi/3$ achieved using a repeating sequence block is shown in Figure 2.40b.

TABLE 2.7

Three-Phase IM Model Parameters

Sl. No.	Parameters	Value	Units
1	RMS line-to-line input supply voltage	260	V
2	Input frequency	50	Hz
3	Power output	4	kV A
4	RMS line-to-line machine voltage	400	V
5	Machine frequency	50	Hz
6	No. of phase	3	–
7	No. of poles	4	–
8	Stator resistance R_s	1.405	Ω
9	Stator leakage inductance L_{ls}	0.005839	H
10	Rotor resistance $R_{r'}$	1.395	Ω
11	Rotor leakage inductance $L_{lr'}$	0.005839	H
12	Mutual inductance L_m	0.1722	H
13	Moment of inertia J	0.0131	kg m^2
14	Damping constant B	0.002985	N m s

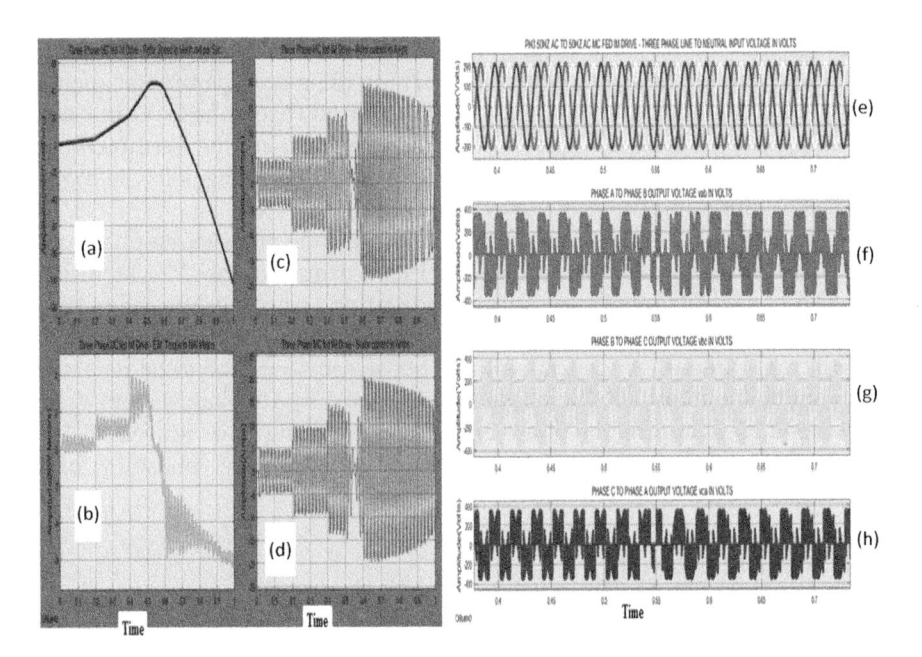

FIGURE 2.39
Three-phase MC-fed IM drive: (a) rotor speed in mechanical radians per second, (b) electro-magnetic torque in Newton-metres, (c) three-phase rotor current in amperes, (d) three-phase stator current in amperes, (e) three-phase line-to-neutral input voltage in volts and (f)–(h) three-phase line-to-line output voltage in volts.

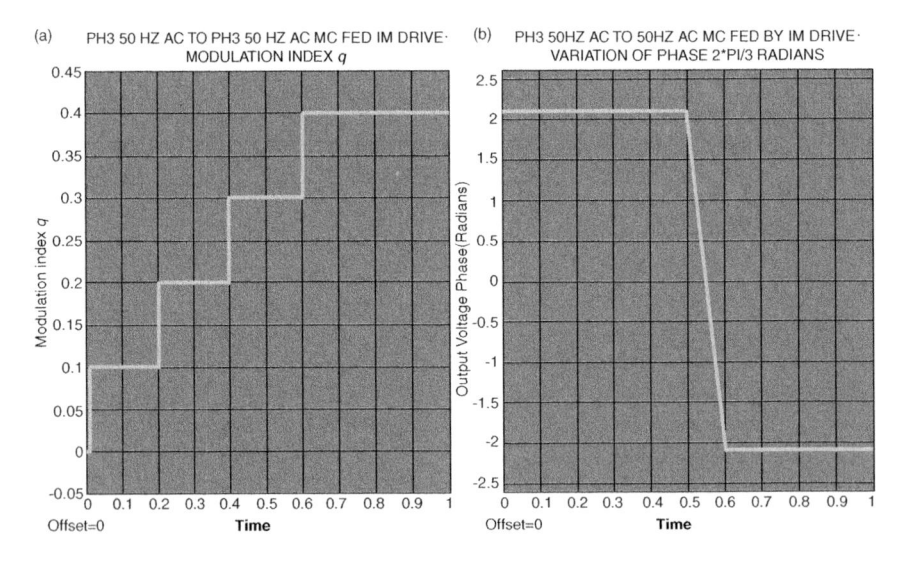

FIGURE 2.40
Three-phase MC-fed IM drive: (a) variation of modulation index q and (b) variation of output voltage phase $2\pi/3$ rad.

2.6.2 Real-Time Implementation

The real-time implementation scheme using SIMULINK model in conjunction with dSPACE hardware controller board is shown in Figure 2.41a–c. The scheme in Figure 2.41a,b is with DC generator and three-phase alternator mode, respectively. Figure 2.41c is the scheme for dSPACE gate drive for MC. Figure 2.41d–f shows the hardware implementation for the saw-tooth carrier generator, output voltage phase value ($2\pi/3$) swapping device and the modulation index q varying device. The hardware realization for saw-tooth carrier generator is using operational amplifier square wave generator, integrator and reset switch. The output voltage phase value swapping device and the modulation index q varying device use operational amplifiers and resistor networks as shown in Figure 2.41e and f, respectively. In Figure 2.41e, switch SS1 can be used to select +5 or –5 V which causes the output voltage phase value to change from $+2\pi/3$ to $-2\pi/3$ rad. In Figure 2.41f, if SS2 is thrown to ground, then q value selected will be zero and the output voltage will be zero. With SS2 in +5 V position, SS3 can be thrown to select contacts from 1 to 9 when the q value varies from 0.1 to 0.5 in steps of 0.05.

2.7 Discussion of Results

Comparing the total harmonic distortion (THD) of line-to-neutral output voltage by the three methods, this value is minimum for the optimum Venturini algorithm and maximum for the advanced modulation algorithm. The THD for line-to-line output voltage and input current are minimum for the Venturini first method and maximum for the advanced modulation algorithm for the former and Venturini second method for the later. The linear modulation ranges for the Venturini first, Venturini second and optimum Venturini algorithms are 0–0.5 for the first two and 0–0.866 for the third. This linear range is 0–0.577 for the advanced modulation algorithm.

Considering the speed control and brake by plugging of three-phase IM fed by MC, it is seen that as the modulation index is zero initially and increased, the output voltage is zero and gradually increases. The stator current, rotor current and speed increase for the positive phase sequence applied voltage. As the phase sequence of output voltage is reversed by swapping the phase value $2\pi/3$ rad at 600 ms, it is seen that braking effect takes place, speed decreases and finally reverses. It is also seen that the electromagnetic torque is positive when the output voltage is positive phase sequence and it is negative when output phase voltage takes negative phase sequence value.

Observation of line-to-line output voltage in Figure 2.39f–h before and after 600 ms clearly reveals phase sequence reversal. A real-time implementation scheme for speed control and brake by plugging of three-phase IM by

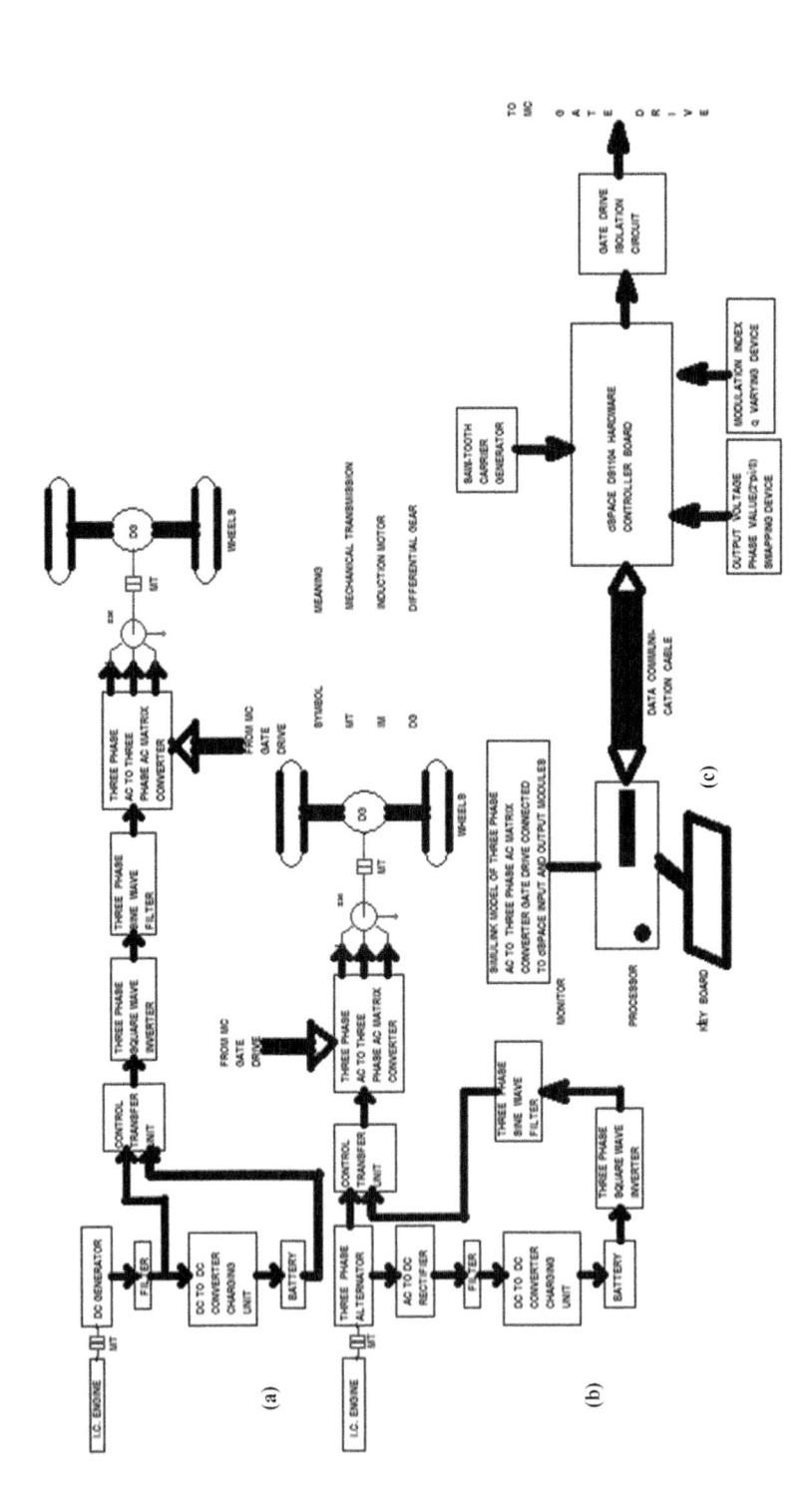

FIGURE 2.41

Real-time implementation of three-phase MC-fed IM drive: (a) using DC generator, (b) using three-phase alternator, (c) dSPACE experimental set up, (d) saw-tooth carrier, (e) output voltage phase value swapping device and (f) modulation index q varying device.

(Continued)

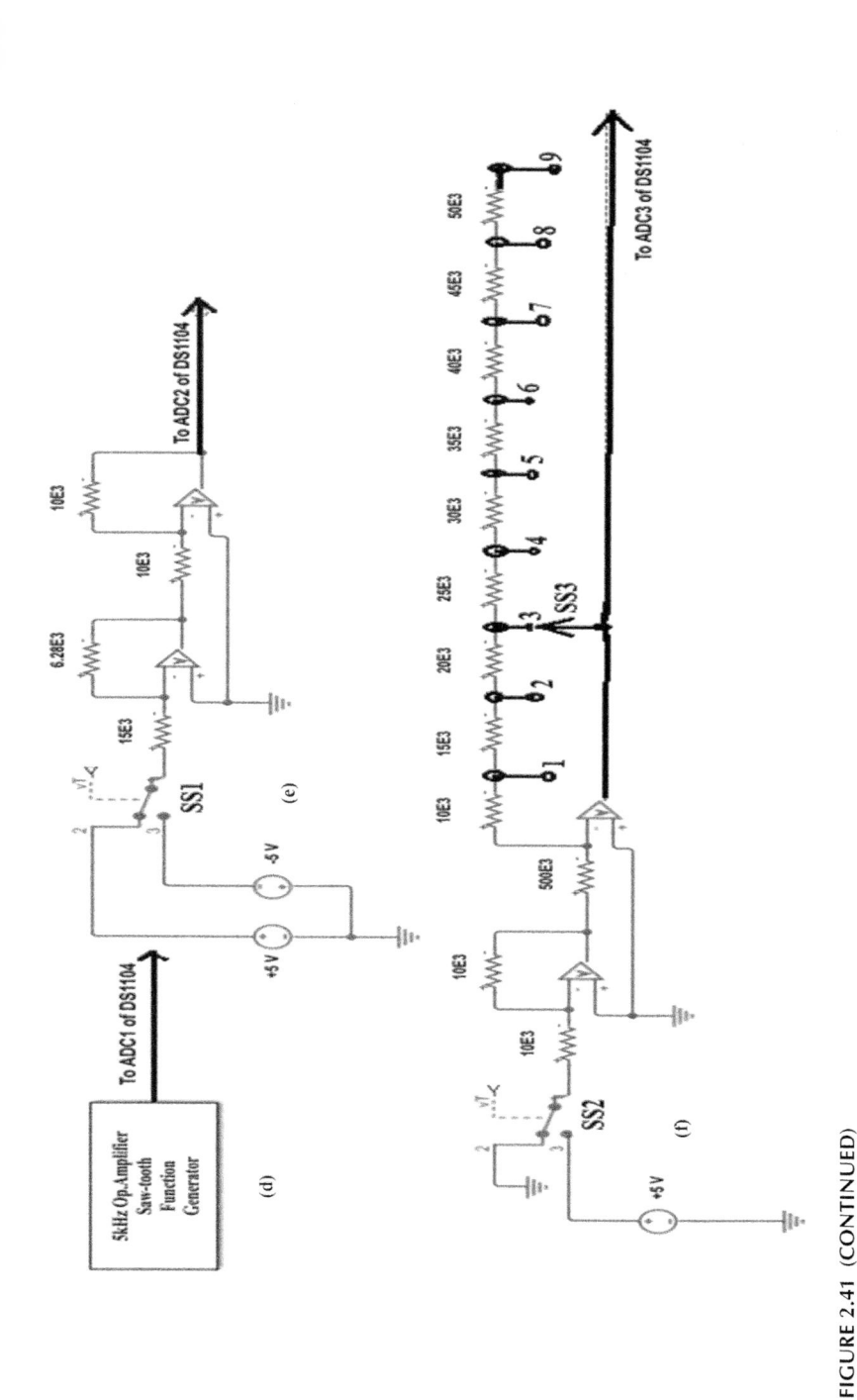

FIGURE 2.41 (CONTINUED)

Real-time implementation of three-phase MC-fed IM drive: (a) using DC generator, (b) using three-phase alternator, (c) dSPACE experimental set up, (d) saw-tooth carrier, (e) output voltage phase value swapping device and (f) modulation index *q* varying device.

using dSPACE DS1104 hardware controller board is presented. This method finds application in HEVs and electric traction.

2.8 Conclusions

The linear modulation range is 0–0.5 for the Venturini first and second methods, 0–0.866 for the optimum Venturini method and 0.577 for the advanced modulation algorithm. As the modulation index in all three cases is within the linear modulation range, the predicted performance is valid for all modulation indices within the linear range. A successful model and its real-time implementation for the speed control and brake by plugging of a three-phase IM fed by MC is presented which has applications in HEVs and electric traction.

References

1. A. Alesina and M.G.B. Venturini: Solid state power conversion: a Fourier analysis approach to generalized transformer synthesis, *IEEE Transactions on Circuits and Systems*, Vol. CAS-28, No. 4, 1981, pp. 319–330.
2. A. Alesina and M. Venturini: Intrinsic amplitude limits and optimum design of 9-switches direct PWM AC-AC converters, *IEEE-PESC*, Kyoto, Japan, 1988, pp. 1284–1291.
3. A. Alesina and M. Venturini: Analysis and design of optimum amplitude nine-switch direct AC-AC converters, *IEEE Transactions on Power Electronics*, Vol. 4, 1989, pp. 101–112.
4. P. Wheeler, J. Rodriguez, J.C. Clare, L. Empringham and A. Weinstein: Matrix converters – a technology review, *IEEE Transactions on Industrial Electronics*, Vol. 49, No. 2, 2002, pp. 276–288.
5. P. Wheeler, J. Clare, L. Empringham, M. Apap and M. Bland: Matrix converters, *Power Engineering Journal*, Vol. 16, No. 6, 2002, pp. 273–282.
6. S. Sunter and J.C. Clare: Development of a matrix converter induction motor drive, *IEEE MELECON*, Antalya, 1994, pp. 833–836.
7. S. Sunter and J.C. Clare: A true four quadrant matrix converter induction motor drive with servo performance, *IEEE PESC*, Italy, 1996, pp. 146–151.
8. S. Sunter and J.C. Clare: Feed forward indirect vector control of a matrix converter-fed induction motor drive, *The International Journal for Computation and Mathematics in Electrical and Electronic Engineering*, Vol. 19, No. 4, 2000, pp. 974–986.
9. A. Zuckerberger, D. Weinstock and A. Alexandrovitz: Simulation of three phase loaded matrix converter, *IEE Proceedings - Electric Power Applications*, Vol. 143, No. 4, 1996, pp. 294–300.

10. M. Imayavaramban, A.V. Krishna Chaithanya and B.G. Fernandez: Analysis and mathematical modeling of matrix converter for adjustable speed AC drives, *IEEE-PSCE*, Atlanta, GA, 2006, pp. 1113–1120.
11. L. Huber and D. Borojevic: Space vector modulated three-phase to three-phase matrix converter with input power factor correction, *IEEE Transactions on Industry Applications*, Vol. 31, No. 6, 1995, pp. 1234–1245.
12. L. Huber, D. Borojevic, X.F. Zhuang and F.C. Lee: Design and implementation of a three phase to three phase matrix converter with input power factor correction, *IEEE Applied Power Electronics Conference and Exposition*, San Diego, CA, 1993, pp. 860–865.
13. L. Huber and D. Borojevic: Space vector modulation with unity input power factor for forced commutated cycloconverters, *IEEE Industry Application Society Annual Meeting*, Dearbom, MI, 1991, pp. 1032–1041.
14. L. Huber and D. Borojevic: Space vector modulator for forced commutated cycloconverters, *IEEE Industry Application Society Annual Meeting*, San Diego, CA, 1989, pp. 871–876.
15. L. Huber, D. Borojevic and N. Burany: Voltage space vector based PWM control of forced commutated cycloconverters, *IEEE – IECON*, Philadelphia, PA, 1989, pp. 106–111.
16. L. Huber, D. Borojevic and N. Burany: Analysis, design and implementation of the space-vector modulator for forced-commutated cycloconverters, *IEE-Proceedings, Part B*, Vol. 139, No. 2, 1992, pp. 103–113.
17. The Mathworks Inc.: www.mathworks.com, MATLAB/SIMULINK user manual, R2017b, 2017.
18. K.K. Mohapatra, P. Jose, A. Drolia, G. Aggarwal, S. Thuta and N. Mohan: A novel carrier-based PWM scheme for matrix converters that is easy to implement, *IEEE-PESC*, Recife, Brazil, 2005, pp. 2410–2414.
19. K.K. Mohapatra, T. Satish and N. Mohan: A speed-sensorless direct torque control scheme for matrix converter driven induction motor, *IEEE-IECON*, Paris, France, 2006, pp. 1435–1440.
20. T. Sathish, K.K. Mohapatra and N. Mohan: Steady state over-modulation of matrix converter using simplified carrier based control, *IEEE-IECON*, Taipei, Taiwan, 2007, pp. 1817–1822.
21. S. Thuta, K.K. Mohapatra and N. Mohan: Matrix converter over-modulation using carrier-based control: maximizing the voltage transfer ratio, *IEEE-PESC*, Rhodes, Greece, 2008, pp. 1727–1733.
22. N.P.R. Iyer: Carrier based modulation technique for three phase matrix converters – state of the art progress, *IEEE Region 8 SIBIRCON-2010*, Irkutsk Listvyanka, Russia, July 11–15, 2010, pp. 659–664.
23. N.P.R. Iyer: Modelling, simulation and real time implementation of a three phase AC to AC matrix converter, Ph.D. thesis, Ch. 3 & Ch. 4, Department of ECE, Curtin University, Perth, WA, Australia, February 2012.
24. N.P.R. Iyer and C.V. Nayar: Performance comparison of a three phase AC to three phase AC matrix converter using different carrier based switching algorithms, *International Journal of Engineering Research & Technology (IJERT)*, Vol. 6, No. 5, 2017, pp. 827–832.

3

Multilevel Matrix Converter

3.1 Introduction

Multilevel technology is a good solution for medium- and high-voltage power conversions [1–5]. There are three kinds of multilevel converters, namely diode clamped [3], capacitor clamped [4,5] and cascade [6]. Among them, capacitor-clamped multilevel converter is the most suitable topology for direct AC to AC conversion [4,5]. This chapter describes the multilevel matrix converter (MMC) with a capacitor clamped or flying capacitor or multicell topology [1,2]. The Venturini algorithm is used for generating gate pulses for MMC, the application of which is different from the conventional single-cell matrix converter. Simulation results using the software SIMULINK are provided.

3.2 Multilevel Matrix Converter with Three Flying Capacitors per Output Phase

The three-phase capacitor-clamped MMC with three flying capacitors per output phase is shown in Figure 3.1 [1,2]. The IGBT switches in Figure 3.1 are bidirectional. The analysis is given below:

Consider switches S_{Aa1}, S_{Aa2}, S_{Ba1} and S_{Ba2} and capacitor C_1. Let all the switches are identical with off resistance R_{off}. With all four switches off, applying Kirchhoff's law to the above four switches with C_1 connected to V_{AB} as shown in Figure 3.1, we have the voltage V_{c1} across C_1 expressed as follows:

$$V_{c1} = \frac{V_{AB}}{4R_{off}} * 2R_{off} = \frac{(V_A - V_B)}{2} \tag{3.1}$$

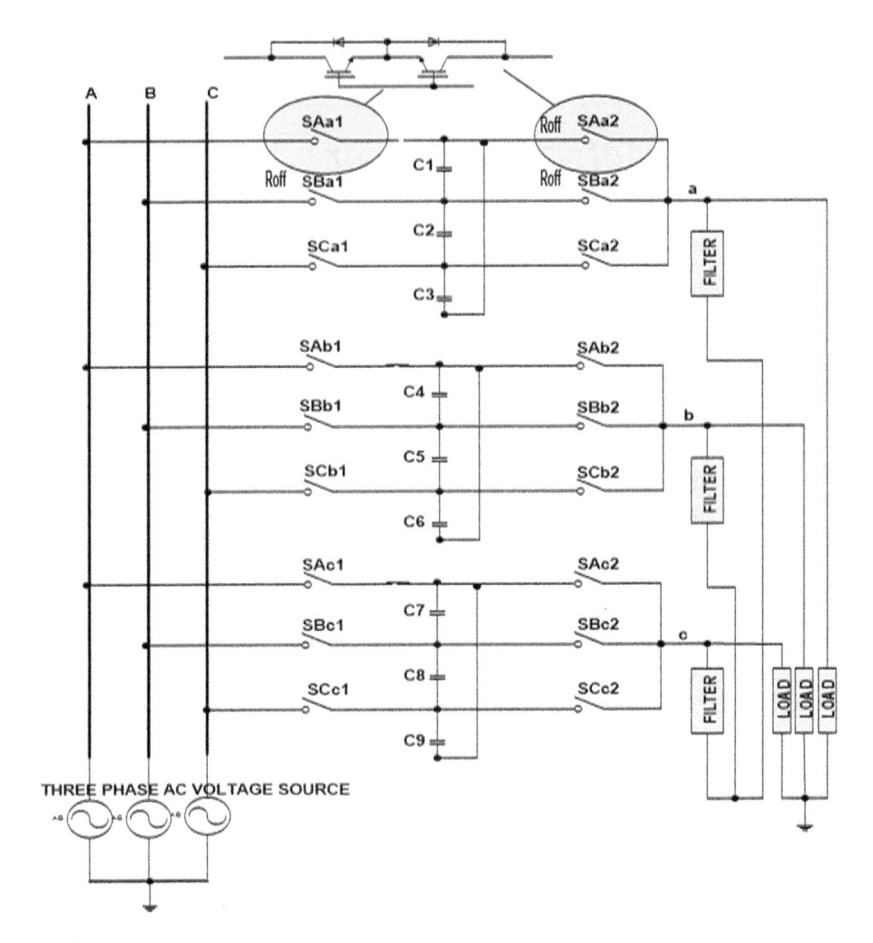

FIGURE 3.1
Three-phase AC to three-phase AC matrix converter with three flying capacitors per output phase.

where V_{c1} has the polarity positive at the top and negative at the bottom plate. The same analysis holds good for other flying capacitors C_2 to C_9. Thus, $V_{c1} = V_{c4} = V_{c7}$, $V_{c2} = V_{c5} = V_{c8}$ and $V_{c3} = V_{c6} = V_{c9}$.

In Figure 3.1, for the above switch combinations, applying Kirchhoff's law, Table 3.1 is valid. A detailed table for output phase a is shown in Table 3.2 [1,2]. This table is illustrated diagrammatically in Figure 3.2a–i. Capacitance of the flying capacitor C_1 is selected using the following equation:

$$C_1 = \frac{I_o}{\Delta V_c * p * f_{sw}} \tag{3.2}$$

TABLE 3.1

PH3 MMC with Three FC – Truth Table

Sl. No.	S_{Aa1}	S_{Aa2}	S_{Ba1}	S_{Ba2}	Output Voltage v_a
1	1	1	0	0	V_A
2	0	0	1	1	V_B
3	1	0	0	1	$(V_A + V_B)/2$
4	0	1	1	0	$(V_A + V_B)/2$

1, switch on; 0, switch off.

TABLE 3.2

PH3 MMC with Three FC – Complete Truth Table

Sl. No.	S_{Aa1}	S_{Aa2}	S_{Ba1}	S_{Ba2}	S_{Ca1}	S_{Ca2}	Output Voltage v_a
1	1	1	0	0	0	0	V_A
2	1	0	0	1	0	0	$(V_A + V_B)/2$
3	0	1	1	0	0	0	$(V_A + V_B)/2$
4	0	0	1	1	0	0	V_B
5	0	0	1	0	0	1	$(V_B + V_C)/2$
6	0	0	0	1	1	0	$(V_B + V_C)/2$
7	0	0	0	0	1	1	V_C
8	0	1	0	0	1	0	$(V_A + V_C)/2$
9	1	0	0	0	0	1	$(V_A + V_C)/2$

1, switch on; 0, switch off.

where I_o is the peak value of load current, p is the number of cells per output phase, f_{sw} is the switching frequency and ΔV_c is the flying capacitor voltage ripple [1,2].

The switching function for the bidirectional switches in Figure 3.1 can be expressed as follows:

$$S_{ijk} = \begin{cases} 0 & \text{when switch open} \\ 1 & \text{when switch closed} \end{cases}$$

$$S_{Ajk} + S_{Bjk} + S_{Cjk} = 1 \tag{3.3}$$

$i \in A, B, C; j \in a, b, c$ and $k =$ switch column count $1, 2$

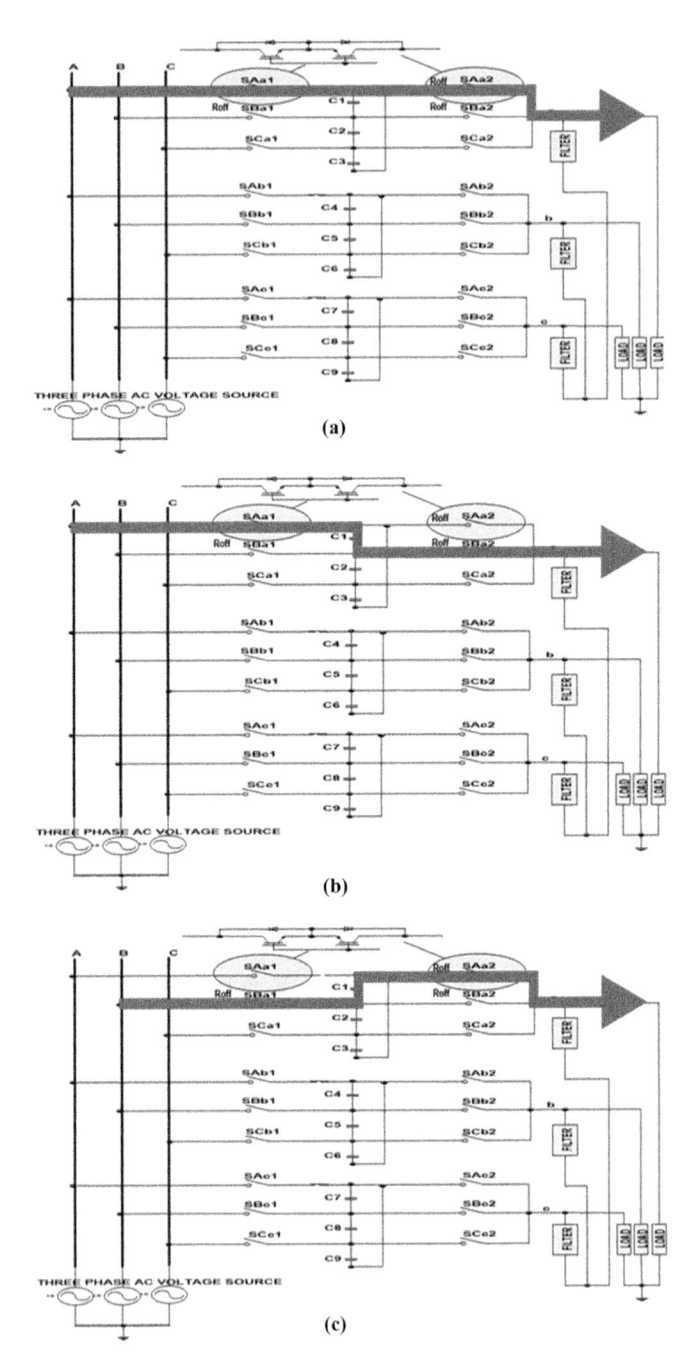

FIGURE 3.2
(a)–(i) Switching combinations of three-phase MMC with three FC.

(*Continued*)

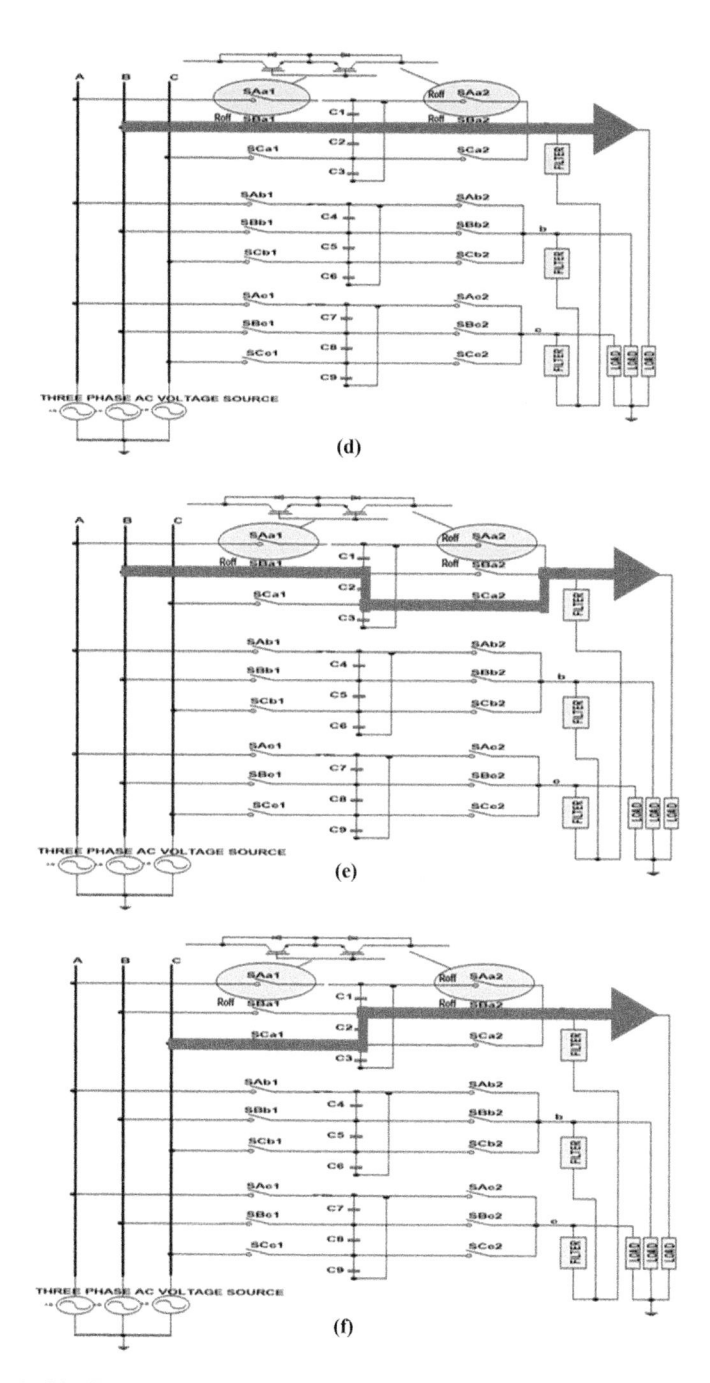

FIGURE 3.2 (CONTINUED)
(a)–(i) Switching combinations of three-phase MMC with three FC.

(*Continued*)

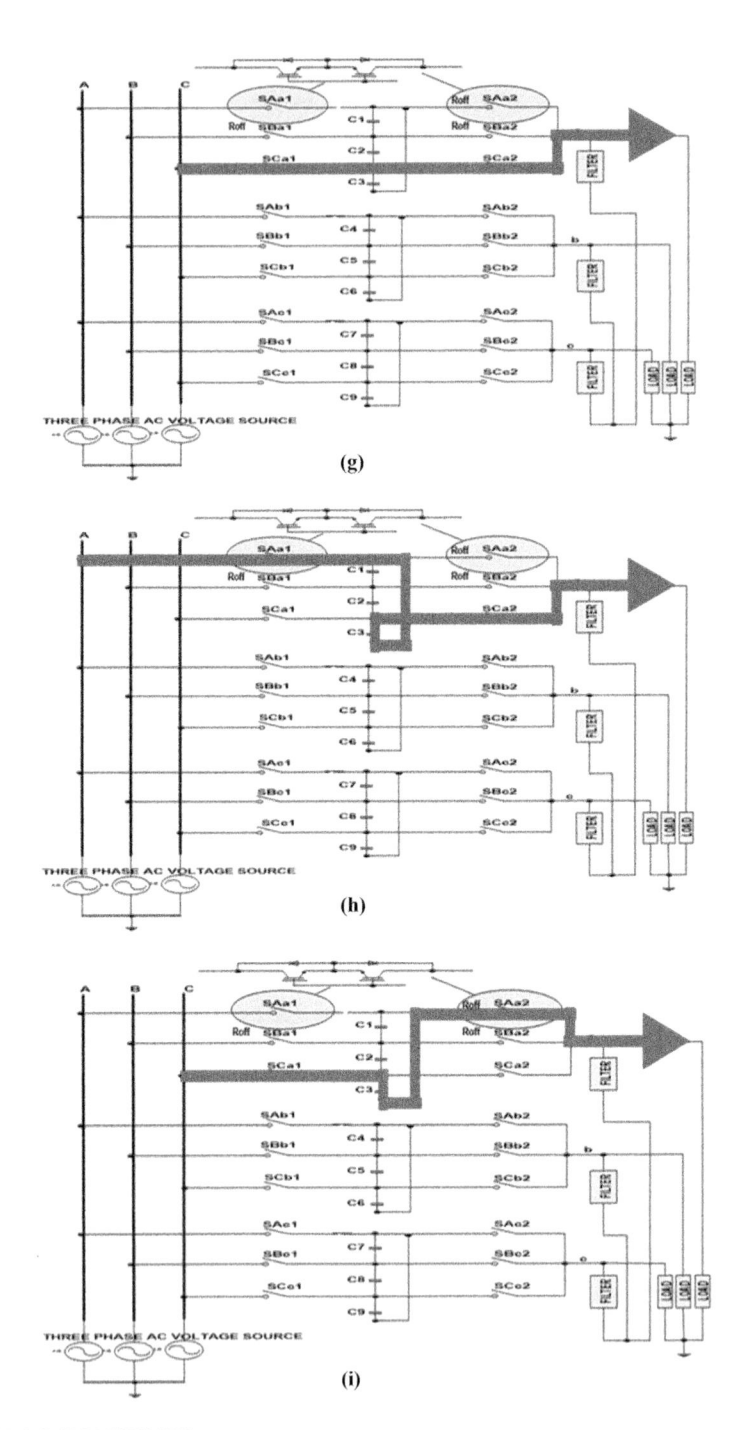

FIGURE 3.2 (CONTINUED)

(a)–(i) Switching combinations of three-phase MMC with three FC.

The modelling equation referring to Table 3.2 for the output voltage can be expressed as follows:

$$\begin{bmatrix} v_a \\ v_b \\ v_c \end{bmatrix} = \begin{bmatrix} S_{Aa} & S_{Ba} & S_{Ca} & S_{C1} & S_{C2} & S_{C3} \\ S_{Ab} & S_{Bb} & S_{Cb} & S_{C4} & S_{C5} & S_{C6} \\ S_{Ac} & S_{Bc} & S_{Cc} & S_{C7} & S_{C8} & S_{C9} \end{bmatrix} * \begin{bmatrix} v_A \\ v_B \\ v_C \\ (v_A + v_B)/2 \\ (v_C + v_B)/2 \\ (v_A + v_C)/2 \end{bmatrix}$$

where

$$S_{Aa} = S_{Aa1} \cap S_{Aa2}\, ; S_{Ba} = S_{Ba1} \cap S_{Ba2}\, ; S_{Ca} = S_{Ca1} \cap S_{Ca2}$$

$$S_{C1} = S_{Aa1} \cap S_{Ba2} \cup S_{Ba1} \cap S_{Aa2}\, ; S_{C2} = S_{Ba1} \cap S_{Ca2} \cup S_{Ca1} \cap S_{Ba2}\, ;$$

$$S_{C3} = S_{Aa1} \cap S_{Ca2} \cup S_{Ca1} \cap S_{Aa2}$$

$$S_{Ab} = S_{Ab1} \cap S_{Ab2}\, ; S_{Bb} = S_{Bb1} \cap S_{Bb2}\, ; S_{Cb} = S_{Cb1} \cap S_{Cb2}$$

$$S_{C4} = S_{Ab1} \cap S_{Bb2} \cup S_{Bb1} \cap S_{Ab2}\, ; S_{C5} = S_{Bb1} \cap S_{Cb2} \cup S_{Cb1} \cap S_{Bb2}\, ;$$

$$S_{C6} = S_{Ab1} \cap S_{Cb2} \cup S_{Cb1} \cap S_{Ab2}$$

$$S_{Ac} = S_{Ac1} \cap S_{Ac2}\, ; S_{Bc} = S_{Bc1} \cap S_{Bc2}\, ; S_{Cc} = S_{Cc1} \cap S_{Cc2}$$

$$S_{C7} = S_{Ac1} \cap S_{Bc2} \cup S_{Bc1} \cap S_{Ac2}\, ; S_{C8} = S_{Bc1} \cap S_{Cc2} \cup S_{Cc1} \cap S_{Bc2}\, ;$$

$$S_{C9} = S_{Ac1} \cap S_{Cc2} \cup S_{Cc1} \cap S_{Ac2}$$

1 = Switch closed

0 = Switch open

$\cap$ = Logical AND operator

$\cup$ = Logical OR operator

$$\tag{3.4}$$

Similarly referring to Figure 3.1, the modelling equation for the three-phase input currents can be expressed as follows:

$$
\begin{bmatrix} i_A \\ i_B \\ i_C \end{bmatrix} = \begin{bmatrix} (S_{Aa} \cup S_{ABa} \cup S_{ACa}) & (S_{Ab} \cup S_{ABb} \cup S_{ACb}) & (S_{Ac} \cup S_{ABc} \cup S_{ACc}) \\ (S_{Ba} \cup S_{BAa} \cup S_{BCa}) & (S_{Bb} \cup S_{BAb} \cup S_{BCb}) & (S_{Bc} \cup S_{BAc} \cup S_{BCc}) \\ (S_{Ca} \cup S_{CAa} \cup S_{CBa}) & (S_{Cb} \cup S_{CAb} \cup S_{CBb}) & (S_{Cc} \cup S_{CAc} \cup S_{CBc}) \end{bmatrix}
$$
$$
* \begin{bmatrix} i_a \\ i_b \\ i_c \end{bmatrix}
$$

where

$$S_{Aa} = S_{Aa1} \cap S_{Aa2} \quad S_{ABa} = S_{Aa1} \cap S_{Ba2} \quad S_{ACa} = S_{Aa1} \cap S_{Ca2}$$

$$S_{Ba} = S_{Ba1} \cap S_{Ba2} \quad S_{BAa} = S_{Ba1} \cap S_{Aa2} \quad S_{BCa} = S_{Ba1} \cap S_{Ca2}$$

$$S_{Ca} = S_{Ca1} \cap S_{Ca2} \quad S_{CAa} = S_{Ca1} \cap S_{Aa2} \quad S_{CBa} = S_{Ca1} \cap S_{Ba2}$$

$$S_{Ab} = S_{Ab1} \cap S_{Ab2} \quad S_{ABb} = S_{Ab1} \cap S_{Bb2} \quad S_{ACb} = S_{Ab1} \cap S_{Cb2}$$

$$S_{Bb} = S_{Bb1} \cap S_{Bb2} \quad S_{BAb} = S_{Bb1} \cap S_{Ab2} \quad S_{BCb} = S_{Bb1} \cap S_{Cb2}$$

$$S_{Cb} = S_{Cb1} \cap S_{Cb2} \quad S_{CAb} = S_{Cb1} \cap S_{Ab2} \quad S_{CBb} = S_{Cb1} \cap S_{Bb2}$$

$$S_{Ac} = S_{Ac1} \cap S_{Ac2} \quad S_{ABc} = S_{Ac1} \cap S_{Bc2} \quad S_{ACc} = S_{Ac1} \cap S_{Cc2}$$

$$S_{Bc} = S_{Bc1} \cap S_{Bc2} \quad S_{BAc} = S_{Bc1} \cap S_{Ac2} \quad S_{BCc} = S_{Bc1} \cap S_{Cc2}$$

$$S_{Cc} = S_{Cc1} \cap S_{Cc2} \quad S_{CAc} = S_{Cc1} \cap S_{Ac2} \quad S_{CBc} = S_{Cc1} \cap S_{Bc2}$$

$1 = $ switch closed

$0 = $ switch open

$\cap = $ Logical AND operator

$\cup = $ Logical OR operator

$$(3.5)$$

3.3 Control of Multilevel Matrix Converter with Three Flying Capacitors per Output Phase by the Venturini Method

Let the input and output voltages be expressed as in equations 2.9 and 2.11, respectively, in Chapter 2. For unity input phase displacement factor, the nine modulation functions for a three-phase AC to three-phase AC conventional matrix converter is given in equation 2.17. Here, the modified modulation function assuming unity input phase displacement factor for the 18 bidirectional switches of the three-phase MMC shown in Figure 3.1 is given below:

$$M_{ijk} = \frac{t_{ijk}}{T_s} = \left[\frac{1}{3} + \frac{2v_i * v_j}{3V_{im}^2} \right]$$

(3.6)

for $i \in A, B, C$ and $j \in a, b, c$ and $k \in 1, 2$.

Signals controlling switches in individual cells of the converter should be shifted with respect to each other by an angle of $2\pi/p$, where p is the number of switching cells which in this case is two. Displacement of carrier signals involved in the control of switches S_{ij1} and S_{ij2} is $T_{sw}/2$, where $T_{sw} = 1/f_{sw}$ is the carrier switching period. Duty cycles for switch group S_{ij1} are by comparison of modulation function with the saw-tooth carrier starting from the origin and that for switch group S_{ij2} are by comparison of the modulation functions with the saw-tooth carrier phase shifted by $T_{sw}/2$ or π rad [2].

3.4 Output Filter

The output filter circuit shown in Figure 3.1 is an RLC circuit, with R_f, L_f and C_f connected in series whose resonant frequency matches with the carrier switching frequency f_{sw} [2], as given below:

$$f_{sw} = \frac{1}{\left(2\pi * \sqrt{L_f C_f} \right)}$$

(3.7)

3.5 Model Development

The model of the three-phase AC to three-Phase AC MMC with three flying capacitors per output phase using SIMULINK [7] is shown in Figure 3.3. The model parameters are shown in Table 3.3. The modulation function shown in equation 3.6 is used to calculate the duty cycle for the bidirectional switches. In Figure 3.3, the Embedded MATLAB Function with the inputs q, V_{im}, f_i, f_o and time module calculates the nine modulation functions for the three-phase input and output voltages defined

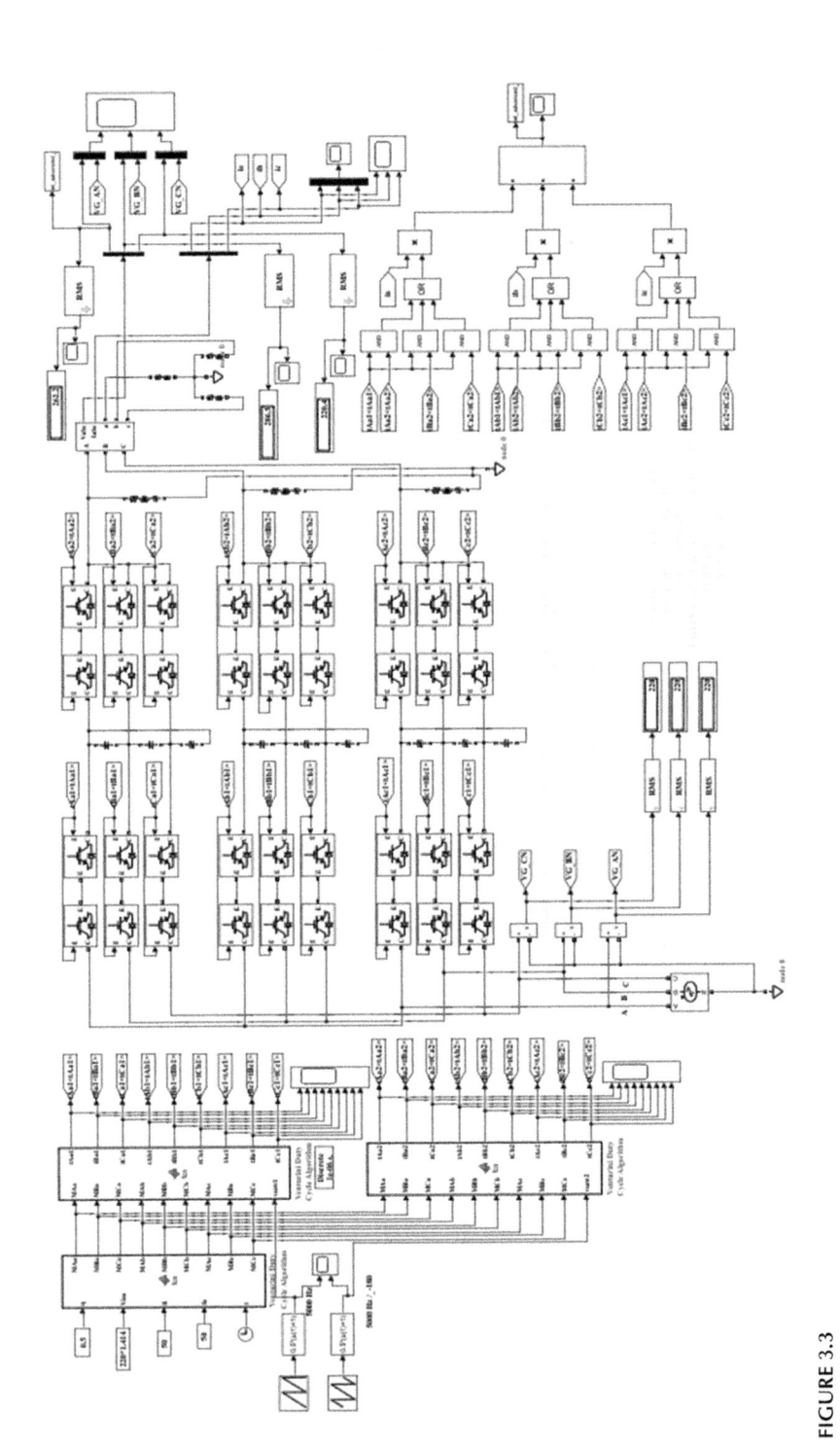

FIGURE 3.3
Model of three-phase AC to three-phase AC multilevel matrix converter with three flying capacitors per output phase.

in equations 2.9 and 2.11 of Chapter 2, each phase voltage corresponding to input and output phase shifted by $\pi/2$ rad. The source code is the same as shown in Program segment 2.3. The switching pulses for the nine switches starting from S_{Aa1} to S_{Cc1} in the first column of Figure 3.1 are obtained by comparison of the respective modulation functions with a 5 kHz saw-tooth waveform V_{saw1} and applying logic operation using a second Embedded MATLAB Function. The same for the other group of bidirectional switches S_{Aa2} to S_{Cc2} in the second column of Figure 3.1 are obtained by comparing the respective modulation functions with another 5 kHz saw-tooth waveform V_{saw2} which is phase shifted by π rad or $T_{sw}/2$ s $(1/(2*f_{sw}))$ from the first saw-tooth waveform V_{saw1} and applying logic operation using a third Embedded MATLAB Function. The source code for the logic operation to generate the gate pulses for the two groups of nine bidirectional switches is the same as in Program segment 2.2. The only difference is that the second Embedded MATLAB Function compares the nine modulation functions with saw-tooth carrier V_{saw1} and performs logic operation to generate gate pulses t_{Aa1} to t_{Cc1} for the nine switches in the first column of Figure 3.1. The third Embedded MATLAB Function compares the nine modulation functions with saw-tooth carrier V_{saw2} and performs logic operation to generate gate pulses t_{Aa2} to t_{Cc2} for the nine switches in the second column of Figure 3.1. The output filter parameter is chosen by assuming a 10 Ω resistance, 2 mH inductance and the output filter capacitor using equation 3.7 for the carrier switching frequency f_{sw} in Table 3.3. The three-phase sine-wave AC voltage can be generated by entering [381.04 0 50] in the box named 'Positive-sequence:' in the three-phase programmable voltage source model. Phase A input current i_A is retrieved by using logic gates, multipliers and summers as shown in Figure 3.3, as per equation 3.5. The flying capacitors C_1 to C_9 in Figure 3.1 are selected by

TABLE 3.3

PH3 MMC with Three FC Model Parameters

Sl. No.	Parameter	Value	Units
1	RMS line-to-neutral input voltage V_i	220	V
2	Input frequency f_i	50	Hz
3	Output frequency f_o	50	Hz
4	Modulation index q	0.5	-
5	Carrier switching frequency f_{sw}	5	kHz
6	Flying capacitor C_1 to C_9	10	μF
7	Series RLC output filter R_f, L_f, C_f	10, 2e-3, 0.50712e-6	Ω, H, F
8	RL load	50, 0.5	Ω, H

assuming a peak value of load current of 2 A, a voltage ripple of ΔV_c of 20% for the carrier switching frequency f_{sw} in Table 3.3.

3.5.1 Simulation Results

The simulation of the three-phase MMC with three FC per output phase for the data in Table 3.3 is carried out using SIMULINK [7]. The fixed-step ode5 (Dormand–Prince) solver is used.

Harmonic spectrum of the line-to-neutral output voltage, phase A input current and line-to-line output voltage are shown in Figures 3.4–3.6, respectively. The simulation results for the three-phase line-to-neutral output voltage, Phase A input current, three-phase line-to-line output voltage and three-phase load current are shown in Figures 3.7–3.10, respectively. The two 5 kHz saw-tooth carriers and the gate pulse for the nine group of bidirectional switches in the first and second columns of Figure 3.1 are shown in Figures 3.11 and 3.12a,b, respectively. The simulation results from the harmonic spectrum are tabulated in Table 3.4.

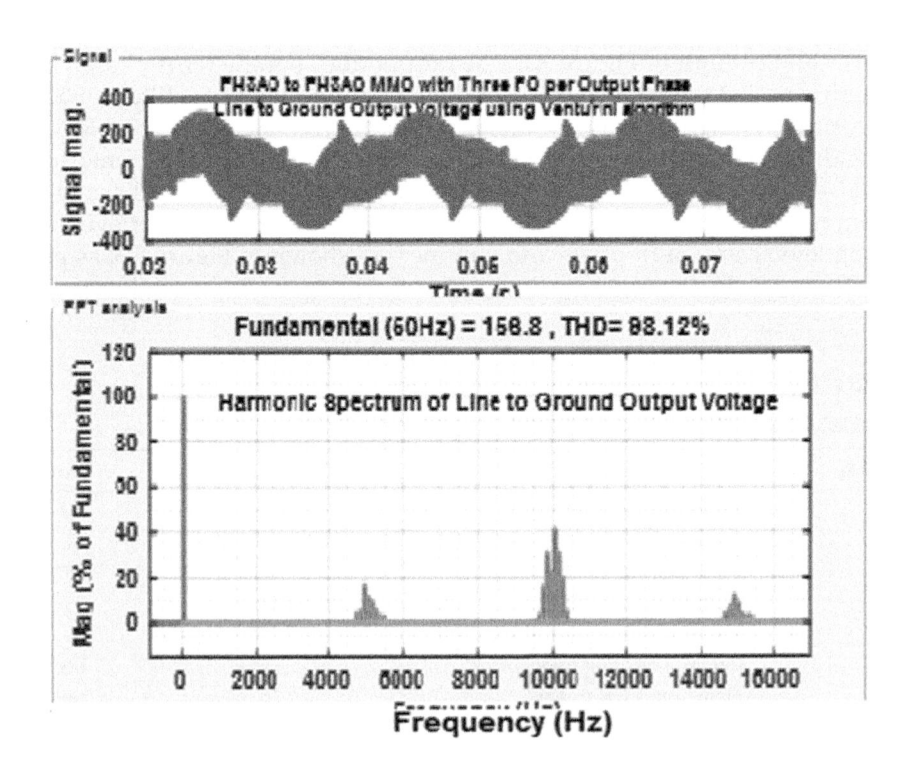

FIGURE 3.4
Three-phase MMC with three FC per output phase – line-to-ground output voltage and harmonic spectrum.

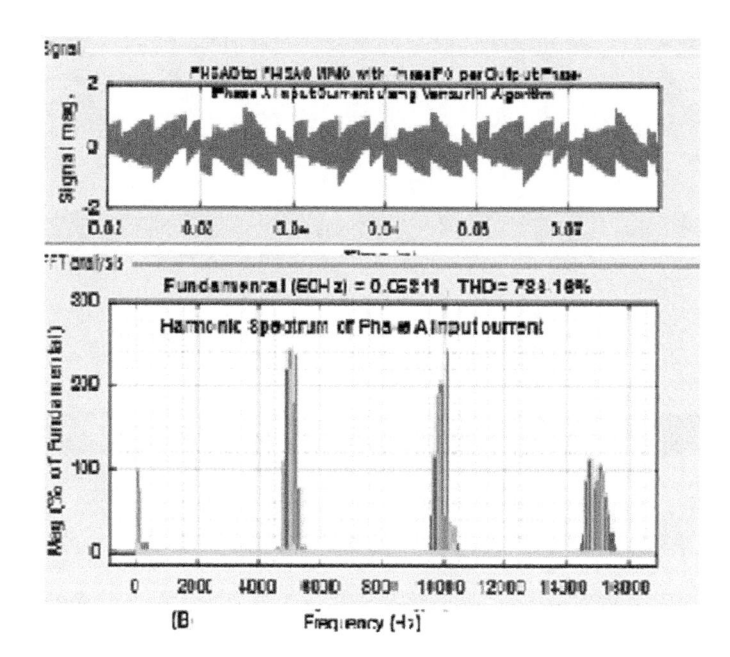

FIGURE 3.5
Three-phase MMC with three FC per output phase – Phase A input current and harmonic spectrum.

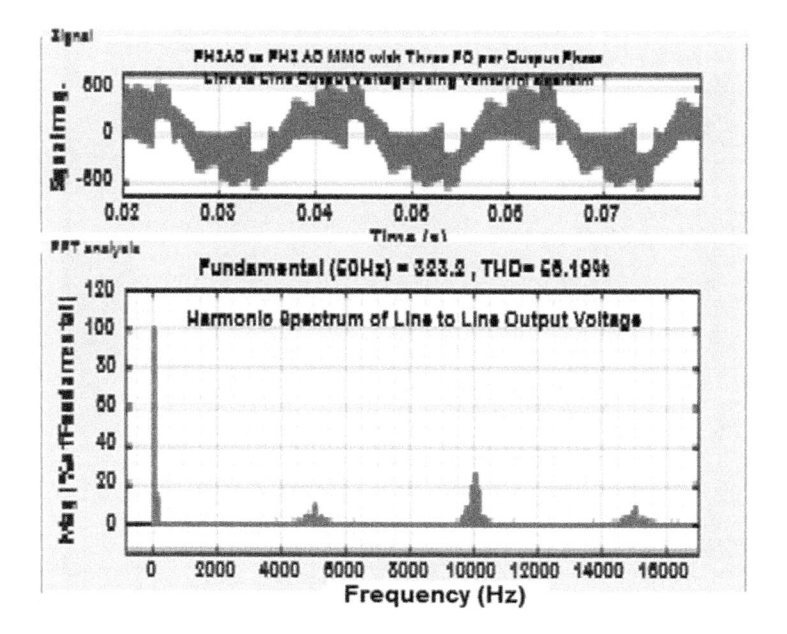

FIGURE 3.6
Three-phase MMC with three FC per output phase – line-to-line output voltage and harmonic spectrum.

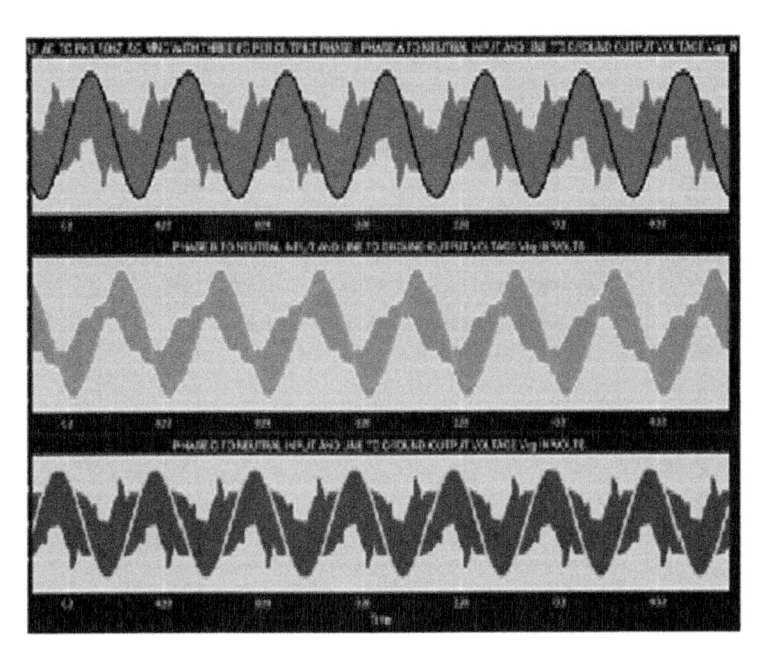

FIGURE 3.7
Three-phase MMC with three FC per output phase – line-to-neutral output voltage.

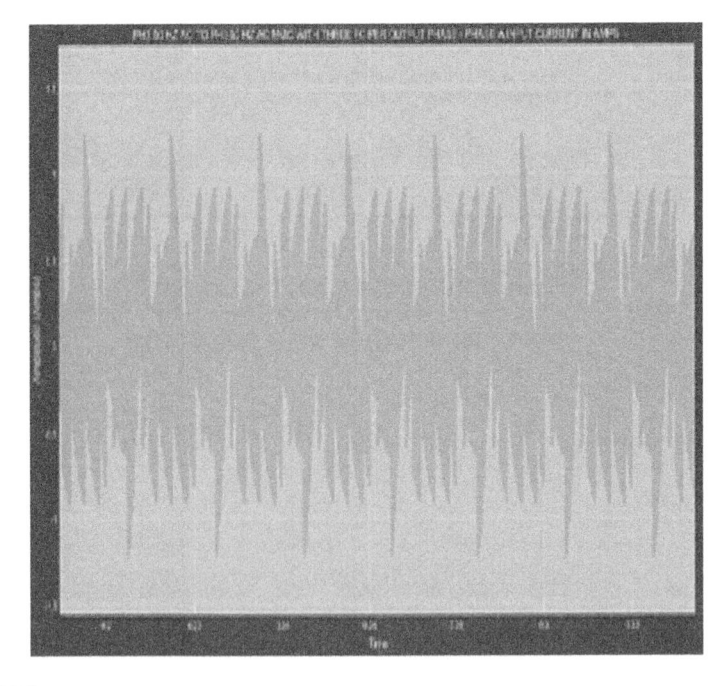

FIGURE 3.8
Three-phase MMC with three FC per output phase – phase A input current.

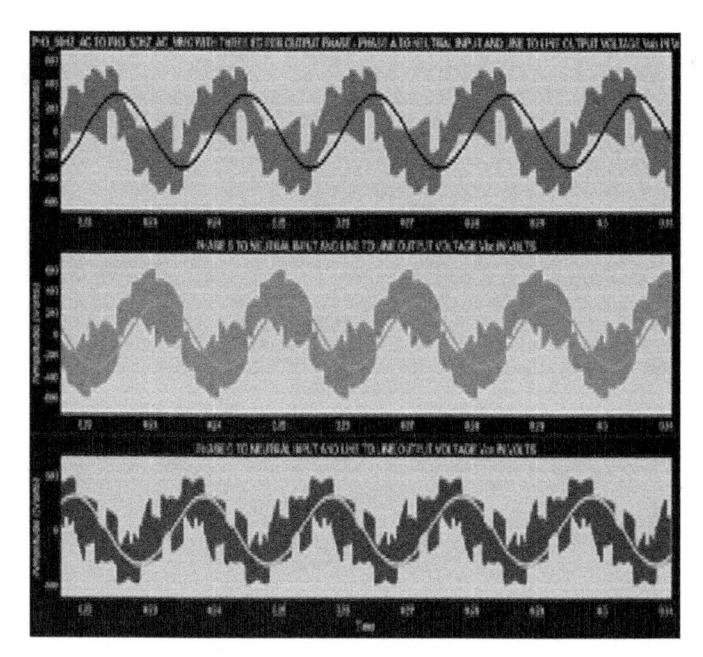

FIGURE 3.9
Three-phase MMC with three FC per output phase – line-to-line output voltage.

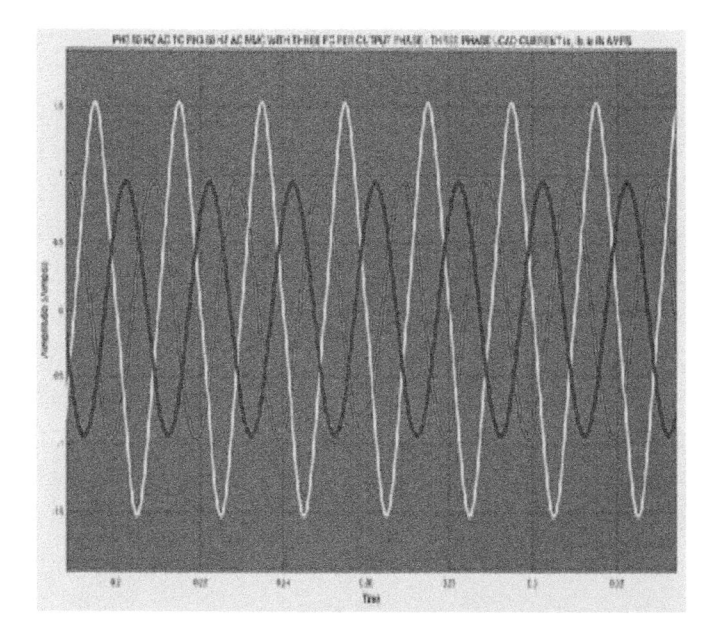

FIGURE 3.10
Three-phase MMC with three FC per output phase – three-phase load current.

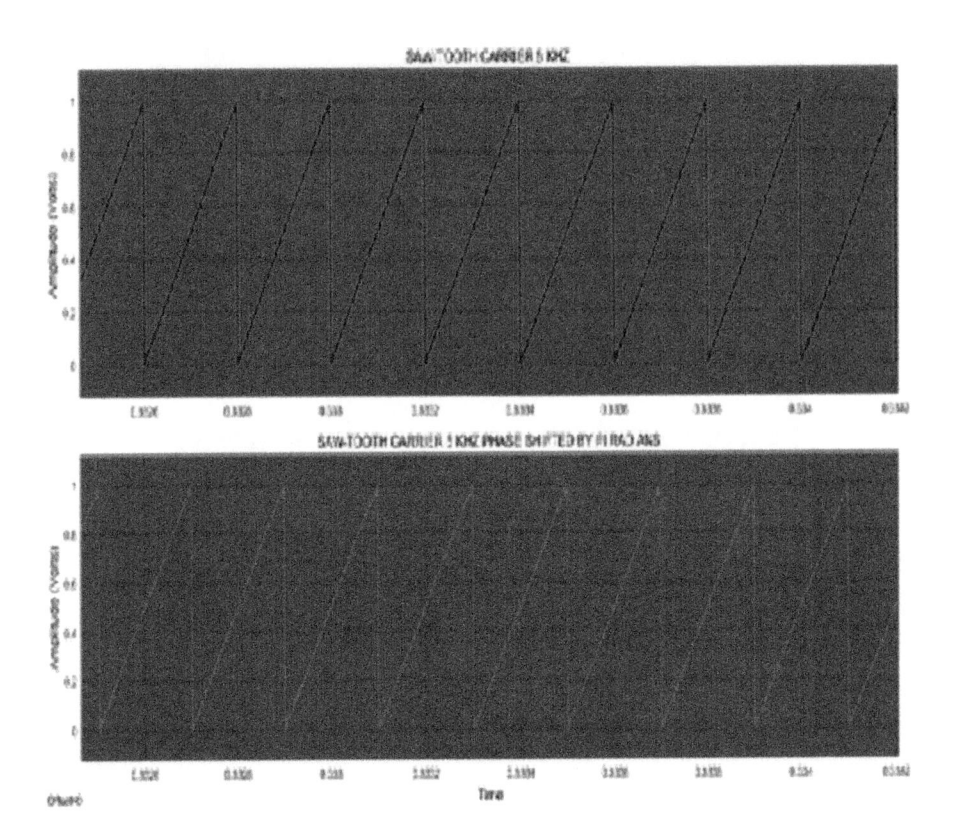

FIGURE 3.11
Two 5-kHz saw-tooth carriers phase shifted by 180°.

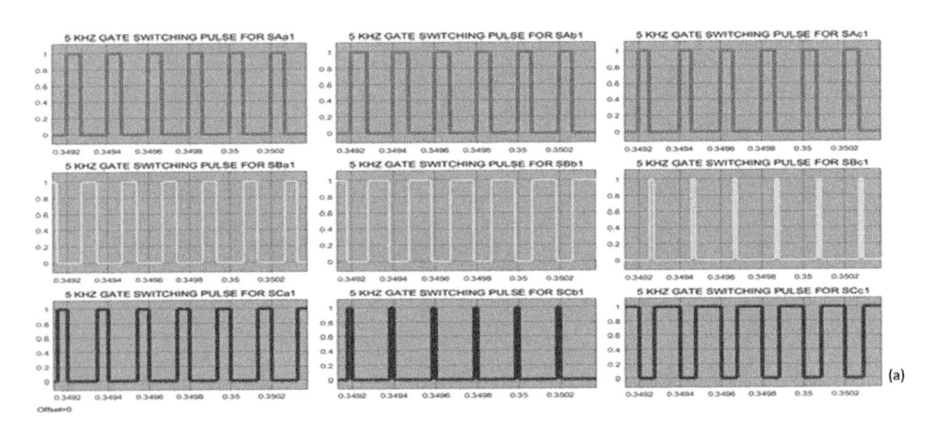

FIGURE 3.12
(a) Gate pulses for the group of nine switches in cell 1 and (b) gate pulses for the group of switches in cell 2 for the three-phase MMC with three FC per output phase.

(Continued)

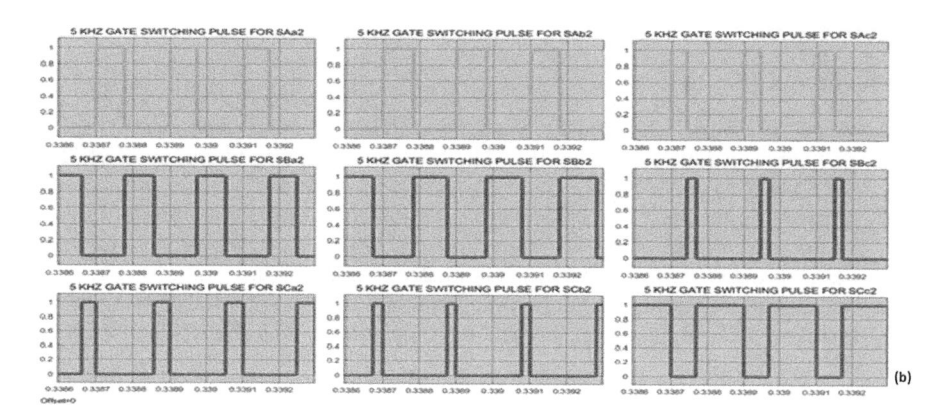

FIGURE 3.12 (CONTINUED)
(a) Gate pulses for the group of nine switches in cell 1 and (b) gate pulses for the group of switches in cell 2 for the three-phase MMC with three FC per output phase.

TABLE 3.4

Three-Phase MMC with Three FC per Output Phase – Simulation Results

			Line-to-Neutral Output Voltage		Line-to-Line Output Voltage		Phase A Input Current	
Sl. No.	Topology	Parameters	Peak Fundamental (V)	T.H.D. (p.u.)	Peak Fundamental (V)	T.H.D. (p.u.)	Peak Fundamental (A)	T.H.D. (p.u.)
1	Three-phase MMC with three FC per output phase	Table 3.3	156.8	0.9812	323.2	0.5619	0.05811	7.86

3.6 Conclusions

The simulation results for the line-to-neutral and line-to-line output voltage indicate severe voltage unbalance for output phase b for the former case and between the phase b and c for the later. This in turn has caused load current unbalance. This problem caused by the three-phase MMC with three FC topology was re-examined using PSCAD, and the simulation results are reported [8,9].

References

1. Y. Shi, X. Yang, Q. He and Z. Wang: Research on a novel capacitor clamped multilevel matrix converter, *IEEE Transactions on Power Electronics*, Vol. 20, No. 5, 2005, pp. 1055–1065.
2. J. Rzasa: Capacitor clamped multilevel matrix converter controlled with Venturini method, *13th International Power Electronics and Motion Control Conference (EPE-PEMC)*, Poznan, Poland, 2008, pp. 357–364.
3. A. Nabae, I. Takahashi and H. Akagi: A new neutral-point-clamped PWM inverter, *IEEE Transactions on Industry Applications*, Vol. IA-17, No. 5, 1981, pp. 518–523.
4. T.A. Meynard and H. Foch: Multilevel conversion: high voltage choppers and voltage-source inverters, *IEEE-Power Electronics Specialists Conference (PESC)*, Toledo, Spain, Vol. 1, July 1992, pp. 397–403.
5. T.A. Meynard, H. Foch, F. Forest, C. Turpin, F. Richardeau, L. Delmas, G. Gateau and E. Lefeuvre: Multicell converters: derived topologies, *IEEE Transactions on Industrial Electronics*, Vol. 49, No. 5, 2002, pp. 978–987.
6. R.H. Baker and L.H. Bannister: Electric power converter, U.S. Patent #3867643, February 18, 1975.
7. The Mathworks Inc.: www.mathworks.com, MATLAB/SIMULINK release notes, MATLAB R2017b, 2017.
8. N.P.R. Iyer: Modelling, simulation and real time implementation of a three phase AC to AC matrix converter, Ph.D. thesis, Ch. 7 and Ch. 14, Department of ECE, Curtin University, Perth, WA, Australia, February 2012.
9. N.P.R. Iyer: Performance comparison of a three-phase multilevel matrix converter with three flying capacitors per output phase with a three-phase conventional matrix converter, *Electrical Engineering*, Vol. 99, No. 2, 2017, pp. 775–789. DOI: 10.1007/s00202-016-0500-4.

4

Direct Space Vector Modulation of Three-Phase Matrix Converter

4.1 Introduction

The most used modulation strategy for matrix converter (MC) is the space vector modulation (SVM) since it is well suitable for digital implementation [1–5]. It provides full exploitation of the input voltage and good load current quality [1–5]. The SVM approach has advantages with respect to the traditional carrier-based modulation technique such as (a) immediate comprehension of the required commutation process, (b) simplified control algorithm, (c) maximum voltage transfer without adding third harmonic components and avoiding a fictitious DC link and (d) no synchronization requirements with input voltage waveform. Moreover, the proposed switching algorithm allows to reduce the number of switching devices involved in a commutation process with respect to conventional switching strategies. The algorithm is based on the instantaneous space vector representation of input current and output voltages. It analyses all the possible switching configurations available in three-phase MCs and does not need the concept of a virtual DC link. This modulation technique is also known as direct space vector modulation (DSVM). In this chapter, the models of the three-phase AC to three-phase AC MC using DSVM algorithm have been developed using SIMULINK and the results obtained by simulation are presented.

Three-phase MCs have received considerable attention in recent years because they may become a good alternative to pulse-width-modulated (PWM) voltage-source inverter topology. This is because the MC provides direct AC to AC power conversion, bidirectional power flow, nearly sinusoidal input/output waveforms and a controllable input power factor. Furthermore, the MC allows a compact design due to the lack of DC-link capacitors for energy storage. The complexity of the topology of MCs makes modulation a difficult task. Two approaches for modulation of MCs, namely modulation duty-cycle matrix (MDCM) [1–3] and SVM, are commonly employed [4,5]. In the Alesina–Venturini method developed in [1–3], full control of output voltage and input power factor is possible and this algorithm provides a maximum voltage transfer ratio of 0.5. The inclusion of third harmonics in the input and output voltage waveforms has been successfully implemented in [3] to increase the maximum voltage transfer ratio to 0.866. In addition, the direct connection between output and input causes a great sensibility to grid distortions. This problem as well as the large number of semiconductor bidirectional switches in the power stage, the complexity of the modulation algorithms or the correct commutation between the bidirectional switches [3] have been overcome in recent years [5–9]. Permanent development of this kind of power converter systems makes it necessary to build reliable simulation models before the physical implementation of the power converter. Therefore, there is a great interest on developing simulation models in order to allow an early design and analysis of the power converter. MATLAB/ SIMULINK is one of the most flexible development environments in order to model this kind of power converter systems because of its capability for simulating complex control and modulation algorithms. It enables modelling of the control stages of the MC with more complex modulation algorithms using Embedded MATLAB Function, MATLAB Function, Math block set, Logic and Bit Operator block set and the power stages from the Power Systems block set. This is why MATLAB/SIMULINK is chosen to simulate the DSVM algorithm for MCs. SVM has become one of the most preferred modulations for MCs. The principal reason for this is the better harmonic performance that can be achieved using different switching strategies on each commutation period. Two versions for SVM are defined: the indirect modulation [4] and the direct one [5–9]. The former proposes an equivalent circuit, very similar to a back-to-back converter with a rectifier and an inverter stage. The latter uses the MC's own space vectors to be used in the modulation. This chapter proposes a MATLAB/SIMULINK simulation model for DSVM MC so that different switching patterns can be simulated and analysed before a physical implementation [10–12].

4.2 Direct Space Vector Modulation Algorithm

The three-phase AC to three-phase AC MC is shown in Figure 2.1. The 21 out of 27 switching states used in DSVM are shown in Table 4.1. The 21 possible states that can be generated by the matrix of bidirectional switches shown in Table 4.1 can fully control both the output voltage vector and input current displacement angle [5–12]. The DSVM technique is explained below [5]. The output voltage and input current space vector hexagon are shown in Figure 4.1a,b. At any sampling instant, the output voltage vector v_o and the input current displacement angle φ_i are known as reference quantities (Figure 4.2a,b). The input line-to-neutral voltage vector v_i is imposed by the source voltages and is known by measurements. Then, the control of φ_i can

TABLE 4.1

DSVM Switching States

Sl. No.	States a b c	Switches On	Output Phase Voltage		Input Current	
			$\|V_o\|$	α_o	$\|I_i\|$	β_i
1	A B B +1	$S_{Aa}\,S_{Bb}\,S_{Bc}$	$\dfrac{2v_{AB}}{3}$	0	$\dfrac{2i_a}{\sqrt{3}}$	$\dfrac{-\pi}{6}$
2	B A A −1	$S_{Ba}\,S_{Ab}\,S_{Ac}$	$\dfrac{-2v_{AB}}{3}$	0	$\dfrac{-2i_a}{\sqrt{3}}$	$\dfrac{-\pi}{6}$
3	B C C +2	$S_{Ba}\,S_{Cb}\,S_{Cc}$	$\dfrac{2v_{BC}}{3}$	0	$\dfrac{2i_a}{\sqrt{3}}$	$\dfrac{\pi}{2}$
4	C B B −2	$S_{Ca}\,S_{Bb}\,S_{Bc}$	$\dfrac{-2v_{BC}}{3}$	0	$\dfrac{-2i_a}{\sqrt{3}}$	$\dfrac{\pi}{2}$
5	C A A +3	$S_{Ca}\,S_{Ab}\,S_{Ac}$	$\dfrac{2v_{CA}}{3}$	0	$\dfrac{2i_a}{\sqrt{3}}$	$\dfrac{7\pi}{6}$
6	A C C −3	$S_{Aa}\,S_{Cb}\,S_{Cc}$	$\dfrac{-2v_{CA}}{3}$	0	$\dfrac{-2i_a}{\sqrt{3}}$	$\dfrac{7\pi}{6}$
7	B A B +4	$S_{Ba}\,S_{Ab}\,S_{Bc}$	$\dfrac{2v_{AB}}{3}$	$\dfrac{2\pi}{3}$	$\dfrac{2i_b}{\sqrt{3}}$	$\dfrac{-\pi}{6}$
8	A B A −4	$S_{Aa}\,S_{Bb}\,S_{Ac}$	$\dfrac{-2v_{AB}}{3}$	$\dfrac{2\pi}{3}$	$\dfrac{-2i_b}{\sqrt{3}}$	$\dfrac{-\pi}{6}$
9	C B C +5	$S_{Ca}\,S_{Bb}\,S_{Cc}$	$\dfrac{2v_{BC}}{3}$	$\dfrac{2\pi}{3}$	$\dfrac{2i_b}{\sqrt{3}}$	$\dfrac{\pi}{2}$
10	B C B −5	$S_{Ba}\,S_{Cb}\,S_{Bc}$	$\dfrac{-2v_{BC}}{3}$	$\dfrac{2\pi}{3}$	$\dfrac{-2i_b}{\sqrt{3}}$	$\dfrac{\pi}{2}$
11	A C A +6	$S_{Aa}\,S_{Cb}\,S_{Ac}$	$\dfrac{2v_{CA}}{3}$	$\dfrac{2\pi}{3}$	$\dfrac{2i_b}{\sqrt{3}}$	$\dfrac{7\pi}{6}$
12	C A C −6	$S_{Ca}\,S_{Ab}\,S_{Cc}$	$\dfrac{-2v_{CA}}{3}$	$\dfrac{2\pi}{3}$	$\dfrac{-2i_b}{\sqrt{3}}$	$\dfrac{7\pi}{6}$
13	B B A +7	$S_{Ba}\,S_{Bb}\,S_{Ac}$	$\dfrac{2v_{AB}}{3}$	$\dfrac{4\pi}{3}$	$\dfrac{2i_c}{\sqrt{3}}$	$\dfrac{-\pi}{6}$
14	A A B −7	$S_{Aa}\,S_{Ab}\,S_{Bc}$	$\dfrac{-2v_{AB}}{3}$	$\dfrac{4\pi}{3}$	$\dfrac{-2i_c}{\sqrt{3}}$	$\dfrac{-\pi}{6}$
15	C C B +8	$S_{Ca}\,S_{Cb}\,S_{Bc}$	$\dfrac{2v_{BC}}{3}$	$\dfrac{4\pi}{3}$	$\dfrac{2i_c}{\sqrt{3}}$	$\dfrac{\pi}{2}$
16	B B C −8	$S_{Ba}\,S_{Bb}\,S_{Cc}$	$\dfrac{-2v_{BC}}{3}$	$\dfrac{4\pi}{3}$	$\dfrac{-2i_c}{\sqrt{3}}$	$\dfrac{\pi}{2}$
17	A A C +9	$S_{Aa}\,S_{Ab}\,S_{Cc}$	$\dfrac{2v_{CA}}{3}$	$\dfrac{4\pi}{3}$	$\dfrac{2i_c}{\sqrt{3}}$	$\dfrac{7\pi}{6}$
18	C C A −9	$S_{Ca}\,S_{Cb}\,S_{Ac}$	$\dfrac{-2v_{CA}}{3}$	$\dfrac{4\pi}{3}$	$\dfrac{-2i_c}{\sqrt{3}}$	$\dfrac{7\pi}{6}$
19	A A A 0_1	$S_{Aa}\,S_{Ab}\,S_{Ac}$	0	0	0	0
20	B B B 0_2	$S_{Ba}\,S_{Bb}\,S_{Bc}$	0	0	0	0
21	C C C 0_3	$S_{Ca}\,S_{Cb}\,S_{Cc}$	0	0	0	0

be achieved by controlling the phase angle β of the input current vector. In principle, the SVM algorithm is based on the selection of four active switching configurations that are applied for suitable time intervals within each sampling period T_s. The zero configurations are applied to complete T_s [5].

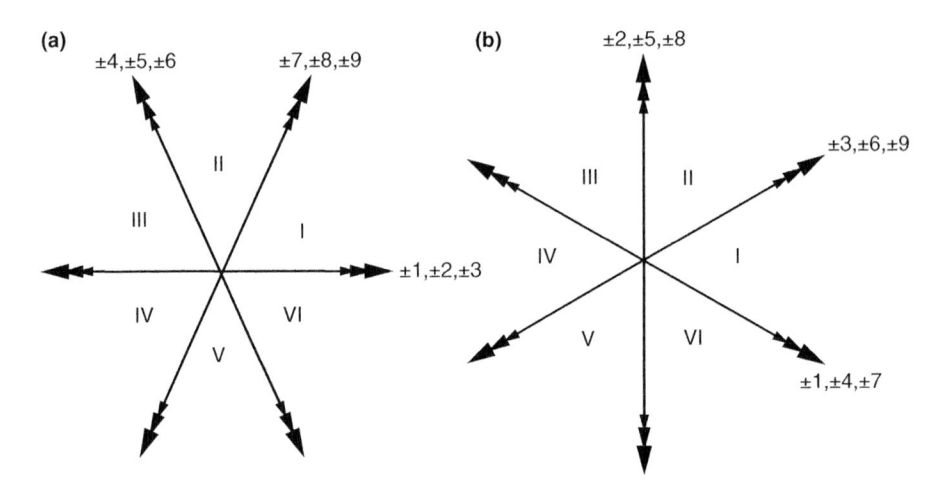

FIGURE 4.1
(a) Output voltage space vector hexagon and (b) input current space vector hexagon.

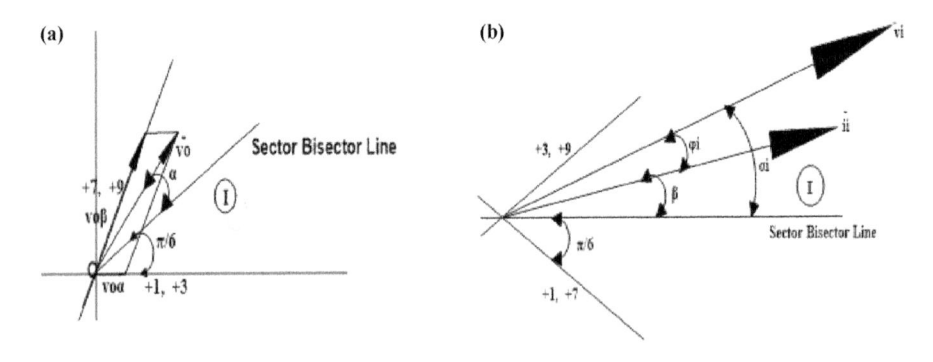

FIGURE 4.2
(a) Output voltage vectors modulation principle and (b) input current vectors modulation principle.

To explain the SVM algorithm, referring to Figure 4.2a,b, both output voltage vector v_o and input current vector i_i are assumed to be in sector I. The reference voltage vector v_o is resolved into the components $v_{o\alpha}$ and $v_{o\beta}$ along the two adjacent vector directions. The component $v_{o\beta}$ can be synthesized using two voltage vectors having the same direction of $v_{o\beta}$. Among the six possible switching configurations, ±7, ±8, ±9 shown in Figure 4.1a, the ones that allow also the modulation of the input current direction must be selected. It is seen that this constraint allows the elimination of two switching configurations, $+8$ and -8 in this case. Among the remaining four switching configurations for sector I, the application of the positive switching configurations $+7$ and $+9$ are assumed which are common to both output voltage and input current sectors. With similar considerations, the switching configurations required

TABLE 4.2

Active Vectors Configuration

Sector of Input Current		Sector of Output Voltage												
		1 or 4				2 or 5				3 or 6				
1 or 4		9	7	3	1	6	4	9	7	3	1	6	4	
2 or 5		8	9	2	3	5	6	8	9	2	3	5	6	
3 or 6		7	8	1	2	4	5	7	8	1	2	4	5	
		I	II	III	IV	I	II	III	IV	I	II	III	IV	

to synthesize the component $v_{o\alpha}$ which are common to both output voltage and input current sectors can be found as +1 and +3, respectively [5].

Using the same procedure, it is possible to determine the four switching configurations related to any possible combination of output voltage and input current sectors, leading to the results summarized in Table 4.2. Four symbols (I, II, III, IV) are also introduced in the last row of Table 4.2 to identify the four general switching configurations, valid for any combination of input current and output voltage sectors [5–12].

Now it is possible to write in a general form, the four basic equations of the SVM algorithm, which satisfy simultaneously the requirements of the reference output voltage vector and input current displacement angle. With reference to the output voltage vector, the following two equations can be written:

$$\overrightarrow{v_{o\beta}} = \overrightarrow{v_o^I} * \delta^I + \overrightarrow{v_o^{II}} * \delta^{II} = \frac{2}{\sqrt{3}} * v_o * \cos\left(\alpha - \frac{\pi}{3}\right) * e^{j\left[(S_v-1)*\frac{\pi}{3}+\frac{\pi}{3}\right]} \tag{4.1}$$

$$\overrightarrow{v_{o\alpha}} = \overrightarrow{v_o^{III}} * \delta^{III} + \overrightarrow{v_o^{IV}} * \delta^{IV} = \frac{2}{\sqrt{3}} * v_o * \cos\left(\alpha + \frac{\pi}{3}\right) * e^{j\left[(S_v-1)*\frac{\pi}{3}\right]} \tag{4.2}$$

With reference to the input current displacement angle, two equations are obtained by imposing on the vectors $\left(\overrightarrow{i_i^I} * \delta^I + \overrightarrow{i_i^{II}} * \delta^{II}\right)$ and $\left(\overrightarrow{i_i^{III}} * \delta^{III} + \overrightarrow{i_i^{IV}} * \delta^{IV}\right)$ the direction defined by β. This can be achieved by imposing a null value on the two-vector component along the direction perpendicular to $e^{j\beta}$ (i.e. $je^{j\beta}$), leading to the following equations:

$$\left(\overrightarrow{i_i^I} * \delta^I + \overrightarrow{i_i^{II}} * \delta^{II}\right) * je^{j\beta} * e^{j(S_i-1)*\frac{\pi}{3}} = 0 \tag{4.3}$$

$$\left(\overrightarrow{i_i^{III}} * \delta^{III} + \overrightarrow{i_i^{IV}} * \delta^{IV}\right) * je^{j\beta} * e^{j(S_i-1)*\frac{\pi}{3}} = 0 \tag{4.4}$$

$\delta^I, \delta^{II}, \delta^{III}$ and δ^{IV} are the duty cycles (i.e. $\delta^I = t^I/T_s$) of the four switching configurations, $S_v = 1,2\ldots6$ represents the output voltage sector, and $S_i = 1,2\ldots6$ represents the input current sector. $\overrightarrow{v_o^I}, \overrightarrow{v_o^{II}}, \overrightarrow{v_o^{III}}, \overrightarrow{v_o^{IV}}$ are the output voltage vectors associated, respectively, with the switching configurations I, II, III and IV given in Table 4.2. The same convention holds good for the input current vectors.

Solving equations 4.1–4.4 leads to the following equations for the duty cycles:

$$\delta^{\mathrm{I}} = (-1)^{(S_v + S_i)} * \frac{2q}{\sqrt{3}} * \frac{\cos\left(\alpha - \dfrac{\pi}{3}\right) * \cos\left(\beta - \dfrac{\pi}{3}\right)}{\cos(\varphi_i)} \tag{4.5}$$

$$\delta^{\mathrm{II}} = (-1)^{(S_v + S_i + 1)} * \frac{2q}{\sqrt{3}} * \frac{\cos\left(\alpha - \dfrac{\pi}{3}\right) * \cos\left(\beta + \dfrac{\pi}{3}\right)}{\cos(\varphi_i)} \tag{4.6}$$

$$\delta^{\mathrm{III}} = (-1)^{(S_v + S_i + 1)} * \frac{2q}{\sqrt{3}} * \frac{\cos\left(\alpha + \dfrac{\pi}{3}\right) * \cos\left(\beta - \dfrac{\pi}{3}\right)}{\cos(\varphi_i)} \tag{4.7}$$

$$\delta^{\mathrm{IV}} = (-1)^{(S_v + S_i)} * \frac{2q}{\sqrt{3}} * \frac{\cos\left(\alpha + \dfrac{\pi}{3}\right) * \cos\left(\beta + \dfrac{\pi}{3}\right)}{\cos(\varphi_i)} \tag{4.8}$$

Furthermore, for the feasibility of the control strategy, the sum of the absolute values of the four duty cycles must be lower than unity:

$$\text{i.e., } \left|\delta^{\mathrm{I}}\right| + \left|\delta^{\mathrm{II}}\right| + \left|\delta^{\mathrm{III}}\right| + \left|\delta^{\mathrm{IV}}\right| \leq 1 \tag{4.9}$$

The zero configurations are applied to complete the sampling period T_s, as shown below:

$$\delta^0 = 1 - \delta^{\mathrm{I}} - \delta^{\mathrm{II}} - \delta^{\mathrm{III}} - \delta^{\mathrm{IV}} \tag{4.10}$$

In equations 4.5–4.8, negative value appears which corresponds to the negative switching state given in Table 4.1.

Switching strategies deal with the switching configuration sequence, that is, the order in which the active and zero vectors are applied along the commutation period. Here, both the asymmetrical SVM (ASVM) and the symmetrical SVM (SSVM) are analysed in detail and a simulation model is developed. The ASVM uses only one of the three zero configurations in the middle of the sequence, so that minimum switch commutations are achieved between one switching state and the next one. Using this technique, the switching commutations are up to eight for each commutation period. This way switching losses are minimized.

Two simple rules are proposed for both ASVM and SSVM which are given below [5–11]:

Rule 1: If (SV + SI) is even, then the order of the switching sequence is: $\delta^{\mathrm{III}}, \delta^{\mathrm{I}}, \delta^{\mathrm{II}}, \delta^{\mathrm{IV}}$.

TABLE 4.3

Zero Configuration for ASVM

I_i Sector	V_o Sector 1, 2, 3, 4, 5 or 6
1 or 4	AAA
2 or 5	CCC
3 or 6	BBB

Rule 2: If (SV + SI) is odd, then the order of the switching sequence is: $\delta^I, \delta^{III}, \delta^{IV}, \delta^{II}$.

Also referring to equations 4.5–4.8, if (SV + SI) is even, δ^{II} and δ^{III} have a negative sign and if (SV + SI) is odd, δ^I and δ^{IV} have a negative sign. This negative sign corresponds to the negative switching state in Table 4.1.

With regard to zero configuration for ASVM, Table 4.3 can be used. For example, considering output voltage vector in sector 1 and input current vector in sector 4 (i.e. SV = 1; SI = 4) within their respective hexagons, it can be seen that (SV + SI) is odd, rule 2 applies with δ^I and δ^{IV} having negative sign. Also using Table 4.3, the following is the only possible double-sided sequences that can be generated for ASVM technique:

−9, +3, −1, +7 which from Table 4.1 is given below:

$$\underbrace{CCA - CAA - AAA - BAA - BBA}_{T_s/2} \Big| \underbrace{BBA - BAA - AAA - CAA - CCA}_{T_s/2}$$

where T_s is the sampling period.

The gate pulse pattern for ASVM using Tables 4.1–4.3, rules 1 and 2 is shown in Table 4.4.

With regard to zero configuration for SSVM, Table 4.5 shown below can be used.

For example, considering output voltage vector in sector 1 and input current vector in sector 4 (i.e. SV = 1; SI = 4) within their respective hexagons, it can be seen that (SV + SI) is odd, rule 2 applies with δ^I and δ^{IV} having negative sign. Also using Table 4.5, the following is the only possible double-sided sequences that can be generated for SSVM technique.

−9, +3, −1, +7 which from Table 4.1 is given below:

$$\underbrace{CCC - CCA - CAA - AAA - BAA - BBA - BBB}_{T_s/2} \Big|$$

$$\underbrace{BBB - BBA - BAA - AAA - CAA - CCA - CCC}_{T_s/2}$$

The gate pulse pattern for SSVM using Tables 4.1, 4.2 and 4.5, rules 1 and 2 is shown in Table 4.6.

TABLE 4.4

ASVM Switching Pattern

Sl. No.	SV, SI	(SV + SI)	ssf_x High	t_a	t_b	t_0	t_c	t_d	S_{Aa} Timing
				Bidirectional Switch Gate Pulse Pattern					
1	1, 1 or 4, 4	Even	ssf1	ACC	AAC	AAA	AAB	ABB	t_{ab0cd}
2	1, 4 or 4, 1	Odd	ssf2	CCA	CAA	AAA	BAA	BBA	t_0
3	1, 5 or 4, 2	Even	ssf3	CBB	CCB	CCC	CCA	CAA	0
4	1, 2 or 4, 5	Odd	ssf4	BBC	BCC	CCC	ACC	AAC	t_{cd}
5	1, 3 or 4, 6	Even	ssf5	BAA	BBA	BBB	BBC	BCC	0
6	1, 6 or 4, 3	Odd	ssf6	AAB	ABB	BBB	CBB	CCB	t_{ab}
7	2, 4 or 5, 1	Even	ssf7	CCA	ACA	AAA	ABA	BBA	t_{b0c}
8	2, 1 or 5, 4	Odd	ssf8	CAC	AAC	AAA	AAB	BAB	t_{b0c}
9	2, 2 or 5, 5	Even	ssf9	BBC	CBC	CCC	CAC	AAC	t_d
10	2, 5 or 5, 2	Odd	ssf10	BCB	CCB	CCC	CCA	ACA	t_d
11	2, 6 or 5, 3	Even	ssf11	AAB	BAB	BBB	BCB	CCB	t_a
12	2, 3 or 5, 6	Odd	ssf12	ABA	BBA	BBB	BBC	CBC	t_a
13	3, 1 or 6, 4	Even	ssf13	CAC	CAA	AAA	BAA	BAB	t_0
14	3, 4 or 6, 1	Odd	ssf14	ACC	ACA	AAA	ABA	ABB	t_{ab0cd}
15	3, 5 or 6, 2	Even	ssf15	BCB	BCC	CCC	ACC	ACA	t_{cd}
16	3, 2 or 6, 5	Odd	ssf16	CBB	CBC	CCC	CAC	CAA	0
17	3, 3 or 6, 6	Even	ssf17	ABA	ABB	BBB	CBB	CBC	t_{ab}
18	3, 6 or 6, 3	Odd	ssf18	BAA	BAB	BBB	BCB	BCC	0

TABLE 4.5

Zero Configuration for SSVM

I_i Sector	V_o Sector 1, 2, 3, 4, 5 or 6
1 or 4	CCC – AAA – BBB
2 or 5	BBB – CCC – AAA
3 or 6	AAA – BBB – CCC

In Tables 4.4 and 4.6, the terms ssf1 to ssf18 represent the sector switch function for sectors 1–18 whose output voltage and input current sector locations are defined in column 2.

4.3 Model of Three-Phase Asymmetrical Space Vector Modulated Matrix Converter

The SIMULINK model of the direct ASVM three-phase AC to three-phase AC MC is shown in Figure 4.3. The power circuit of the model is developed using

TABLE 4.6

SSVM Switching Pattern

Sl. No.	SV, SI	(SV+SI)	ssf_x High	Bidirectional Switch Gate Pulse Pattern							S_{Aa} Timing
				t_{01}	t_a	t_b	t_{02}	t_c	t_d	t_{03}	
1	1, 1 or 4, 4	Even	ssf1	CCC	ACC	AAC	AAA	AAB	ABB	BBB	t_{ab02cd}
2	1, 4 or 4, 1	Odd	ssf2	CCC	CCA	CAA	AAA	BAA	BBA	BBB	t_{02}
3	1, 5 or 4, 2	Even	ssf3	BBB	CBB	CCB	CCC	CCA	CAA	AAA	t_{03}
4	1, 2 or 4, 5	Odd	ssf4	BBB	BBC	BCC	CCC	ACC	AAC	AAA	t_{cd03}
5	1, 3 or 4, 6	Even	ssf5	AAA	BAA	BBA	BBB	BBC	BCC	CCC	t_{01}
6	1, 6 or 4, 3	Odd	ssf6	AAA	AAB	ABB	BBB	CBB	CCB	CCC	t_{01ab}
7	2, 4 or 5, 1	Even	ssf7	CCC	CCA	ACA	AAA	ABA	BBA	BBB	t_{b02c}
8	2, 1 or 5, 4	Odd	ssf8	CCC	CAC	AAC	AAA	AAB	BAB	BBB	t_{b02c}
9	2, 2 or 5, 5	Even	ssf9	BBB	BBC	CBC	CCC	CAC	AAC	AAA	t_{d03}
10	2, 5 or 5, 2	Odd	ssf10	BBB	BCB	CCB	CCC	CCA	ACA	AAA	t_{d03}
11	2, 6 or 5, 3	Even	ssf11	AAA	AAB	BAB	BBB	BCB	CCB	CCC	t_{01a}
12	2, 3 or 5, 6	Odd	ssf12	AAA	ABA	BBA	BBB	BBC	CBC	CCC	t_{01a}
13	3, 1 or 6, 4	Even	ssf13	CCC	CAC	CAA	AAA	BAA	BAB	BBB	t_{02}
14	3, 4 or 6, 1	Odd	ssf14	CCC	ACC	ACA	AAA	ABA	ABB	BBB	t_{ab02cd}
15	3, 5 or 6, 2	Even	ssf15	BBB	BCB	BCC	CCC	ACC	ACA	AAA	t_{cd03}
16	3, 2 or 6, 5	Odd	ssf16	BBB	CBB	CBC	CCC	CAC	CAA	AAA	t_{03}
17	3, 3 or 6, 6	Even	ssf17	AAA	ABA	ABB	BBB	CBB	CBC	CCC	t_{01ab}
18	3, 6 or 6, 3	Odd	ssf18	AAA	BAA	BAB	BBB	BCB	BCC	CCC	t_{01}

the Power Systems block set in SIMULINK [13,15]. This mainly consists of three-phase AC voltage source (Figure 4.3), bidirectional switch matrix (Figure 2.2), output filter and R–L load (Figure 4.3). The modulation algorithm is developed in several sub-units using Embedded MATLAB Function, MATLAB Function, Math Function, Logical and Bit Operator and using Sources block set in SIMULINK [13]. The various sub-units are explained below.

4.3.1 Duty-Cycle Sequence and Sector Switch Function Generator

This is developed using Embedded MATLAB Function block in SIMULINK, as shown in Figure 4.3. With peak line-to-neutral input voltage, desired peak output phase voltage, input frequency, output frequency, time and angular frequency of reference frame ω_c as input parameters, the three-phase output voltage can be resolved into dq-axis component voltages [14] and the absolute value of the angle of the output phase voltage, vout_angle can be calculated using the function atan2(voq, vod).

Using input voltage and input phase displacement angle, a similar procedure is used to calculate the absolute value of the input current angle, i_in_angle. The value of reference frame frequency ω_c is zero. This is illustrated in Program segment 4.1.

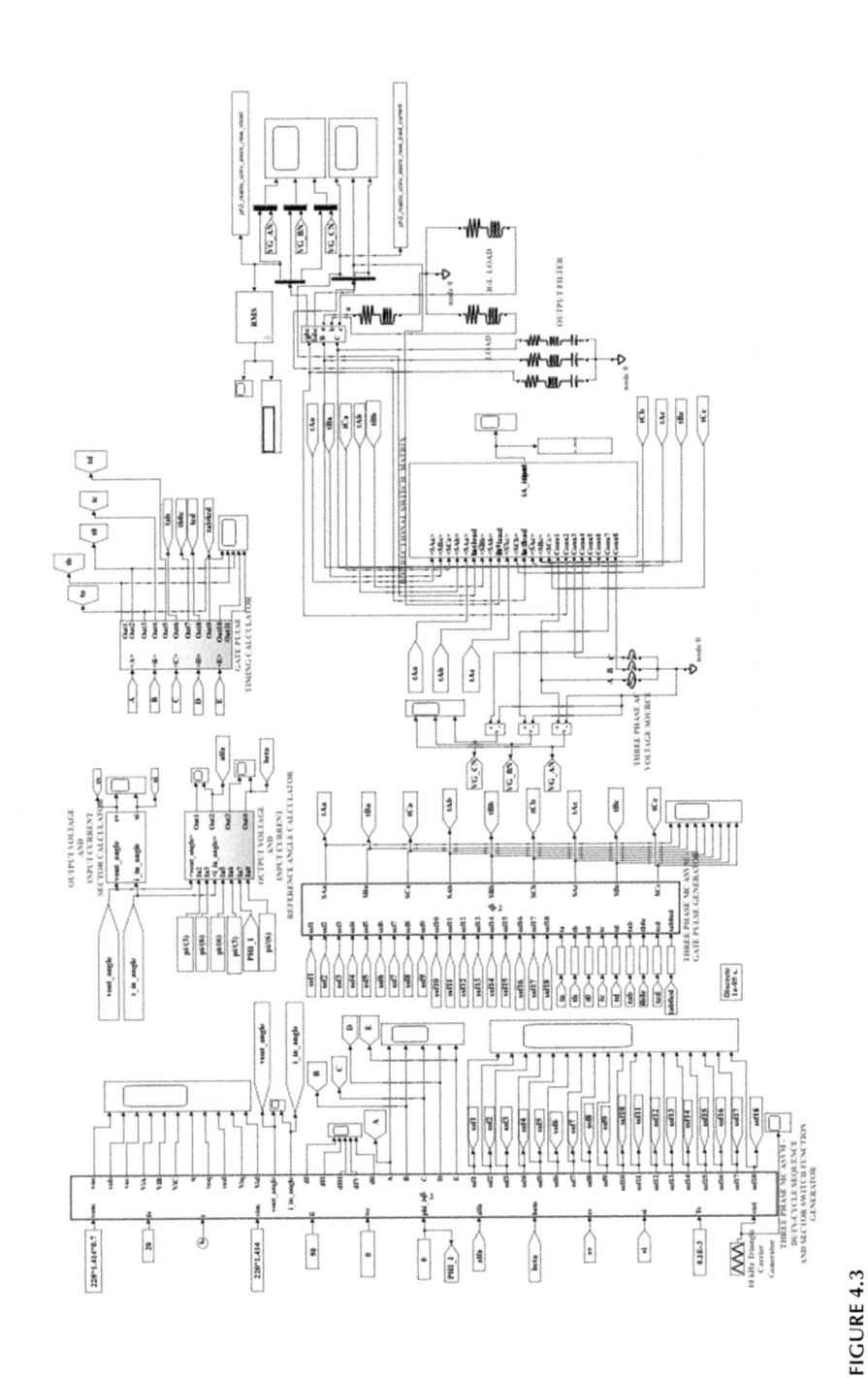

FIGURE 4.3

SIMULINK model of three-phase AC to three-phase AC direct ASVM matrix converter.

Program Segment 4.1

```
%% Dr.Narayanaswamy. P.R. Iyer
function [voa,vob,voc,ViA,ViB,ViC,q,voq,vod,Viq,Vid,vout_
angle,i_in_angle,....] = fcn(vom,fo,t,vim,fi,wc,phi_i,....)
voa = vom*sin(2*pi*fo*t);
vob = vom*sin(2*pi*fo*t - 2*pi/(3));
voc = vom*sin(2*pi*fo*t + 2*pi/(3));
ViA = vim*sin(2*pi*fi*t);
ViB = vim*sin(2*pi*fi*t - 2*pi/(3));
ViC = vim*sin(2*pi*fi*t + 2*pi/(3));
%%phi_i is the input p.f. i.e.Iin_A = k*vim*sin(2*pi*fi*t - phi_i)
%%q is the ratio of vo/vi or vom/vim
q = vom/vim;
voq = (2/(3))*(voa*cos(wc*t)+ vob*cos(wc*t - (2*pi/(3))) +
voc*cos(wc*t + (2*pi/(3)))) ;
vod = (2/(3))*(voa*sin(wc*t)+ vob*sin(wc*t - (2*pi/(3))) +
voc*sin(wc*t + (2*pi/(3)))) ;
Viq = (2/(3))*(ViA*cos(wc*t)+ ViB*cos(wc*t - (2*pi/(3))) +
ViC*cos(wc*t + (2*pi/(3)))) ;
Vid = (2/(3))*(ViA*sin(wc*t)+ ViB*sin(wc*t - (2*pi/(3))) +
ViC*sin(wc*t + (2*pi/(3)))) ;
vout_angle = atan2(voq,vod);
i_in_angle = atan2(Viq,Vid);
```

These two values of vout_angle and i_in_angle are used to calculate the output voltage sector s_v, input current sector s_i, α and β values for output voltage and input current using MATLAB Function, REM functions and SUBTRACT modules [15].

Program segment 4.2 illustrates the MATLAB code for duty-cycle sequencing and sector switch function generation. Duty cycles dI, dII, dIII, dIV and d0 are calculated using equations 4.5–4.8 and 4.10 assuming unity input power factor. With (sv + si) even and odd, rules 1 and 2 for duty cycle are followed with zero vector duty cycle d0 introduced at the middle of the four duty cycles dI, dII, dIII and dIV. This is used to calculate the timings A, B, C, D and E, by comparison of the cumulative sector timing with a triangle carrier V_{tri} as shown in Program segment 4.2. This triangle carrier V_{tri} has a period T_s, peak value $T_s/2$ V and minimum value zero, as shown in Figure 4.4. Cumulative sector timing is obtained by appropriately adding the value of duty ratio and multiplying by $T_s/2$, as per rules 1 and 2 given above. Sector switch functions ssf1 to ssf18 are determined using if-then-else statement. For example, sector switch function ssf1 is HIGH only when sv = si = 1 or sv = si = 4 and is LOW otherwise [15].

4.3.2 Output Voltage and Input Current Sector Calculator

This is developed using MATLAB Function, a mux and a demux as shown in Figure 4.5. The inputs to the mux are vout_angle and i_in_angle. The

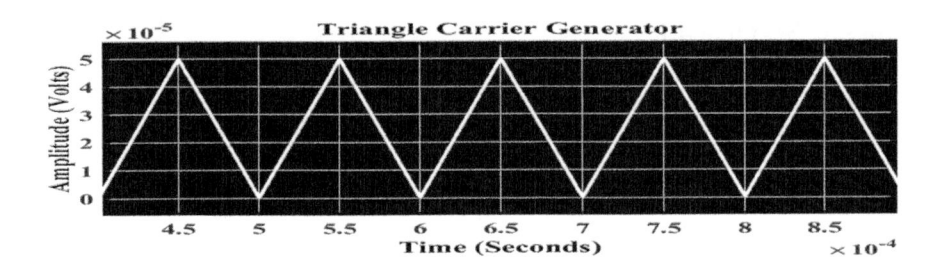

FIGURE 4.4
Triangle carrier waveform.

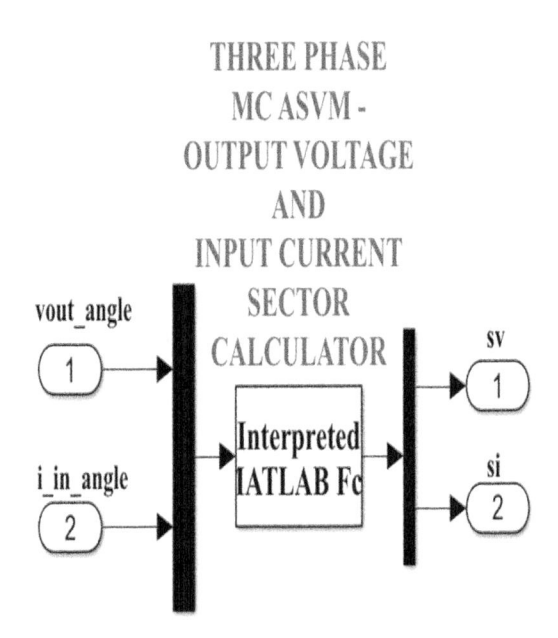

FIGURE 4.5
Output voltage and input current sector calculator.

outputs from the demux are SV and SI. The source code to determine SV and SI is shown in Program segment 4.3 [15].

Program Segment 4.2

```
%% Dr.Narayanaswamy. P.R. Iyer
function(q,dI,dII,dIII,dIV,d0,A,B,C,D,E,ssf1,ssf2,ssf3,...ssf18)
= fcn(Vom,fo,t,Vim,fi,wc,ph_i,alfa,beta,sv,si,Ts,vtri)
q = Vom/Vim  %%Duty cycle calculations
dI = (2*q*cos(alfa - pi/(3))*cos(beta - pi/(3))/
(sqrt(3)*cos(phi_i)));
dII = (2*q*cos(alfa - pi/(3))*cos(beta + pi/(3))/
(sqrt(3)*cos(phi_i)));
```

```
dIII = (2*q*cos(alfa+ pi/(3))*cos(beta - pi/(3))/
(sqrt(3)*cos(phi_i)));
dIV =(2*q*cos(alfa+pi/(3))*cos(beta + pi/(3))/
(sqrt(3)*cos(phi_i)));
d0 = (1 - dI - dII - dIII - dIV);
if (rem((sv+si),2) == 0) %%even sequence
    if ( vtri <= dIII*Ts/(2))
    A = 1;     else
    A = 0;
    end
    if ( vtri <= (dIII + dI)*Ts/(2))
    B = 1;     else
    B = 0;
    end
    if ( vtri <= (dIII + dI + d0)*Ts/(2))
    C = 1;     else
    C = 0;
    end
    if ( vtri <= (dIII + dI + d0 + dII)*Ts/(2))
    D = 1;     else
    D = 0;
    end
    if ( vtri <= (dIII + dI + d0 + dII + dIV)*Ts/(2))
    E = 1;     else
    E = 0;
    end
else
    if ( vtri <= dI*Ts/(2)) %%odd sequence
    A = 1;     else
    A = 0;
    end
    if ( vtri <= (dIII + dI)*Ts/(2))
    B = 1;     else
    B = 0;
    end
    if ( vtri <= (dIII + dI + d0)*Ts/(2))
    C = 1;     else
    C = 0;
    end
    if ( vtri <= (dIII + dI + d0 + dIV)*Ts/(2))
    D = 1;     else
    D = 0;
    end
    if ( vtri <= (dIII + dI + d0 + dII + dIV)*Ts/(2))
    E = 1;     else
    E = 0;
    end
end
if (sv == 1 && si == 1 || sv == 4 && si == 4)
    ssf1 = 1; else
```

```
    ssf1 = 0;
end
if ( sv == 4 && si == 1  ||  sv == 1 && si == 4 )
    ssf2 = 1; else
    ssf2 = 0;
end
%%similar statements for ssf3 to ssf18.
.
.
.
if ( sv == 3 && si == 6  ||  sv == 6 && si == 3 )
    ssf18 = 1; else
    ssf18 = 0;
end
```

Program Segment 4.3

```
function [y] = fcn(u)
%% Dr.Narayanaswamy. P.R. Iyer
vout_angle = u(1);
i_in_angle = u(2);
if (vout_angle > -pi && vout_angle <= -2*pi/(3))
    y(1) = 4;
else if (vout_angle > -2*pi/(3) && vout_angle <= -pi/(3))
        y(1) = 5;
else if (vout_angle > -pi/(3) && vout_angle <= 0)
            y(1) = 6;
else if (vout_angle > 0 && vout_angle <= pi/(3))
                y(1) = 1;
else if (vout_angle > pi/(3) && vout_angle <= 2*pi/(3))
                    y(1) = 2;
else if (vout_angle > 2*pi/(3) && vout_angle <= pi)
                        y(1) = 3;
                    end
                end
            end
        end
    end
end
if (i_in_angle > -pi && i_in_angle <= -2*pi/(3))
    y(2) = 4;
else if (i_in_angle > -2*pi/(3) && i_in_angle <=
 -pi/(3))
        y(2) = 5;
else if (i_in_angle > -pi/(3) && i_in_angle <= 0)
            y(2) = 6;
else if (i_in_angle > 0 && i_in_angle <= pi/(3))
                y(2) = 1;
else if (i_in_angle > pi/(3) && i_in_angle <= 2*pi/(3))
                    y(2) = 2
```

```
else if (i_in_angle > 2*pi/(3) && i_in_angle <= pi)
                    y(2) = 3;
                  end
              end
          end
      end
   end
end
sv = y(1);
si = y(2);
```

4.3.3 Output Voltage and Input Current Reference Angle Calculator

This is shown in Figure 4.6. This uses two REM function blocks and SUBTRACT modules. The first REM block has the inputs v_out_angle and π/(3). The second REM block has the inputs i_in_angle and π/(3). The output of the first REM block is subtracted from π/(6) to obtain angle alfa defined in Figure 4.2a. The output of the second REM block is subtracted from π/(6), input p.f. angle phi_i and once again from π/(6) to obtain beta as per the configuration defined in Figure 4.2b.

4.3.4 Gate Pulse Timing Calculator

This is shown in Figure 4.7. The inputs are A, B, C, D and E, respectively. The individual gate pulse timings t_a, t_b, t_0, t_c and t_d are, respectively, dIII*T_s/(2),

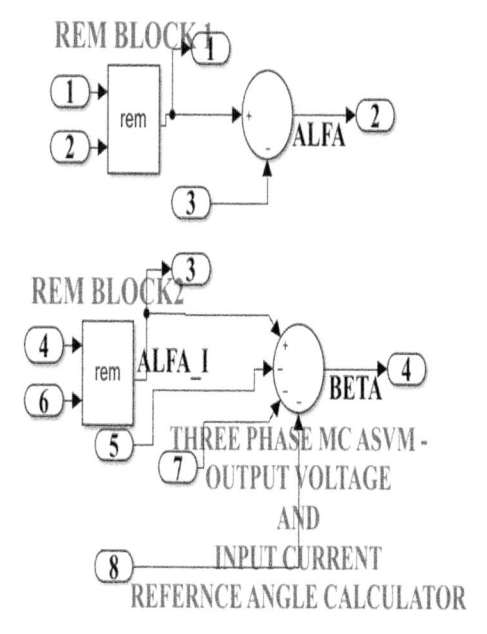

FIGURE 4.6
Output voltage and input current reference angle calculator.

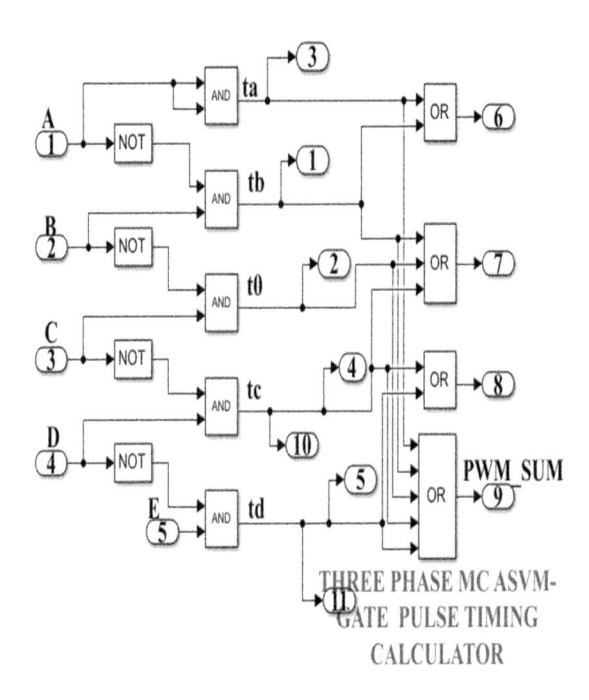

FIGURE 4.7
Gate pulse timing calculator.

$dI^*T_s/(2)$, $d0^*T_s/(2)$, $dII^*T_s/(2)$ and $dIV^*T_s/(2)$ when $(SV + SI)$ is even and $dI^*T_s/(2)$, $dIII^*T_s/(2)$, $d0^*T_s/(2)$, $dIV^*T_s/(2)$ and $dII^*T_s/(2)$ when $(SV + SI)$ is odd. This individual gate pulse duration is obtained as follows:

$$t_a = (A \& A)$$
$$t_b = (\sim A \& B)$$
$$t_0 = (\sim B \& C) \qquad (4.11)$$
$$t_c = (\sim C \& D)$$
$$t_d = (\sim D \& E)$$

where symbol (&) represents logical AND and symbol (~) represents NOT operation. To easily generate the MC gate pulse, the timing pulse in equation 4.11 is given to OR gates to obtain the following additional timing pulse:

$$t_{ab} = (t_a \| t_b)$$
$$t_{b0c} = (t_b \| t_0 \| t_c)$$
$$t_{cd} = (t_c \| t_d) \qquad (4.12)$$
$$t_{ab0cd} = (t_a \| t_b \| t_0 \| t_c \| t_d)$$

where symbol (‖) represents logical OR operation.

The gate pulse timing for the bidirectional switch S_{Aa} using direct ASVM algorithm is given in the last column marked S_{Aa} timing in Table 4.4 [15].

4.3.5 Gate Pulse Generator

The gate pulse generator for the MC is developed using Embedded MATLAB, as shown in Figure 4.3. The sector switch functions ssf1 to ssf18 and gate pulse timing calculator outputs defined in equations 4.11 and 4.12 form the input modules, and the outputs are the gate pulses for the nine bidirectional switches of the MC. In Table 4.4, the gate timing for the bidirectional switch S_{Aa} is shown. The gate timing for switch S_{Aa} can be combined by logically ANDing the respective value with sector switch function ssf_x and ORing the value for each sector. Thus, the gate timing pulse for switch S_{Aa} can be expressed as

$$S_{Aa} = \text{ssf1} \,\&\, (t_{ab0cd}) \,\|\, \text{ssf2} \,\&\, (t_0) \,\|\, \cdots \,\|\, \text{ssf18} \,\&\, (0) \qquad (4.13)$$

Program segment 4.4 gives the method of generating the gate timing pulses for the nine bidirectional switches calculated using Table 4.4 [15].

Program Segment 4.4

```
function [SAa,SBa,SCa,SAb,SBb,SCb,SAc,SBc,SCc] = fcn(ssf1,ssf2,
ssf3,ssf4,ssf5,ssf6,ssf7,ssf8,ssf9,ssf10,ssf11,ssf12,ssf1
3,ssf14,ssf15,ssf16,ssf17,ssf18,ta, tb,t0,tc,td,tab,tb0c,
tcd,tab0cd)
%% Dr.Narayanaswamy. P.R.Iyer.
SAa = ssf1&(tab0cd)||ssf2&(t0)||ssf3&(0)||ssf4&(tcd)||ssf5&(0)
||ssf6&(tab)||ssf7&(tb0c)||ssf8&(tb0c)||ssf9& (td)||ssf10&(td)
||ssf11&(ta)||ssf12&(ta)||ssf13&(t0)||ssf14&(tab0cd)||ssf15&(t
cd)||ssf16&(0)||ssf17& (tab)||ssf18&(0);
SBa = ssf1&(0)||ssf2&(tcd)||ssf3&(0)||ssf4&(tab)||ssf5&(tab0cd)
||ssf6&(t0)||ssf7&(td)||ssf8&(td)||ssf9&(ta)|| ssf10&(ta)||ssf
11&(tb0c)||ssf12&(tb0c)||ssf13&(tcd)||ssf14&(0)||ssf15&(tab)||
ssf16&(0)||ssf17&(t0)|| ssf18&(tab0cd);
SCa = ssf1&(0)||ssf2&(tab)||ssf3&(tab0cd)||ssf4&(t0)||ssf5&(0)
||ssf6&(tcd)||ssf7&(ta)||ssf8&(ta)||ssf9&(tb0c) ||ssf10&(tb0c)
||ssf11&(td)||ssf12&(td)||ssf13&(tab)||ssf14&(0)||ssf15&(t0)||
ssf16&(tab0cd)||ssf17& (tcd)||ssf18&(0);
SAb = ssf1&(tb0c)||ssf2&(tb0c)||ssf3&(td)||ssf4&(td)||ssf5&(ta)
||ssf6&(ta)||ssf7&(t0)||ssf8&(tab0cd)||ssf9& (tcd)||ssf10&(0)|
|ssf11&(tab)||ssf12&(0)||ssf13&(tab0cd)||ssf14&(t0)||ssf15&(0)
||ssf16&(tcd)||ssf17& (0)||ssf18&(tab);
SBb = ssf1&(td)||ssf2&(td)||ssf3&(ta)||ssf4&(ta)||ssf5&(tb0c)|
|ssf6&(tb0c)||ssf7&(tcd)||ssf8&(0)||ssf9&(tab)|| ssf10&(0)||
ssf11&(t0)||ssf12&(tab0cd)||ssf13&(0)||ssf14&(tcd)||ssf15&(0)||
ssf16&(tab)||ssf17& (tab0cd)||ssf18&(t0);
```

```
SCb = ssf1&(ta)||ssf2&(ta)||ssf3&(tb0c)||ssf4&(tb0c)||ssf5&(td)
||ssf6&(td)||ssf7&(tab)||ssf8&(0)||ssf9&(t0)|| ssf10&(tab0cd)
||ssf11&(tcd)||ssf12&(0)||ssf13&(0)||ssf14&(tab)||ssf15&(tab0cd)
||ssf16&(t0)||ssf17& (0)||ssf18&(tcd);
SAc = ssf1&(t0)||ssf2&(tab0cd)||ssf3&(tcd)||ssf4&(0)||ssf5&(tab)
||ssf6&(0)||ssf7&(tab0cd)||ssf8&(t0)||ssf9& (0)||ssf10&(tcd)||
ssf11&(0)||ssf12&(tab)||ssf13&(tb0c)||ssf14&(tb0c)||ssf15&(td)
||ssf16&(td)||ssf17& (ta)||ssf18&(ta);
SBc = ssf1&(tcd)||ssf2&(0)||ssf3&(tab)||ssf4&(0)||ssf5&(t0)||
ssf6&(tab0cd)||ssf7&(0)||ssf8&(tcd)||ssf9&(0)|| ssf10&(tab)||
ssf11&(tab0cd)||ssf12&(t0)||ssf13&(td)||ssf14&(td)||ssf15&(ta)
||ssf16&(ta)||ssf17&(tb0c)||ssf18&(tb0c);
SCc = ssf1&(tab)||ssf2&(0)||ssf3&(t0)||ssf4&(tab0cd)||ssf5&(tcd)
||ssf6&(0)||ssf7&(0)||ssf8&(tab)||ssf9& (tab0cd)||ssf10&(t0)||
ssf11&(0)||ssf12&(tcd)||ssf13&(ta)||ssf14&(ta)||ssf15&(tb0c)||
ssf16&(tb0c)|| ssf17&(td)||ssf18&(td);
```

Given here is the method to calculate the dwell time for the bidirectional switch S_{Aa}. In Table 4.4, for serial number 1 corresponding to ssf1 HIGH, the input phase 'A' occupies the position of output phase 'a' in all columns corresponding to t_a, t_b, t_0, t_c and t_d. Thus, when ssf1 is HIGH dwell time for S_{Aa} is $(t_a \parallel t_b \parallel t_0 \parallel t_c \parallel t_d)$ which is by definition tab0cd. Similarly, when ssf2 is HIGH this dwell time for S_{Aa} is t_0. When ssf3 is HIGH, 'A' is never seen in the position of output phase 'a' in any timing columns from t_a to t_d and S_{Aa} dwell time is zero. Output phase a, b and c must be considered from left to right in order in the timing columns from t_a to t_d. In this way, the dwell time for all the nine bidirectional switches of the MC can be calculated for all 18 sector switch functions. This is shown in Program segment 4.4 [15].

4.4 Simulation Results

The simulation of the three-phase ASVM MC was carried out in SIMULINK [13]. The parameters used for simulation are shown in Table 4.7. The ode15S(Stiff/NDF) solver is used. Simulation of the above three-phase AC to three-phase AC direct ASVM of MC was carried out for two different output frequencies with all other parameters constant [15]. The simulation results of the harmonic spectrum of line-to-neutral output voltage, input current, load current and line-to-line output voltage for a 50 Hz output frequency are shown in Figure 4.8a–d and the oscilloscope waveform of the above in order is shown in Figure 4.9a–d. Similarly, the harmonic spectrum of the above in order for a 20 Hz output frequency is shown in Figure 4.10a–d and the oscilloscope waveform of the above in order is shown in Figure 4.11a–d. The simulation results are tabulated in Table 4.8.

TABLE 4.7

Model Parameters

Sl. No.	Parameter	Value	Units
1	RMS line-to-neutral input voltage	220	V
2	Input frequency f_i	50	Hz
3	Modulation index q	0.4	–
4	Output frequency f_o	50, 20	Hz
5	Carrier sampling frequency f_{sw}	10	kHz
6	R–L–C output filter	10, 0.01e-3, 25.356e-6	Ω, H, F
7	R–L load	100, 500e-3	Ω, H
8	Input phase displacement φ_i	0	rad

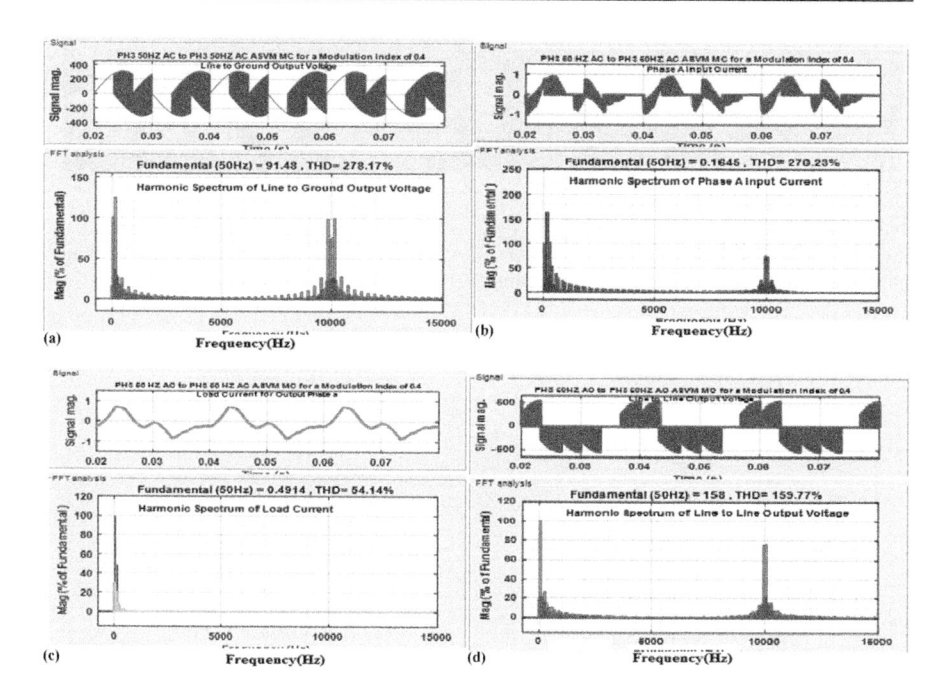

FIGURE 4.8
Harmonic spectrum of three-phase 50 Hz AC to 50 Hz AC ASVM MC: (a) line-to-neutral output voltage, (b) phase A input current, (c) load current and (d) line-to-line output voltage.

4.5 Model of Direct Symmetrical Space Vector Modulated Three-Phase Matrix Converter

The SIMULINK model of the Direct Symmetrical Space Vector Modulated (DSSVM) three-phase AC to three-phase AC MC is shown in Figure 4.12. The power circuit of the model is developed using the Power Systems block

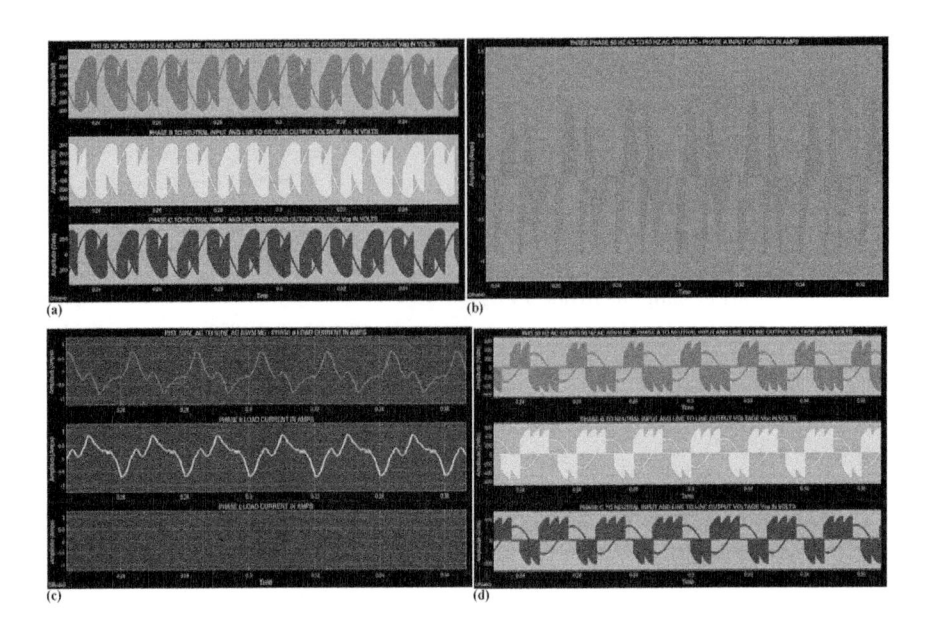

FIGURE 4.9
Three-phase 50 HZ AC to three-phase 50 HZ AC ASVM MC: (a) line-to-ground output voltage,
(b) phase A input current, (c) load current and (d) line-to-line output voltage.

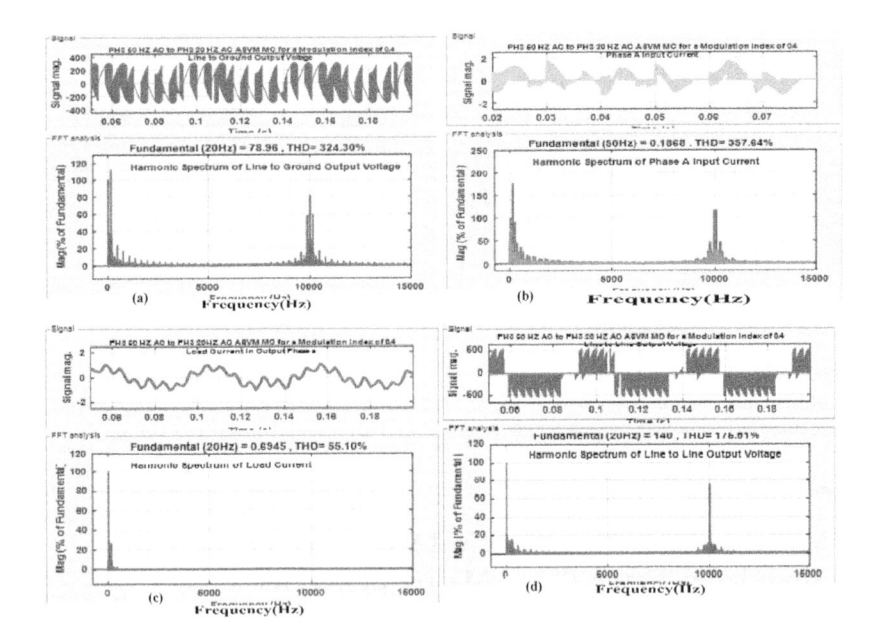

FIGURE 4.10
Harmonic spectrum of three-phase 50 Hz AC to 20 Hz AC ASVM MC: (a) line-to-neutral output
voltage, (b) phase A input current, (c) load current and (d) line-to-line output voltage.

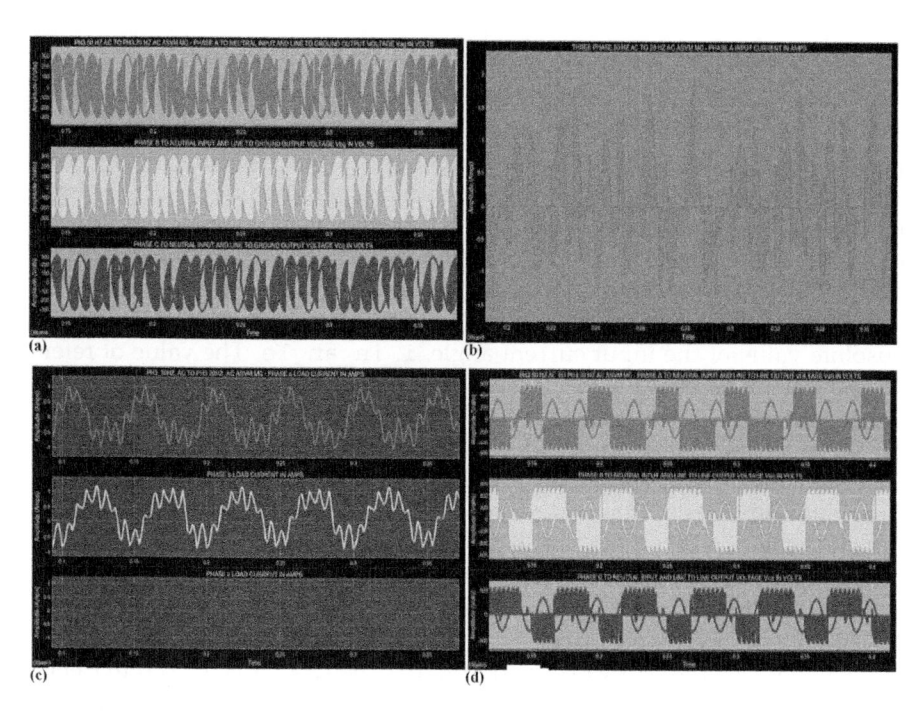

FIGURE 4.11
Three-phase 50 Hz AC to three-phase 20 Hz AC ASVM MC: (a) line-to-ground output voltage, (b) phase A input current, (c) load current and (d) line-to-line output voltage.

TABLE 4.8

Three-Phase ASVM Matrix Converter – Simulation Results

Sl. No.	Three-Phase ASVM MC Input – Output Frequency (Hz)	Modulation Index q	Line-to-Line Output Voltage THD (p.u.)	Line-to-Neutral Output Voltage THD (p.u.)	Input Current THD (p.u.)	Load Current THD (p.u.)
1	50 – 50	0.4	1.597	2.78	2.70	0.54
2	50 – 20	0.4	0.9559	1.1765	1.4765	0.3678

set in SIMULINK [13,15]. This mainly consists of three-phase AC voltage source, bidirectional switch matrix, output filter and R–L load. The arrangement of the bidirectional switch matrix using IGBTs is the same as shown in Figure 2.2 of Chapter 2.

The modulation algorithm is developed in several sub-units using Embedded MATLAB Function, MATLAB Function, Math Function, Logical and Bit Operator and using Sources block set in SIMULINK [13,15]. The various sub-units are explained below.

4.5.1 Duty-Cycle Sequence and Sector Switch Function Generator

This is developed using Embedded MATLAB Function block in SIMULINK, as shown in Figure 4.12. With peak line-to-neutral input voltage, desired peak output phase voltage, input frequency, output frequency, time, and angular frequency of reference frame ω_c as input parameters, the three-phase output voltage can be resolved into dq-axis component voltages [14] and the absolute value of the angle of the output phase voltage, vout_angle can be calculated using the function atan2(voq, vod). Using input voltage and input phase displacement angle, a similar procedure is used to calculate the absolute value of the input current angle, i_in_angle. The value of reference frame frequency ω_c is zero. This is illustrated in Program segment 4.1. These two values of the angles are used to calculate the output voltage sector sv, input current sector si, α and β values for output voltage and input current using MATLAB Function, REM functions and SUBTRACT modules. Program segment 4.5 illustrates the MATLAB code for duty-cycle sequencing and sector switch function generation. Duty cycles dI, dII, dIII, dIV and d0 are calculated using equations 4.5–4.8 and 4.10 assuming unity input power factor. With (sv + si) even and odd, rules 1 and 2 for duty cycle are followed with zero vector duty cycle (d0/3) introduced at the start, middle and at the end of the four duty cycles dI, dII, dIII and dIV. This is used to calculate the timings A, B, C, D, E, F and G, by comparison of the cumulative sector timing with a triangle carrier. This triangle carrier has a period T_s, peak value $T_s/2$ V and minimum value zero, as shown in Figure 4.4. Cumulative sector timing is obtained by appropriately adding the value of duty ratio and multiplying by $T_s/2$, as per rules 1 and 2 given above. Sector switch functions ssf1 to ssf18 are determined using if-then-else statement. For example, sector switch function ssf1 is HIGH only when sv = si = 1 or sv = si = 4 and is LOW otherwise [15].

4.5.2 Output Voltage and Input Current Sector Calculator

This is developed using MATLAB Function, a mux and a demux as shown in Figure 4.5. The inputs to the mux are vout_angle and i_in_angle. The outputs from the demux are SV and SI. The source code to determine SV and SI is shown in Program segment 4.3 [15].

Program Segment 4.5

```
%% Dr.NARAYANASWAMY. P.R. IYER
function(q,dI,dII,dIII,dIV,d0,A,B,C,D,E,ssf1,ssf2,ssf3,...
ssf18)= fcn(Vom,fo,t,Vim,fi,wc,ph_i,alfa,beta,sv,si,Ts,vtri)
q = Vom/Vim  %%Duty cycle calculations
dI = (2*q*cos(alfa - pi/(3))*cos(beta - pi/(3))/
(sqrt(3)*cos(phi_i)));
```

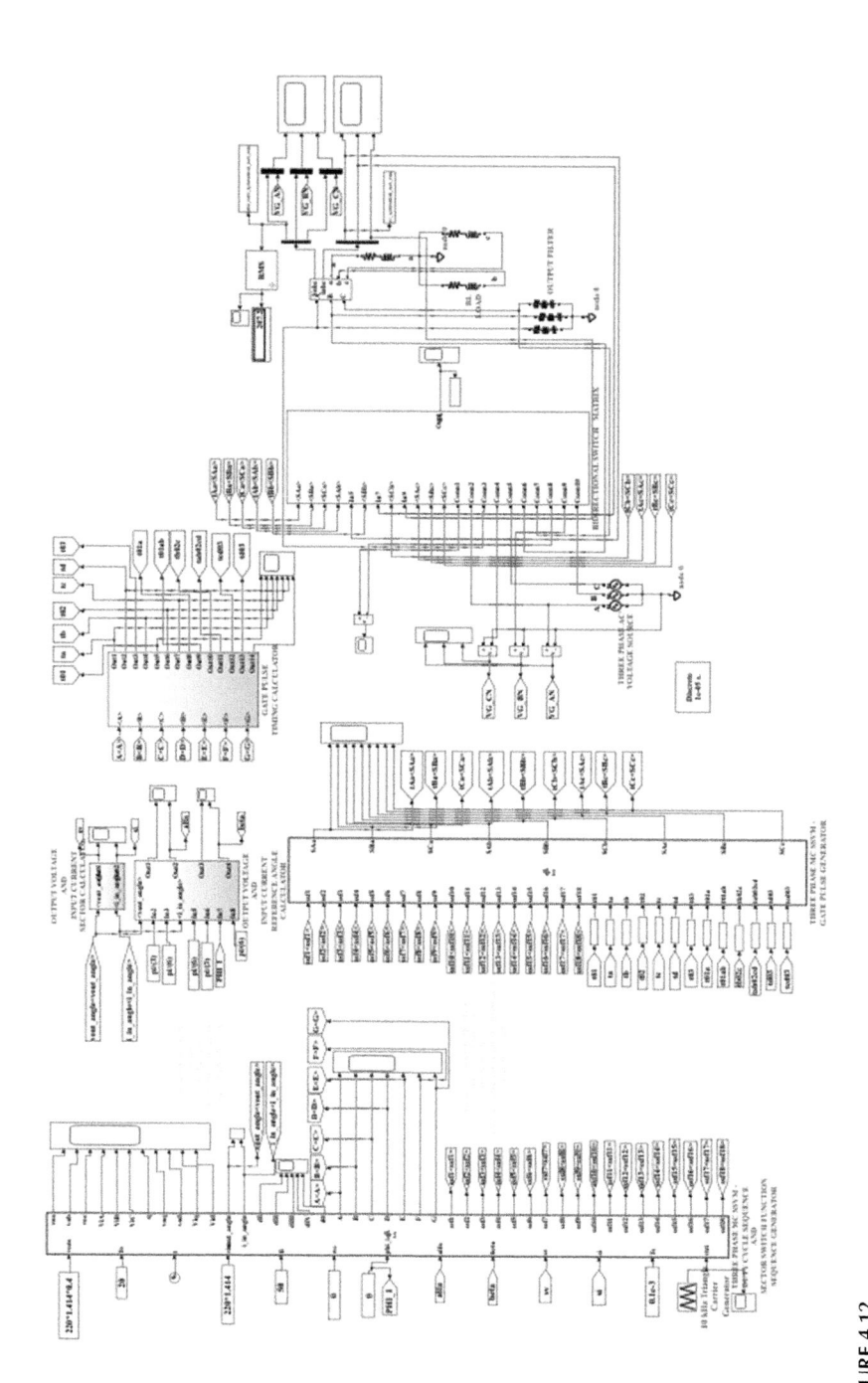

FIGURE 4.12
SIMULINK model of three-phase AC to three-phase AC SSVM matrix converter.

```
dII = (2*q*cos(alfa - pi/(3))*cos(beta + pi/(3))/
(sqrt(3)*cos(phi_i)));
dIII = (2*q*cos(alfa + pi/(3))*cos(beta - pi/(3))/
(sqrt(3)*cos(phi_i)));
dIV = (2*q*cos(alfa + pi/(3))*cos(beta + pi/(3))/
(sqrt(3)*cos(phi_i)));
d0 = (1 - dI - dII - dIII - dIV);
if (rem((sv+si),2) == 0)
    %%even sequence
if ( vtri <= (d0/(3))*Ts/(2))
A = 1;     else
A = 0;
end
if ( vtri <= (dIII + d0/(3))*Ts/(2))
B = 1;     else
B = 0;
end
if ( vtri <= (dIII + dI + d0/(3))*Ts/(2))
C = 1;     else
C = 0;
end
if ( vtri <= (dIII + dI + d0/(3) + d0/(3))*Ts/(2))
D = 1;     else
D = 0;
end
if ( vtri <= (dIII + dI + d0/(3) + d0/(3) + dII)*Ts/(2))
E = 1;     else
E = 0;
end
if ( vtri <= (dIII + dI + d0/(3) + d0/(3) + dII +
dIV)*Ts/(2))
F = 1;      else
F = 0;
end
if ( vtri <= (dIII + dI + d0/(3) + d0/(3) + d0/(3) + dII +
dIV)*Ts/(2))
G = 1;     else
G = 0;
end
else
%%odd sequence
if ( vtri <= (d0/(3))*Ts/(2))
A = 1;     else
A = 0;
end
if ( vtri <= (dI + d0/(3))*Ts/(2))
B = 1;     else
B = 0;
end
if ( vtri <= (dIII + dI + d0/(3))*Ts/(2))
```

```
C = 1;        else
C = 0;
end
if ( vtri <= (dIII + dI + d0/(3) + d0/(3))*Ts/(2))
D = 1;        else
D = 0;
end
if ( vtri <= (dIII + dI + d0/(3) + d0/(3) + dIV)*Ts/(2))
E = 1;         else
E = 0;
end
if ( vtri <= (dIII + dI + d0/(3) + d0/(3) + dII +
dIV)*Ts/(2))
F = 1;    else
F = 0;
end
if ( vtri <= (dIII + dI + d0/(3) + d0/(3) + d0/(3) + dII +
dIV)*Ts/(2))
G = 1;    else
G = 0;
end
end
if (sv == 1 && si == 1 || sv == 4 && si == 4)
ssf1 = 1;    else
ssf1 = 0;
end
if ( sv == 4 && si == 1   ||   sv == 1 && si == 4 )
ssf2 = 1;    else
ssf2 = 0;
end
%%similar statements for ssf3 to ssf18.
.
.
.
if ( sv == 3 && si == 6   ||   sv == 6 && si == 3 )
ssf18 = 1; else
ssf18 = 0;
end
```

4.5.3 Output Voltage and Input Current Reference Angle Calculator

This is shown in Figure 4.6. This uses two REM function blocks and SUBTRACT modules. The first REM block has the inputs v_out_angle and $\pi/(3)$. The second REM block has the inputs i_in_angle and $\pi/(3)$. The output of the first REM block is subtracted from $\pi/(6)$ to obtain angle alfa defined in Figure 4.2a. The output of the second REM block is subtracted from $\pi/(6)$, input p.f. angle phi_i and once again from $\pi/(6)$ to obtain beta as per the configuration defined in Figure 4.2b [15].

4.5.4 Gate Pulse Timing Calculator

This is shown in Figure 4.13. The inputs are A, B, C, D, E, F and G, respectively. The individual gate pulse timings t_{01}, t_a, t_b, t_{02}, t_c, t_d and t_{03} are, respectively, $(d0/3)^*T_s/2$, $dIII^*T_s/(2)$, $dI^*T_s/(2)$, $(d0/3)^*T_s/(2)$, $dII^*T_s/(2)$, $dIV^*T_s/(2)$ and $(d0/3)^*T_s/2$ when (SV + SI) is even and $(d0/3)^*T_s/2$, $dI^*T_s/(2)$, $dIII^*T_s/(2)$, $(d0/3)^*T_s/2$, $dIV^*T_s/(2)$, $dII^*T_s/(2)$ and $(d0/3)^*T_s/2$ when (SV + SI) is odd. This individual gate pulse duration is obtained as follows:

$$t_{01} = (A \,\&\, A)$$
$$t_a = (\sim A \,\&\, B)$$
$$t_b = (\sim B \,\&\, C)$$
$$t_{02} = (\sim C \,\&\, D) \tag{4.14}$$
$$t_c = (\sim D \,\&\, E)$$
$$t_d = (\sim E \,\&\, F)$$
$$t_{03} = (\sim F \,\&\, G)$$

where symbol (&) represents logical AND and symbol (~) represents NOT operation. To easily generate the MC gate pulse, the timing pulse in equation 4.14 is given to OR gates to obtain the following additional timing pulse:

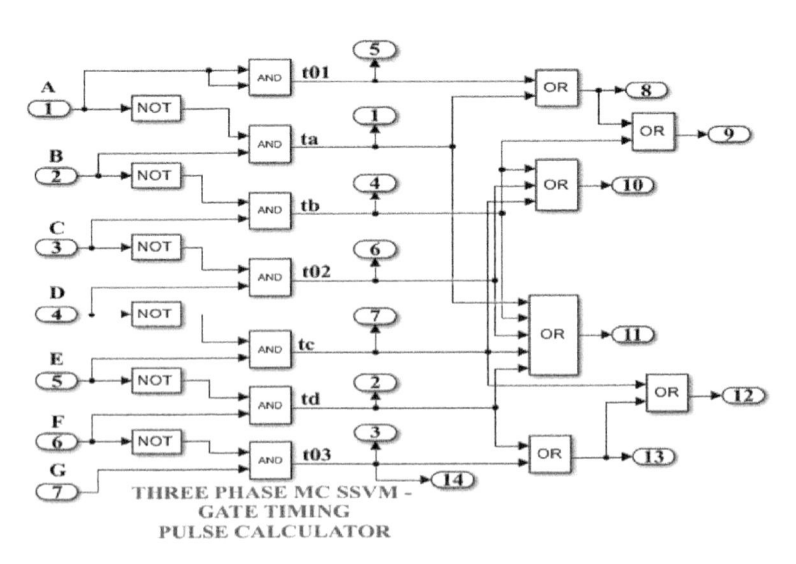

FIGURE 4.13
Gate pulse timing calculator.

$$t_{01a} = \left(t_a \parallel t_{01} \right)$$

$$t_{01ab} = \left(t_b \parallel t_{01} \parallel t_a \right)$$

$$t_{b02c} = \left(t_c \parallel t_{02} \parallel t_b \right)$$

$$t_{ab02cd} = \left(t_a \parallel t_b \parallel t_{02} \parallel t_c \parallel t_d \right)$$

$$t_{d03} = \left(t_d \parallel t_{03} \right)$$

$$t_{cd03} = \left(t_c \parallel t_d \parallel t_{03} \right)$$

(4.15)

where symbol (∥) represents logical OR operation. The gate timing for the bidirectional switch S_{Aa} using DSSVM algorithm is shown in the last column marked S_{Aa} timing in Table 4.6 [15].

4.5.5 Gate Pulse Generator

The gate pulse generator for the MC is developed using Embedded MATLAB Function as shown in Figure 4.12 [15]. The sector switch functions ssf1 to ssf18 and gate pulse timing calculator outputs defined in equations 4.14 and 4.15 form the input modules and the outputs are the gate pulses for the nine bidirectional switches of the MC. The gate timing for switch S_{Aa} can be combined by logically ANDing the respective value with sector switch function ssf_x and ORing the value for each sector. Thus, the gate timing pulse for switch S_{Aa} can be expressed as

$$S_{Aa} = \text{ssf1} \,\&\, \left(t_{ab02cd} \right) \parallel \text{ssf2} \,\&\, \left(t_{02} \right) \parallel \text{ssf3} \,\&\, \left(t_{03} \right) \parallel \ldots \parallel \text{ssf18} \,\&\, \left(t_{01} \right) \quad (4.16)$$

Program segment 4.6 gives the method of generating the gate timing pulses for the nine bidirectional switches using the calculated sector timings from Table 4.6 [15].

Program Segment 4.6

```
function [SAa,SBa,SCa,SAb,SBb,SCb,SAc,SBc,SCc] = fcn(ssf1,ssf2,
ssf3,ssf4,ssf5,ssf6,ssf7,ssf8,ssf9,ssf10,ssf11,ssf12,ssf13,ssf
14,ssf15,ssf16,ssf17,ssf18,t01,ta,tb,t02,tc,td,t03,t01a,
t01ab,tb02c,tab02cd,td03,tcd03)
%% Dr.Narayanaswamy. P.R.Iyer
 SAa = ssf1&(tab02cd)||ssf2&(t02)||ssf3&(t03)||ssf4&(tcd03)
||ssf5&(t01)||ssf6&(t01ab)||ssf7&(tb02c)||ssf8&(tb02c)||ssf9&
(td03)||
 ssf10&(td03)||ssf11&(t01a)||ssf12&(t01a)||ssf13&(t02)||ssf14&
(tab02cd)||ssf15&(tcd03)||ssf16&(t03)||ssf17&(t01ab)||  ssf18&
(t01);
 SBa = ssf1&(t03)||ssf2&(tcd03)||ssf3&(t01)||ssf4&(t01ab)||
ssf5&(tab02cd)||ssf6&(t02)||ssf7&(td03)||ssf8&(td03)||ssf9&
```

```
(t01a) || ssf10& (t01a) ||ssf11&(tb02c) ||ssf12&(tb02c) ||ssf13&
(tcd03) ||ssf14&(t03) ||ssf15&(t01ab) ||ssf16&(t01) ||ssf17&
(t02) || ssf18& (tab02cd);
 SCa = ssf1&(t01) ||ssf2&(t01ab) ||ssf3&(tab02cd) ||ssf4&(t02) ||
ssf5&(t03) ||ssf6&(tcd03) ||ssf7&(t01a) ||ssf8&(t01a) ||ssf9&
(tb02c) || ssf10&(tb02c) ||ssf11&(td03) ||ssf12&(td03) ||ssf13&
(t01ab) ||ssf14&(t01) ||ssf15&(t02) ||ssf16&(tab02cd) ||ssf17&
(tcd03) || ssf18& (t03);
 SAb = ssf1&(tb02c) ||ssf2&(tb02c) ||ssf3&(td03) ||ssf4&(td03) ||
ssf5&(t01a) ||ssf6&(t01a) ||ssf7&(t02) ||ssf8&(tab02cd) ||ssf9&
(tcd03) ||
 ssf10&(t03) ||ssf11&(t01ab) ||ssf12&(t01) ||ssf13&(tab02cd) ||
ssf14&(t02) ||ssf15&(t03) ||ssf16&(tcd03) ||ssf17&(t01) ||
ssf18& (t01ab);
 SBb = ssf1&(td03) ||ssf2&(td03) ||ssf3&(t01a) ||ssf4&
(t01a) ||ssf5&(tb02c) ||ssf6&(tb02c) ||ssf7&(tcd03) ||ssf8&(t03) ||
ssf9&(t01ab) || ssf10&(t01) ||ssf11&(t02) ||ssf12&(tab02cd) ||
ssf13&(t03) ||ssf14&(tcd03) ||ssf15&(t01) ||ssf16&(t01ab) ||ssf17&
(tab02cd) || ssf18& (t02);
 SCb = ssf1&(t01a) ||ssf2&(t01a) ||ssf3&(tb02c) ||ssf4&(tb02c) ||
ssf5&(td03) ||ssf6&(td03) ||ssf7&(t01ab) ||ssf8&(t01) ||ssf9&
(t02) ||ssf10&(tab02cd) ||ssf11&(tcd03) ||ssf12&(t03) ||ssf13&(t01)
||ssf14&(t01ab) ||ssf15&(tab02cd) ||ssf16&(t02) ||ssf17&(t03) ||
ssf18&(tcd03);
 SAc = ssf1&(t02) ||ssf2&(tab02cd) ||ssf3&(tcd03) ||ssf4&(t03) ||
ssf5&(t01ab) ||ssf6&(t01) ||ssf7&(tab02cd) ||ssf8&(t02) ||ssf9&
(t03) || ssf10&(tcd03) ||ssf11&(t01) ||ssf12&(t01ab) ||ssf13&
(tb02c) ||ssf14&(tb02c) ||ssf15&(td03) ||ssf16&(td03) ||ssf17&
(t01a) || ssf18&(t01a);
 SBc = ssf1&(tcd03) ||ssf2&(t03) ||ssf3&(t01ab) ||ssf4&(t01) ||
ssf5&(t02) ||ssf6&(tab02cd) ||ssf7&(t03) ||ssf8&(tcd03) ||ssf9&
(t01) || ssf10&(t01ab) ||ssf11&(tab02cd) ||ssf12&(t02) ||ssf13&
(td03) ||ssf14&(td03) ||ssf15&(t01a) ||ssf16&(t01a) ||ssf17&
(tb02c) || ssf18&(tb02c);
 SCc = ssf1&(t01ab) ||ssf2&(t01) ||ssf3&(t02) ||ssf4&(tab02cd) ||
ssf5&(tcd03) ||ssf6&(t03) ||ssf7&(t01) ||ssf8&(t01ab) ||ssf9&
(tab02cd) ||
 ssf10&(t02) ||ssf11&(t03) ||ssf12&(tcd03) ||ssf13&(t01a) ||ssf14&
(t01a) ||ssf15&(tb02c) ||ssf16&(tb02c) ||ssf17&(td03) ||ssf18&
(td03);
```

Given here is the method to calculate the dwell time for the bidirectional switch S_{Aa}. In Table 4.6, for serial number 1 corresponding to ssf1 HIGH, the input phase 'A' occupies the position of output phase 'a' in columns corresponding to t_a, t_b, $t_{02,}$ t_c and t_d. Thus, when ssf1 is HIGH dwell time for S_{Aa} is $(t_a \parallel t_b \parallel t_{02} \parallel t_c \parallel t_d)$ which is by definition t_{ab02cd}. Similarly, when ssf2 is HIGH this dwell time for S_{Aa} is t_{02}. When ssf3 is HIGH, 'A' is seen in the position of output phase 'a' in the timing columns t_{03} and dwell time for S_{Aa} is t_{03}. Output phase a, b and c must be considered from left to right in order in the timing columns from t_{01} to t_{03}. In this way, the dwell time for all the nine bidirectional switches of the MC can be calculated for all 18 sector switch functions. This is shown in Program segment 4.6 [15].

4.6 Simulation Results

The simulation of the three-phase SSVM MC was carried out in SIMULINK [13]. The parameters used for simulation are shown in Table 4.7. The ode15S(Stiff/NDF) solver is used. Simulation of the above three-phase AC to three-phase AC DSSVM of MC was carried out for two different output frequencies with all other parameters constant [15]. The simulation results of the harmonic spectrum of line-to-neutral output voltage, input current, load current and line-to-line output voltage for a 50 Hz output frequency are shown in Figure 4.14a–d and the oscilloscope waveform of the above in order is shown in Figure 4.15a–d. Similarly, the harmonic spectrum of the above in order for a 20 Hz output frequency is shown in Figure 4.16a–d, and the

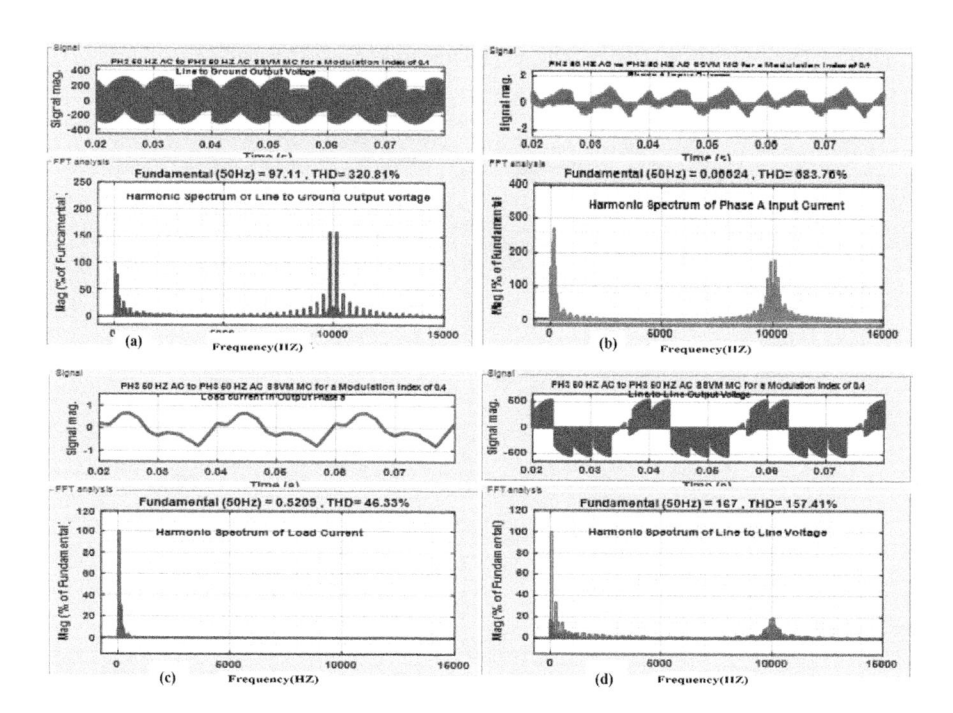

FIGURE 4.14
Harmonic spectrum of three-phase 50 Hz AC to three-phase 50 Hz AC SSVM MC: (a) line-to-neutral output voltage, (b) phase A input current, (c) load current and (d) line-to-line output voltage.

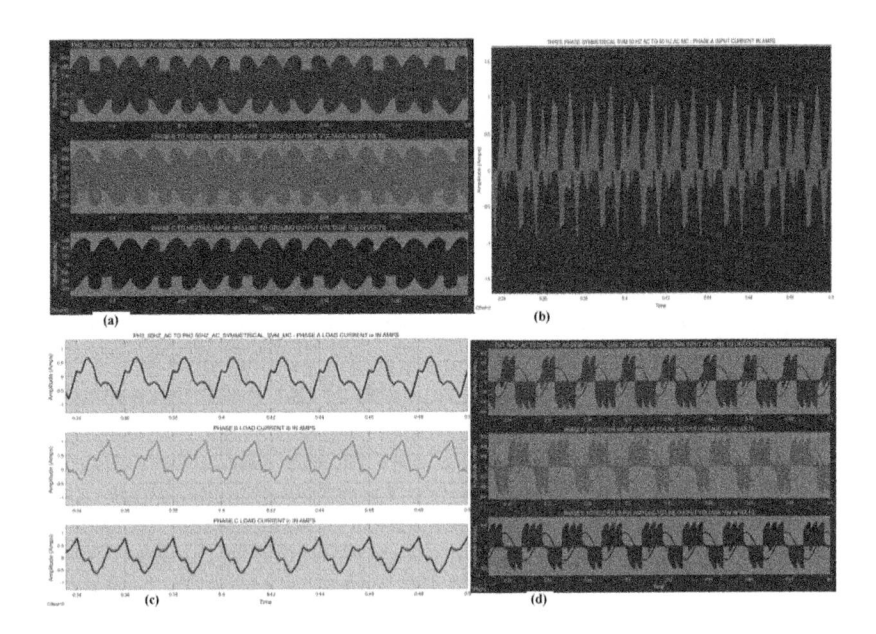

FIGURE 4.15
Three-phase 50 Hz AC to three-phase 50 Hz AC SSVM MC: (a) line-to-ground output voltage, (b) phase A input current, (c) load current and (d) line-to-line output voltage.

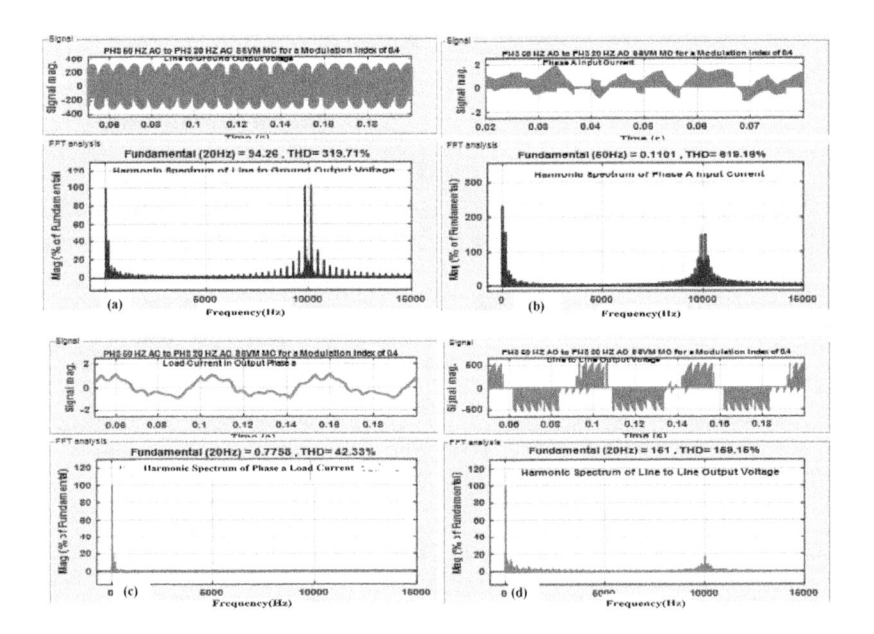

FIGURE 4.16
Harmonic spectrum of three-phase 50 Hz AC to three-phase 20 Hz AC SSVM MC: (a) line-to-ground output voltage, (b) phase A input current, (c) load current and (d) line-to-line output voltage.

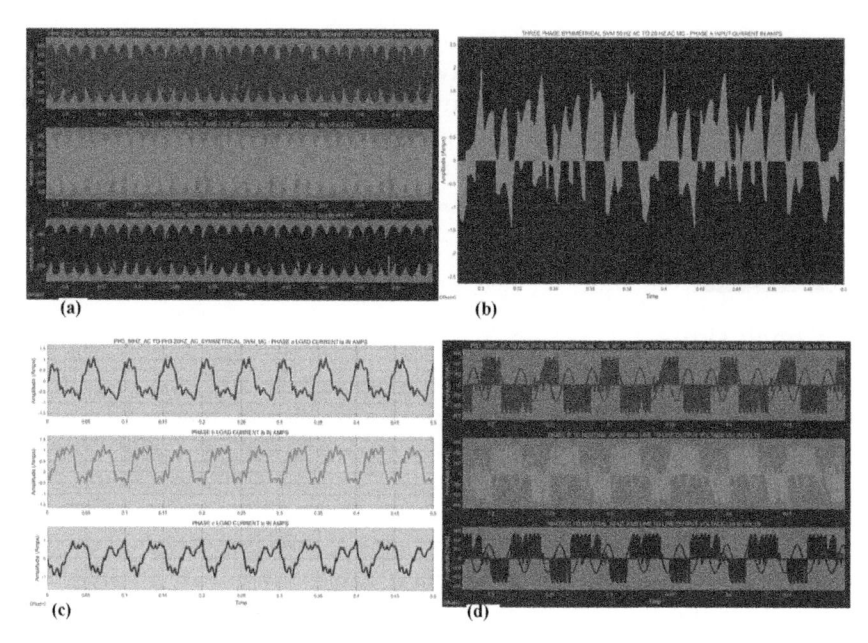

FIGURE 4.17

Three-phase 50 Hz AC to three-phase 20 Hz AC SSVM MC: (a) line-to-ground output voltage (b) phase A input current (c) load current and (d) line-to-line output voltage.

TABLE 4.9

Three-Phase SSVM MC – Simulation Results

Sl. No.	Three-Phase SSVM MC Input – Output Frequency (Hz)	Modulation Index q	THD of Line-to-Line Output Voltage (p.u.)	THD of Line-to-Neutral Output Voltage (p.u.)	THD of Input Current (p.u.)	THD of Load Current (p.u.)
1	50 – 50	0.4	1.574	3.2	6.83	0.4633
2	50 – 20	0.4	1.59	3.19	6.19	0.42

oscilloscope waveform of the above in order is shown in Figure 4.17a–d. The simulation results are tabulated in Table 4.9.

4.7 Discussion of Results

The simulation results for the three-phase ASVM and SSVM MC are tabulated in Tables 4.8 and 4.9, respectively. The THD of line-to-neutral, line-to-line output voltage, input current and load current decreases as the output frequency is reduced for ASVM technique. The input power factor is assumed to be unity in this analysis. In all the two SVM techniques, it is seen from the simulation results that all the harmonic components are centred around the integral multiples of carrier switching frequency. The ASVM has 8 commutations, whereas SSVM has 12 commutations per carrier switching period.

4.8 Conclusions

The model of the three-phase AC to three-phase AC MC has been developed in SIMULINK using the duty-cycle space vector approach. All possible switching combinations are represented by space vectors. The SVM technique for three-phase MC is considered an optimal solution. Owing to its intrinsic two degrees of freedom, SVM technique represents the general solution of the MC modulation problem and can be considered the best solution for achieving the highest voltage transfer ratio and optimizing the switching pattern through a suitable use of the zero configurations. With SVM technique modulation ratio as high as 0.866 can be achieved. The ASVM technique is found to be better from the point of view of harmonic performance such as THD of line-to-neutral output voltage and input current.

References

1. A. Alesina and M. Venturini: Solid-state power conversion: a Fourier analysis approach to generalized transformer synthesis, *IEEE Transactions on Circuits and Systems*, Vol. CAS-28, 1981, pp. 319–330.
2. A. Alesina and M.G.B. Venturini: Analysis and design of optimum amplitude nine-switch direct AC–AC converters, *IEEE Transactions on Power Electronics*, Vol. 4, pp. 101–112, 1989.
3. P.W. Wheeler, J.C. Clare, L. Empringham, M. Bland and K.G. Kerris: Matrix converters, *IEEE Industry Applications Magazine*, Vol. 10, 2004, pp. 59–65.
4. L. Huber and D. Borojevic: Space vector modulated three-phase to three-phase matrix converter with input power factor correction, *IEEE Transactions on Industry Applications*, Vol. 31, pp. 1234–1246, 1995.
5. D. Casadei, G. Serra, A. Tani and L. Zarri: Matrix converter modulation strategies: a new general approach based on space-vector representation of the switching state, *IEEE Transactions on Industrial Electronics*, Vol. 49, No. 2, 2002, pp. 370–381.
6. D. Casadei, G. Grandi, G. Serra and A. Tani: Space vector control of matrix converters with unity input power factor and sinusoidal input/output waveforms, *European Power Electronics Conference*, Vol. 7, 1993, Brighton, Great Britain, pp. 170–175.

7. D. Casadei, G. Serra, A. Tani and L. Zarri: A space vector modulation strategy for matrix converters minimizing the RMS value of the load current ripple, *IEEE-IECON*, 2006, Paris, France, pp. 2757–2762.
8. D. Casadei, G. Serra, A. Trentin, L. Zarri and M. Calvini: Experimental analysis of a matrix converter prototype based on new IGBT modules, *IEEE-ISIE*, Croatia, June 2005, pp. 559–564.
9. D. Casadei, G. Serra and A. Tani: Reduction of the input current harmonic content in matrix converters under input/output unbalance, *IEEE Transactions on Industrial Electronics*, Vol. 45, No. 3, 1998, pp. 401–411.
10. J. Vadillo, J.M. Echeverria, L. Fontan, M. Martinez-Iturralde and I. Elosegui: Modeling and simulation of a direct space vector modulated matrix converter using different switching strategies, *SPEEDAM*, 2008, Ischia, Italy, pp. 944–949.
11. J. Vadillo, J.M. Echeverria, A. Galarza and L. Fontan: Modelling and simulation of space vector modulation techniques for matrix converters: analysis of different switching strategies, *International Conference on Electrical Machines and Systems*, 2008, Wuhan, China, pp. 1299–1304.
12. K. You and M.F. Rahman: Modulations for voltage source rectification and voltage source inversion based on general direct space vector modulation approach of AC-AC matrix converter theory, *IEEE-ISIE*, July 9–13, 2006, Montreal, Canada, pp. 943–948.
13. The Mathworks Inc.: www.mathworks.com, MATLAB/SIMULINK user manual, MATLAB R2017b, 2017.
14. P.C. Krause, O. Wasynczuk and S.D. Sudhoff: *Analysis of Electrical Machinery*, IEEE Press, New York, 1995.
15. N.P.R. Iyer: Modelling, simulation and real time implementation of a three-phase AC to AC matrix converter, Ph.D. thesis, Ch. 8, Department of ECE, Curtin University, Perth, WA, Australia, February 2012.

5

Indirect Space Vector Modulation of Three-Phase Matrix Converter

5.1 Introduction

Three-phase AC to three-phase AC matrix converter (MC) has advantages over the conventional three-phase rectifier-inverter frequency converters in the sense that the former directly converts AC voltage at any given frequency to AC output voltage of any other magnitude and frequency without the need for a DC-link capacitor storage element and with a regeneration capability [1–6]. In the proposed voltage space vector pulse-width modulation (SVPWM) control of MCs, the indirect transfer function (ITF) approach is used where the input AC voltage is first rectified to create a fictitious DC voltage which is then inverted at the required output frequency [1–6]. For low harmonic distortion, the inverter operation is achieved by space vector modulation (SVM) simultaneously with input current SVM for rectification [1–6]. The simultaneous output voltage–input current SVM algorithm is reviewed here and the model of the 3×3 MC is derived based on this algorithm using SIMULINK. The simulation results are presented.

5.2 Principle of Indirect Space Vector Modulation

The three-phase AC to three-phase AC conventional matrix converter (CMC) is shown in Figure 5.1 with the load represented as a current source connected in delta. The equivalent circuit of this CMC is shown in Figure 5.2. The CMC can be considered equivalent to a three-phase rectifier forming a virtual DC-link output voltage which is then inverted using a three-phase inverter to get the required AC output voltage and frequency [1–6]. The model equivalence of Figures 5.1 and 5.2 can be expressed as in equation 5.1.

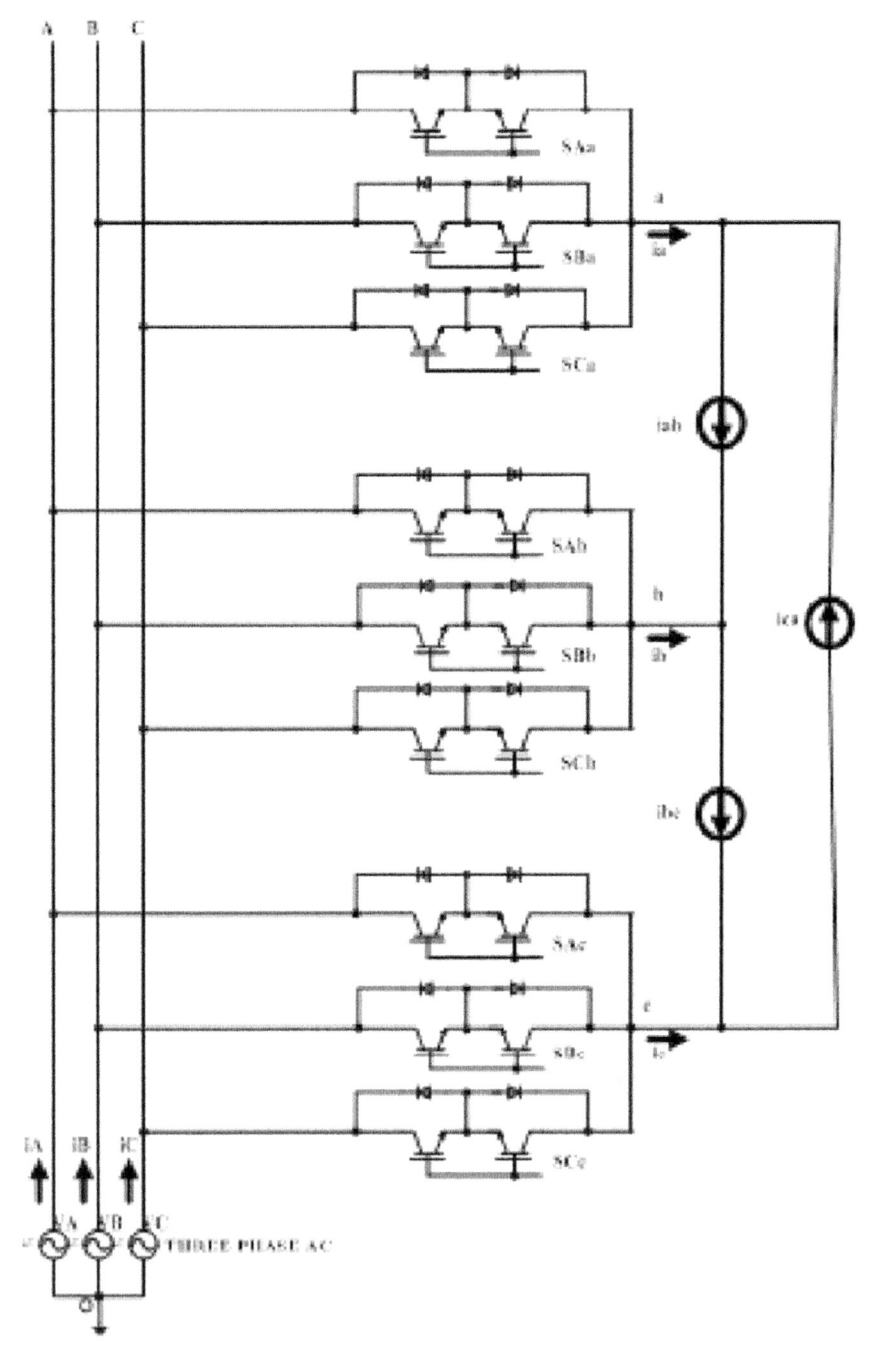

FIGURE 5.1
Three-phase AC to three-phase AC conventional matrix converter.

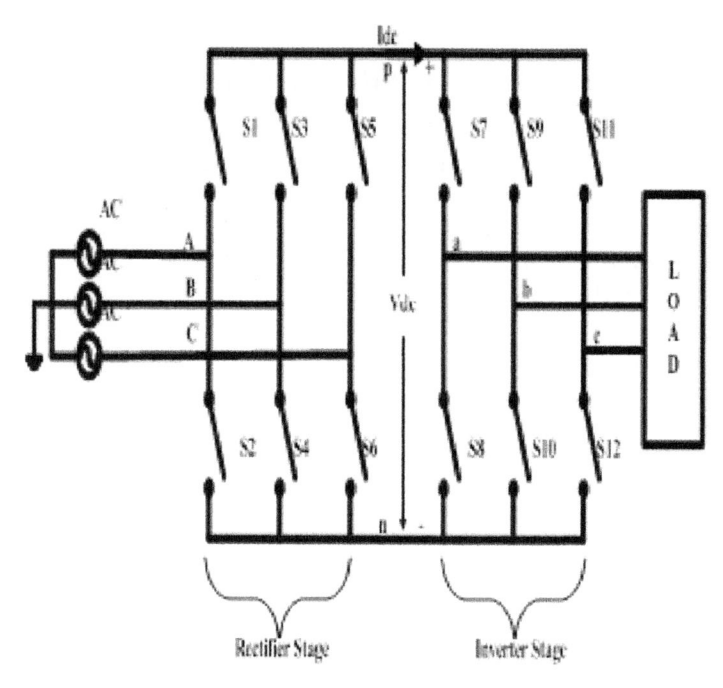

FIGURE 5.2
Matrix converter equivalent circuit.

The basic idea of the indirect SVM technique is to decouple the control of the input current and the control of the output voltage. Modelling the MC in this way enables the well-known space vector PWM to be applied for a rectifier as well as an inverter stage independently [1–7]:

$$
\begin{bmatrix} S_{Aa} & S_{Ba} & S_{Ca} \\ S_{Ab} & S_{Bb} & S_{Cb} \\ S_{Ac} & S_{Bc} & S_{Cc} \end{bmatrix} = \begin{bmatrix} S_7 & S_8 \\ S_9 & S_{10} \\ S_{11} & S_{12} \end{bmatrix} \cap \begin{bmatrix} S_1 & S_3 & S_5 \\ S_2 & S_4 & S_6 \end{bmatrix}
$$

$$
= \begin{bmatrix} (S_7 \cap S_1) \cup (S_8 \cap S_2) & (S_7 \cap S_3) \cup (S_8 \cap S_4) & (S_7 \cap S_5) \cup (S_8 \cap S_6) \\ (S_9 \cap S_1) \cup (S_{10} \cap S_2) & (S_9 \cap S_3) \cup (S_{10} \cap S_4) & (S_9 \cap S_5) \cup (S_{10} \cap S_6) \\ (S_{11} \cap S_1) \cup (S_{12} \cap S_2) & (S_{11} \cap S_3) \cup (S_{12} \cap S_4) & (S_{11} \cap S_5) \cup (S_{12} \cap S_6) \end{bmatrix}
$$

$$(5.1)$$

where $\cap$ = Logical AND; $\cup$ = Logical OR.

5.3 Indirect Space Vector Modulation Algorithm

The 3×3 MC shown in Figure 5.1 connects the three-phase AC source to the three-phase load. The switching function for a 3×3 MC can be defined as in equations 2.1 and 2.2 under Section 2.2 in Chapter 2. Also the 27 switching states of the 3×3 MC is presented in Table 1.1 in Chapter 1. Using this Table 1.1 and Figure 5.1, the following expressions for the output line voltage and input phase currents are obtained [1–6]:

$$\begin{bmatrix} v_{ab} \\ v_{bc} \\ v_{ca} \end{bmatrix} = \begin{bmatrix} (S_{Aa} - S_{Ab}) & (S_{Ba} - S_{Bb}) & (S_{Ca} - S_{Cb}) \\ (S_{Ab} - S_{Ac}) & (S_{Bb} - S_{Bc}) & (S_{Cb} - S_{Cc}) \\ (S_{Ac} - S_{Aa}) & (S_{Bc} - S_{Ba}) & (S_{Cc} - S_{Ca}) \end{bmatrix} * \begin{bmatrix} v_{AO} \\ v_{BO} \\ v_{CO} \end{bmatrix} \tag{5.2}$$

$$\text{i.e., } v_{oL} = T_{phl} * v_{iph} \tag{5.3}$$

$$\text{and } i_{iph} = \begin{bmatrix} i_A \\ i_B \\ i_C \end{bmatrix} = \left[T_{phl} \right]^T * \begin{bmatrix} i_{ab} \\ i_{bc} \\ i_{ca} \end{bmatrix} = \left[T_{phl} \right]^T * i_{oL} \tag{5.4}$$

T_{phl} is the instantaneous input phase to output line Transfer Function matrix of the three-phase MC:

$$v_{oph} = \begin{bmatrix} v_{ao} \\ v_{bo} \\ v_{co} \end{bmatrix} = \begin{bmatrix} S_{Aa} & S_{Ba} & S_{Ca} \\ S_{Ab} & S_{Bb} & S_{Cb} \\ S_{Ac} & S_{Bc} & S_{Cc} \end{bmatrix} * \begin{bmatrix} v_{AO} \\ v_{BO} \\ v_{CO} \end{bmatrix} = T_{phph} * v_{iph} \tag{5.5}$$

and

$$i_{iph} = \begin{bmatrix} i_A \\ i_B \\ i_C \end{bmatrix} = \left[T_{phph} \right]^T * \begin{bmatrix} i_a \\ i_b \\ i_c \end{bmatrix} = \left[T_{phph} \right]^T * i_{oph} \tag{5.6}$$

T_{phph} is the instantaneous input phase to output phase conversion matrix. To use the high frequency (HF) synthesis, the switching frequency must be higher than the frequency of the input voltages and output currents. The switching function S_{Kj} is the duty cycle of the switch S_{Kj} and is denoted by M_{Kj}. The low frequency (LF) equivalents are given below:

$$0 \le M_{Kj} \le 1 \text{ where } K \in A, B, C \text{ and } j \in a, b, c. \tag{5.7}$$

$$M_{Aj} + M_{Bj} + M_{Cj} = 1 \text{ for } j \in a,b,c \qquad (5.8)$$

The remaining LF equivalents are given in equations 5.4–5.7. Equations for T_{phl} and T_{phph} can now be rewritten as follows:

$$T_{phl} = \begin{bmatrix} (M_{Aa} - M_{Ab}) & (M_{Ba} - M_{Bb}) & (M_{Ca} - M_{Cb}) \\ (M_{Ab} - M_{Ac}) & (M_{Bb} - M_{Bc}) & (M_{Cb} - M_{Cc}) \\ (M_{Ac} - M_{Aa}) & (M_{Bc} - M_{Ba}) & (M_{Cc} - M_{Ca}) \end{bmatrix} \qquad (5.9)$$

$$\text{and } T_{phph} = \begin{bmatrix} M_{Aa} & M_{Ba} & M_{Ca} \\ M_{Ab} & M_{Bb} & M_{Cb} \\ M_{Ac} & M_{Bc} & M_{Cc} \end{bmatrix} \qquad (5.10)$$

Let the input phase voltage be expressed as follows:

$$v_{iph} = V_{im} * \begin{bmatrix} \cos(\omega_i t) \\ \cos(\omega_i t - 120) \\ \cos(\omega_i t + 120) \end{bmatrix} \qquad (5.11)$$

The averaged output line voltage is expressed as follows:

$$v_{oL} = \begin{bmatrix} v_{ab} \\ v_{bc} \\ v_{ca} \end{bmatrix} = \sqrt{3} V_{om} * \begin{bmatrix} \cos(\omega_o t - \varphi_o + 30) \\ \cos(\omega_o t - \varphi_o + 30 - 120) \\ \cos(\omega_o t - \varphi_o + 30 + 120) \end{bmatrix} \qquad (5.12)$$

where φ_o is the output phase displacement angle. The input phase to output line transfer matrix is chosen as follows:

$$T_{phl} = m * \begin{bmatrix} \cos(\omega_o t - \varphi_o + 30) \\ \cos(\omega_o t - \varphi_o + 30 - 120) \\ \cos(\omega_o t - \varphi_o + 30 + 120) \end{bmatrix} * \begin{bmatrix} \cos(\omega_i t - \varphi_i) \\ \cos(\omega_i t - \varphi_i - 120) \\ \cos(\omega_i t - \varphi_i + 120) \end{bmatrix}^T \qquad (5.13)$$

where $0 \le m \le 1$ is the modulation index and φ_i is the input phase displacement angle. Equations 5.12–5.14 satisfy equation 5.3 for value of V_{om} given below:

$$V_{om} = \frac{\sqrt{3}}{2} V_{im} * m * \cos(\varphi_i) \qquad (5.14)$$

The output line current is assumed sinusoidal and is given below:

$$i_{oL} = \begin{bmatrix} i_{ab} \\ i_{bc} \\ i_{ca} \end{bmatrix} = \frac{I_{om}}{\sqrt{3}} * \begin{bmatrix} \cos(\omega_o t - \varphi_o - \varphi_L + 30) \\ \cos(\omega_o t - \varphi_o - \varphi_L + 30 - 120) \\ \cos(\omega_o t - \varphi_o - \varphi_L + 30 + 120) \end{bmatrix} \tag{5.15}$$

where φ_L is the load displacement angle at the output frequency f_o Hz. If equations 5.14 and 5.15 are substituted in equation 5.4, the local averaged input currents are obtained as follows:

$$i_{iph} = \begin{bmatrix} i_A \\ i_B \\ i_C \end{bmatrix} = I_{im} * \begin{bmatrix} \cos(\omega_i t) \\ \cos(\omega_i t - 120) \\ \cos(\omega_i t + 120) \end{bmatrix} \tag{5.16}$$

where

$$I_{im} = \frac{\sqrt{3}}{2} I_{om} * m * \cos(\varphi_L) \tag{5.17}$$

Unity input displacement factor is obtained by letting φ_i equal to zero which from equation 5.14 results in a voltage gain of 0.866 for unity modulation index. The HF synthesis requires determination of the value of duty cycle M_{Kj} and the position of the switching pulses such that the constraint given by equations 5.7 and 5.8 are satisfied and the required transfer matrix given by equation 5.13 is implemented. For the control of three-phase AC to three-phase AC MC, indirect and direct Transfer Function approaches are used. Here, ITF approach is discussed below:

Equation 5.13 represents the ITF approach. The transfer matrix T_{phl} may be expressed as follows:

$$T_{phl} = \left[T_{VSI}(\omega_o) \right] * \left[T_{VSR}(\omega_i) \right]^T \tag{5.18}$$

Multiplying $\left[T_{VSR}(\omega_i) \right]^T$ with input phase voltage vector given by equation 5.11, the following is obtained:

$$\left[T_{VSR}(\omega_i) \right]^T * v_{iph} = \frac{3}{2} V_{im} * \cos(\varphi_i) \tag{5.19}$$

Equations 5.14, 5.17 and 5.19 are derived in the Appendix A. This constant voltage given by equation 5.19 represents the operation of a voltage source rectifier (VSR). Multiplying equation 5.19 with $T_{VSI}(\omega_o)$, the operation of a voltage source inverter (VSI) is obtained. Therefore, ITF approach emulates a VSR–VSI combination as shown in Figure 5.2. From this figure, it is clear that the three-phase AC to three-phase AC MC falls to one of the six possible sub-topologies with $V_{dc} \in \left[V_{ab}, V_{bc}, V_{ca}, V_{ba}, V_{cb}, V_{ac}\right]$. It can be seen that for each allowed switching combination in the figure, there is only one allowed switching combination in Table 1.1 of Chapter 1 that results in the same output voltage and input current. The ITF approach enables application of well-known VSR–VSI PWM technique for three-phase AC to three-phase AC MC control.

Here based on the ITF approach, SVM is simultaneously employed for both VSR and VSI part of the MC. The procedure for VSI SVM and VSR SVM are reviewed and their LF transfer functions are derived. The steps for HF synthesis of the simultaneous output voltage and input current SVM are derived and the representation of the modulation process in the complex plane is given.

5.3.1 Voltage Source Inverter Output Voltage SVM

Consider the VSI part of Figure 5.2, supplied by a DC voltage source $v_{pn} = V_{dc}$.

The VSI can assume six non-zero and two zero output voltage values. The resulting output line voltage space vector is defined as follows [1–6]:

$$v_{oL} = \frac{2}{3}\left(v_{ab} + v_{bc}e^{+j120} + v_{ca}e^{-j120}\right) \tag{5.20}$$

The output line voltage space vector can assume seven discrete voltage switching space vectors (SSVs) $V_o - V_6$ shown in Figure 5.3. The desired output line voltage space vector is given below:

$$v_{oL} = \sqrt{3}V_{om} * e^{j(\omega_o t - \varphi_o + 30)} \tag{5.21}$$

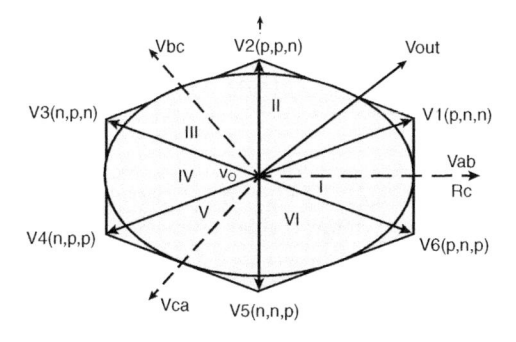

FIGURE 5.3
Output voltage hexagon.

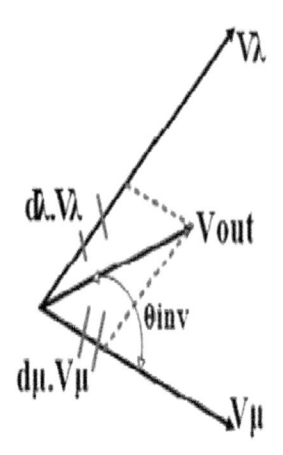

FIGURE 5.4
Output voltage space vector.

The line voltage space vector can be resolved into V_μ, V_λ and V_o using PWM as shown in Figure 5.4. In this figure, V_{out} is the sampled value of v_{oL} at any instant within the switching period T_s. The duty cycles of the two SSVs are given below:

$$d_\mu = \frac{T_\mu}{T_s} = m_v * \sin\left(\frac{\pi}{3} - \theta_{inv}\right)$$ (5.22)

$$d_\lambda = \frac{T_\lambda}{T_s} = m_v * \sin(\theta_{inv})$$ (5.23)

$$d_{ov} = \frac{T_{ov}}{T_s} = \left(1 - d_\mu - d_\lambda\right)$$ (5.24)

where m_v is the voltage modulation index whose range is defined below:

$$0 \le m_v \le \frac{\left(\sqrt{3}V_{om}\right)}{V_{dc}} \le 1$$ (5.25)

The sectors of the VSI hexagon in Figure 5.3 correspond to the six 60° segments within a period of the desired three-phase output line voltages shown in Figure 5.5. In Figure 5.6, the line voltage during the period T_μ corresponds $V_6(p,n,p)$, T_λ corresponds to $V_1(p,n,n)$ and T_o corresponds to zero voltage vector in Figure 5.3. When the switching in Figure 5.2 corresponds to $V_6(p,n,p)$, the line-to-line output voltages V_{ab}, V_{bc} and V_{ca} are $+V_{dc}$, $-V_{dc}$ and zero and when this switching is $V_1(p,n,n)$ the corresponding line-to-line output voltages are

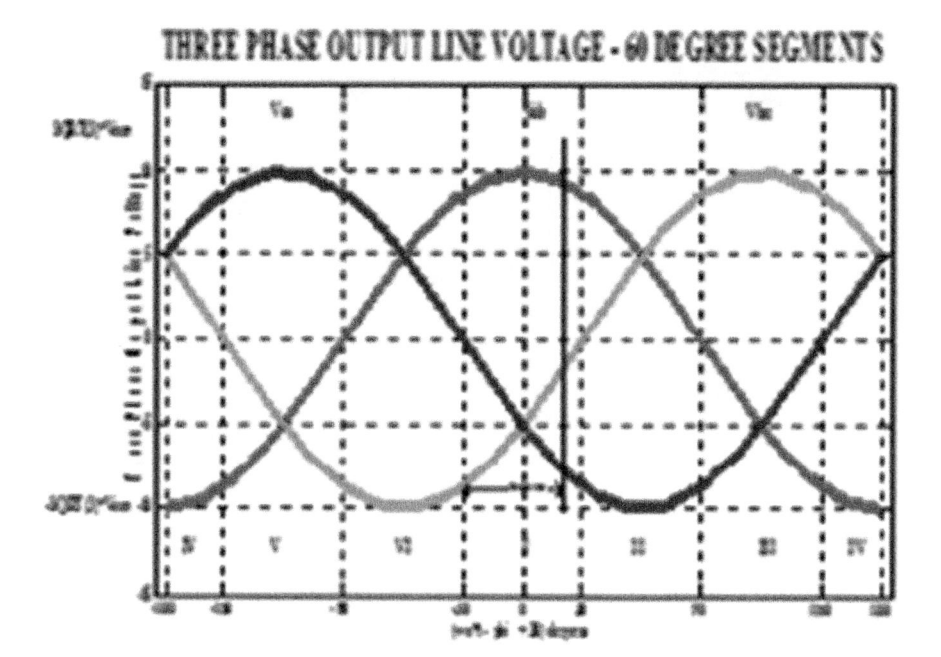

FIGURE 5.5
Three-phase line voltage = 60° segments.

$+V_{dc}$, 0 and $-V_{dc}$, respectively. These values are shown in Figure 5.6. The local averaged output line voltages are given below:

$$
\begin{bmatrix} v_{ab} \\ v_{bc} \\ v_{ca} \end{bmatrix} = \begin{bmatrix} d_\mu + d_\lambda \\ -d_\mu \\ -d_\lambda \end{bmatrix} * V_{dc} \tag{5.26}
$$

Using equations 5.22–5.23 in 5.26 gives

$$
\begin{bmatrix} v_{ab} \\ v_{bc} \\ v_{ca} \end{bmatrix} = m_v * \begin{bmatrix} \cos(\theta_{inv} - 30) \\ -\sin(60 - \theta_{inv}) \\ -\sin(\theta_{inv}) \end{bmatrix} * V_{dc} \tag{5.27}
$$

For the first 60° segment,

$$
-30 \le (\omega_o t - \varphi_o + 30) \le 30
$$

$$
\theta_{inv} = (\omega_o t - \varphi_o + 30) + 30 \tag{5.28}
$$

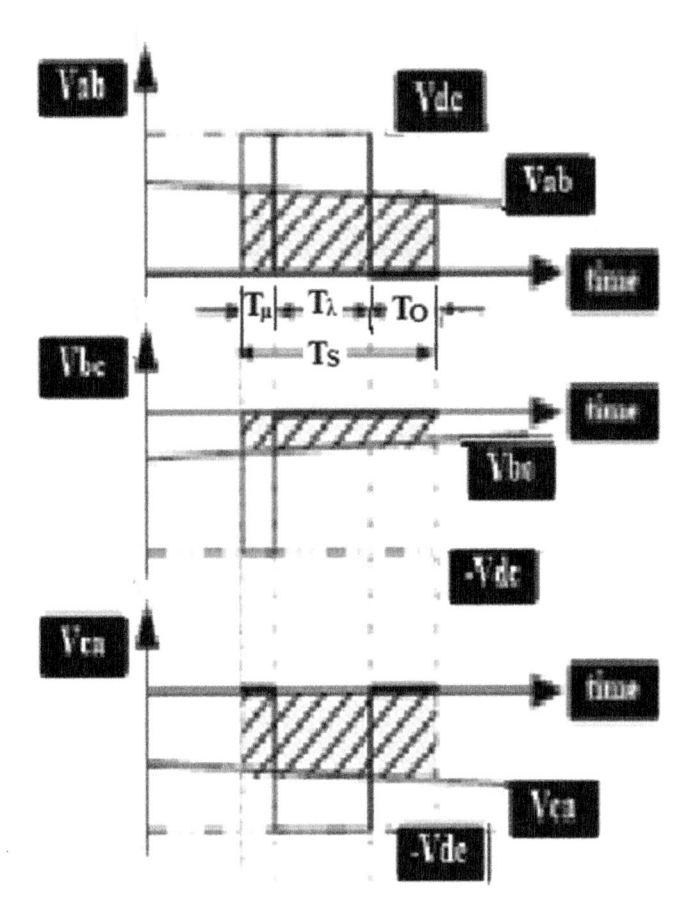

FIGURE 5.6
Synthesis of VSI output line voltage.

Using equation 5.28 in equation 5.27 gives

$$
\begin{bmatrix} v_{ab} \\ v_{bc} \\ v_{ca} \end{bmatrix} = m_v * \begin{bmatrix} \cos(\omega_o t - \varphi_o + 30) \\ \cos(\omega_o t - \varphi_o + 30 - 120) \\ \cos(\omega_o t - \varphi_o + 30 + 120) \end{bmatrix} * V_{dc} = T_{\text{VSI}} * V_{dc} \qquad (5.29)
$$

where T_{VSI} is the LF transfer matrix of VSI. Using equation 5.25 in 5.29, equation 5.12 is obtained. The local averaged input current is defined as follows:

$$
i_{dc} = [T_{\text{VSI}}]^T * i_{oL} = \frac{\sqrt{3}}{2} I_{om} * m_v * \cos(\varphi_L) \qquad (5.30)
$$

5.3.2 Voltage Source Rectifier Input Current SVM

Consider the VSR part of Figure 5.2 as a standalone VSR loaded by a DC current generator I_{dc}. The VSR input current hexagon is similar to VSI output voltage hexagon except that the subscripts μ, λ and inv are replaced by α, β and rect, respectively. The VSR hexagon is shown in Figure 5.7. The input current space vector is shown in Figure 5.8. The VSR duty cycles are defined as follows [1–6]:

$$d_\alpha = \frac{T_\alpha}{T_s} = m_c * \sin\left(\frac{\pi}{3} - \theta_{rect}\right) \tag{5.31}$$

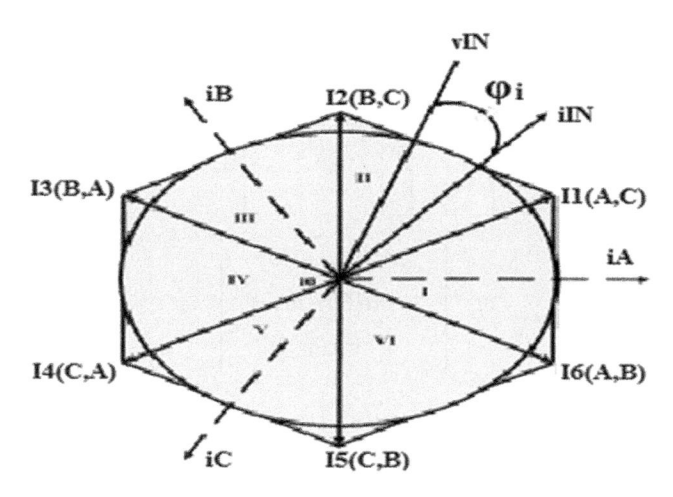

FIGURE 5.7
Input current vector hexagon.

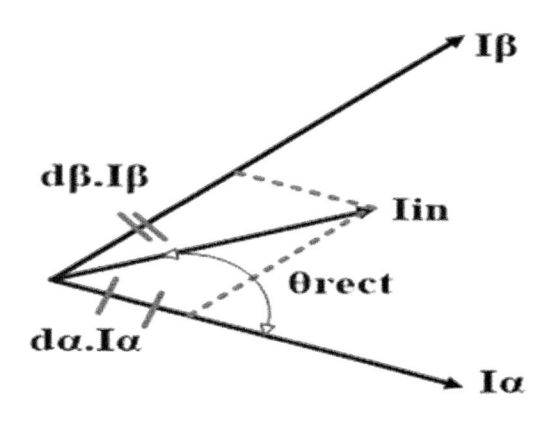

FIGURE 5.8
Input current space vector.

$$d_\beta = \frac{T_\beta}{T_s} = m_c * \sin(\theta_{rect}) \tag{5.32}$$

$$d_{ov} = \frac{T_{oc}}{T_s} = (1 - d_\alpha - d_\beta) \tag{5.33}$$

where m_c is the VSR modulation index whose range is defined below:

$$0 \le m_c \le \frac{I_{im}}{I_{dc}} \le 1 \tag{5.34}$$

The local averaged input phase currents in the first sector of VSR hexagon are given below:

$$\begin{bmatrix} i_A \\ i_B \\ i_C \end{bmatrix} = \begin{bmatrix} d_\alpha + d_\beta \\ -d_\alpha \\ -d_\beta \end{bmatrix} * I_{dc} \tag{5.35}$$

$$= m_c * \begin{bmatrix} \cos(\theta_{rect} - 30) \\ -\sin(60 - \theta_{rect}) \\ -\sin(\theta_{rect}) \end{bmatrix} * I_{dc} \tag{5.36}$$

But θ_{rect} in the first sector of input current hexagon can be expressed as follows:

$$\theta_{rect} = (\omega_i t - \varphi_i + 30)$$
$$-30 \le (\omega_i t - \varphi_i) \le +30 \tag{5.37}$$

Using equation 5.37 in 5.36, the LF transfer matrix T_{VSR} for the first sector of input current hexagon is defined as follows:

$$\begin{bmatrix} i_A \\ i_B \\ i_C \end{bmatrix} = m_c * \begin{bmatrix} \cos(\omega_i t - \varphi_i) \\ \cos(\omega_i t - \varphi_i - 120) \\ \cos(\omega_i t - \varphi_i + 120) \end{bmatrix} * I_{dc} = T_{VSR} * I_{dc} \tag{5.38}$$

Using equation 5.34 in 5.38, equation 5.16 is obtained. The VSR local averaged output voltage $v_{pn} = V_{dc}$ in Figure 5.2 is obtained as follows:

$$v_{pn} = V_{dc} = [T_{VSR}]^T * v_{iph} = \frac{3}{2} V_{im} * m_c * \cos(\varphi_i) \tag{5.39}$$

5.3.3 Matrix Converter Output Voltage and Input Current SVM

The local averaged output voltage of the SVM VSR given in equation 5.39 and that for the input current SVM VSI given in equation 5.30 are constants and hence can be directly interconnected. Also, it is seen that the product $[T_{\text{VSR}}] * [T_{\text{VSI}}]^T$ from equations 5.38 and 5.29 corresponds to T_{phl} in equation 5.13, where m corresponds to the product $m_v * m_c$. For simplicity, m_c is chosen as one and m_v is chosen to be the value of m.

There are six sectors each for the VSI and VSR hexagons, and hence 36 operating modes are possible. Consider the instant when the first 60° segment for both the output voltage and the input current is active. Then, using equations 5.27 and 5.36, the following expression for LF transfer matrix is obtained [1–6]:

$$T_{phl} = m * \begin{bmatrix} \cos(\theta_{\text{inv}} - 30) \\ -\sin(60 - \theta_{\text{inv}}) \\ -\sin(\theta_{\text{inv}}) \end{bmatrix} * \begin{bmatrix} \cos(\theta_{\text{rect}} - 30) \\ -\sin(60 - \theta_{\text{rect}}) \\ -\sin(\theta_{\text{rect}}) \end{bmatrix}^T \tag{5.40}$$

where $m = m_v * m_c$. Using equations 5.26 and 5.35 in equation 5.3, we have the following:

$$\begin{bmatrix} v_{ab} \\ v_{bc} \\ v_{ca} \end{bmatrix} = \begin{bmatrix} d_\mu + d_\lambda \\ -d_\mu \\ -d_\lambda \end{bmatrix} * \begin{bmatrix} d_\alpha + d_\beta \\ -d_\alpha \\ -d_\beta \end{bmatrix}^T * \begin{bmatrix} v_{AO} \\ v_{BO} \\ v_{CO} \end{bmatrix} \tag{5.41}$$

Equation 5.41 simplifies to the following:

$$\begin{bmatrix} v_{ab} \\ v_{bc} \\ v_{ca} \end{bmatrix} = \begin{bmatrix} (d_{\alpha\mu} + d_{\alpha\lambda}) & (d_{\beta\mu} + d_{\beta\lambda}) \\ -d_{\alpha\mu} & -d_{\beta\mu} \\ -d_{\alpha\lambda} & -d_{\beta\lambda} \end{bmatrix} * \begin{bmatrix} v_{AB} \\ v_{AC} \end{bmatrix} \tag{5.42}$$

where

$$v_{AB} = v_{AO} - v_{BO} \tag{5.43}$$

$$v_{AC} = v_{AO} - v_{CO} \tag{5.44}$$

$$d_{\alpha\mu} = d_\alpha d_\mu = m * \sin\left(\frac{\pi}{3} - \theta_{\text{rect}}\right) * \sin\left(\frac{\pi}{3} - \theta_{\text{inv}}\right) = \frac{T_{\alpha\mu}}{T_s} \tag{5.45}$$

$$d_{\alpha\lambda} = d_\alpha d_\lambda = m * \sin\left(\frac{\pi}{3} - \theta_{\text{rect}}\right) * \sin(\theta_{\text{inv}}) = \frac{T_{\alpha\lambda}}{T_s} \tag{5.46}$$

$$d_{\beta\mu} = d_\beta d_\mu = m * \sin(\theta_{\text{rect}}) * \sin\left(\frac{\pi}{3} - \theta_{\text{inv}}\right) = \frac{T_{\beta\mu}}{T_s} \tag{5.47}$$

$$d_{\beta\lambda} = d_{\beta}d_{\lambda} = m^* \sin(\theta_{\text{rect}})^* \sin(\theta_{\text{inv}}) = \frac{T_{\beta\lambda}}{T_s} \qquad (5.48)$$

From equation 5.42 (Appendix A), it is clear that the standard output voltage and input current SVM can be implemented in two VSI sub-topologies. When the VSI SVM sub-topology corresponds to $v_{pn} = V_{dc} = v_{AB,}$ then the two duty cycles of the two adjacent voltage vectors are $d_{\alpha\mu}$ and $d_{\alpha\lambda}$. When the VSI SVM sub-topology corresponds to $v_{pn} = V_{dc} = v_{ac,}$ then the two duty cycles of the adjacent output voltage vectors are $d_{\beta\mu}$ and $d_{\beta\lambda}$. During the remaining part of the switching period, the output line voltage is zero and the duty cycle corresponds to the following:

$$d_0 = 1 - d_{\alpha\mu} - d_{\alpha\lambda} - d_{\beta\mu} - d_{\beta\lambda} = \frac{T_o}{T_s} \qquad (5.49)$$

Finally, it is the selection of the zero switching state vector from Group III of Table 1.1 in Chapter 1 and the order of arrangement of the five SSVs. The pattern followed for arrangement of the state switching vectors is: $d_{\alpha\mu} \rightarrow d_{\beta\mu} \rightarrow d_{\beta\lambda} \rightarrow d_{\alpha\lambda} \rightarrow d_0$. The optimum zero switching state vector is used for each of the 36 switching combinations.

The 36 switching combinations are tabulated as shown in Tables 5.1–5.6, each with six sub-tables. In Tables 5.1–5.6, SI and SV correspond to the input current and output voltage hexagon sector number. Also, in Tables 5.1–5.6, serial numbers 1–5 correspond to the timing $T_{\alpha\mu} \rightarrow T_{\beta\mu} \rightarrow T_{\beta\lambda} \rightarrow T_{\alpha\lambda} \rightarrow T_0$. Also, it is seen from Tables 5.1–5.6 that with reference to VSI output voltage switching combination, there are 18 pair of identical sub-tables.

Referring to Table 5.1, assume that the reference output voltage and input current vectors are in quadrant six and one, respectively, so that Sv = VI and SI = I. From Figures 5.3, 5.4, 5.7 and 5.8, the SSV pair using the above duty-cycle sequence will be I_6–V_5, I_1–V_5, I_1–V_6, I_6–V_6 and I_o–V_o. From Figures 5.7 and 5.8, for the time interval $T_{\alpha\mu}$, I_α corresponds to I_6. Thus, A is connected to positive p terminal and B is connected to negative n terminal of rectifier shown in Figure 5.1. Similarly, referring to Figures 5.3 and 5.4, V_μ corresponds to V_5 and V_λ corresponds to V_6. Thus, the switch combination for the inverter of Figure 5.1 is B B A. Similarly, for the time interval $T_{\beta\mu}$, I_β corresponds to I_1 and hence A is connected to p terminal and C is connected to n terminal of rectifier and the switch combination for the inverter output is C C A. For the time interval $T_{\beta\lambda}$, V_λ corresponds to V_6 and hence the switch combination for the inverter output is A C A. For the time interval $T_{\alpha\lambda}$, the switch combination for the inverter output is A B A. For the time interval T_0, positive terminal corresponding to A is common to both I_α and I_β and the switch combination of the inverter is A A A. The complete Tables 5.1–5.6 are filled in this way. The method of completing Tables 5.1–5.6 is explained in the Appendix A.

TABLE 5.1 TO 5.6

Three-Phase ISVM MC Switching Combinations

TABLE 5.1

SI = I & Sv = I		(A)				SI = I & Sv = II		(B)				SI = I & Sv = III		(C)			
		Switching Combinations						Switching Combinations						Switching Combinations			
Sl. No.	SSV Pair	p	n	a	b	c	Sl. No.	SSV Pair	p	n	a	b	c	Sl. No.	SSV Pair	p	n	a	b	c
1	$I_6 - V_6$	A	B	A	B	A	1	$I_6 - V_1$	A	B	A	B	B	1	$I_6 - V_2$	A	B	A	A	B
2	$I_1 - V_6$	A	C	A	C	A	2	$I_1 - V_1$	A	C	A	C	C	2	$I_1 - V_2$	A	C	A	A	C
3	$I_1 - V_1$	A	C	A	C	C	3	$I_1 - V_2$	A	C	A	A	C	3	$I_1 - V_3$	A	C	C	A	C
4	$I_6 - V_1$	A	B	A	B	B	4	$I_6 - V_2$	A	B	A	A	B	4	$I_6 - V_3$	A	B	B	A	B
5	$I_0 - V_0$	A	A	A	A	A	5	$I_0 - V_0$	A	A	A	A	A	5	$I_0 - V_0$	A	A	A	A	A

SI = I & Sv = IV		(D)				SI = I & Sv = V		(E)				SI = I & Sv = VI		(F)			
		Switching Combinations						Switching Combinations						Switching Combinations			
Sl. No.	SSV Pair	p	n	a	b	c	Sl. No.	SSV Pair	p	n	a	b	c	Sl. No.	SSV Pair	p	n	a	b	c
1	$I_6 - V_3$	A	B	B	A	B	1	$I_6 - V_4$	A	B	B	A	A	1	$I_6 - V_5$	A	B	B	B	A
2	$I_1 - V_3$	A	C	C	A	C	2	$I_1 - V_4$	A	C	C	A	A	2	$I_1 - V_5$	A	C	C	C	A
3	$I_1 - V_4$	A	C	C	A	A	3	$I_1 - V_5$	A	C	C	C	A	3	$I_1 - V_6$	A	C	A	C	A
4	$I_6 - V_4$	A	B	B	A	A	4	$I_6 - V_5$	A	B	B	B	A	4	$I_6 - V_6$	A	B	A	B	A
5	$I_0 - V_0$	A	A	A	A	A	5	$I_0 - V_0$	A	A	A	A	A	5	$I_0 - V_0$	A	A	A	A	A

(Continued)

TABLE 5.1 TO 5.6 (*Continued*)

TABLE 5.2

SI = IV & Sv = I		(A)					SI = IV & Sv = II		(B)					SI = IV & Sv = III		(C)			
		Switching Combinations							Switching Combinations							Switching Combinations			
Sl. No.	SSV Pair	p	n	a	b	c	Sl. No.	SSV Pair	p	n	a	b	c	Sl. No.	SSV Pair	p	n	a	b	c
1	I_3-V_6	B	A	B	A	B	1	I_3-V_1	B	A	B	A	A	1	I_3-V_2	B	A	B	B	A
2	I_4-V_6	C	A	C	A	C	2	I_4-V_1	C	A	C	A	A	2	I_4-V_2	C	A	C	C	A
3	I_4-V_1	C	A	C	A	A	3	I_4-V_2	C	A	C	C	A	3	I_4-V_3	C	A	A	C	A
4	I_3-V_1	B	A	B	A	A	4	I_3-V_2	B	A	B	B	A	4	I_3-V_3	B	A	A	B	A
5	I_o-V_o	A	A	A	A	A	5	I_o-V_o	A	A	A	A	A	5	I_o-V_o	A	A	A	A	A

SI = IV & Sv = IV		(D)					SI = IV & Sv = V		(E)					SI = IV & Sv = VI		(F)			
		Switching Combinations							Switching Combinations							Switching Combinations			
Sl. No.	SSV Pair	p	n	a	b	c	Sl. No.	SSV Pair	p	n	a	b	c	Sl. No.	SSV Pair	p	n	a	b	c
1	I_3-V_3	B	A	A	B	A	1	I_3-V_4	B	A	A	B	B	1	I_3-V_5	B	A	A	A	B
2	I_4-V_3	C	A	A	C	A	2	I_4-V_4	C	A	A	C	C	2	I_4-V_5	C	A	A	A	C
3	I_4-V_4	C	A	A	C	C	3	I_4-V_5	C	A	A	A	C	3	I_4-V_6	C	A	C	A	C
4	I_3-V_4	B	A	A	B	B	4	I_3-V_5	B	A	A	A	B	4	I_3-V_6	B	A	B	A	B
5	I_o-V_o	A	A	A	A	A	5	I_o-V_o	A	A	A	A	A	5	I_o-V_o	A	A	A	A	A

(Continued)

TABLE 5.1 TO 5.6 (*Continued*)

TABLE 5.3

SI = II & Sv = I		(A)					SI = II & Sv = II		(B)					SI = II & Sv = III		(C)				
		Switching Combinations							Switching Combinations							Switching Combinations				
Sl. No.	SSV Pair	p	n	a	b	c	Sl. No.	SSV Pair	p	n	a	b	c	Sl. No.	SSV Pair	p	n	a	b	c
1	I_1-V_6	A	C	A	C	A	1	I_1-V_1	A	C	A	C	C	1	I_1-V_2	A	C	A	A	C
2	I_2-V_6	B	C	B	C	B	2	I_2-V_1	B	C	B	C	C	2	I_2-V_2	B	C	B	B	C
3	I_2-V_1	B	C	B	C	C	3	I_2-V_2	B	C	B	B	C	3	I_2-V_3	B	C	C	B	C
4	I_1-V_1	A	C	A	C	C	4	I_1-V_2	A	C	A	A	C	4	I_1-V_3	A	C	C	A	C
5	I_0-V_0	C	C	C	C	C	5	I_0-V_0	C	C	C	C	C	5	I_0-V_0	C	C	C	C	C

SI = II & Sv = IV		(D)					SI = II & Sv = V		(E)					SI= II & Sv = VI		(F)				
		Switching Combinations							Switching Combinations							Switching Combinations				
Sl. No.	SSV Pair	p	n	a	b	c	Sl. No.	SSV Pair	p	n	a	b	c	Sl. No.	SSV Pair	p	n	a	b	c
1	I_1-V_3	A	C	C	A	C	1	I_1-V_4	A	C	C	A	A	1	I_1-V_5	A	C	C	C	A
2	I_2-V_3	B	C	C	B	C	2	I_2-V_4	B	C	C	B	B	2	I_2-V_5	B	C	C	C	B
3	I_2-V_4	B	C	C	B	B	3	I_2-V_5	B	C	C	C	B	3	I_2-V_6	B	C	B	C	B
4	I_1-V_4	A	C	C	A	A	4	I_1-V_5	A	C	C	C	A	4	I_1-V_6	A	C	A	C	A
5	I_0-V_0	C	C	C	C	C	5	I_0-V_0	C	C	C	C	C	5	I_0-V_0	C	C	C	C	C

(*Continued*)

TABLE 5.1 TO 5.6 (*Continued*)

TABLE 5.4

SI = V & Sv = I		(A)					SI = V & Sv = II		(B)					SI = V & Sv = III		(C)				
		Switching Combinations							Switching Combinations							Switching Combinations				
Sl. No.	SSV Pair	p	n	a	b	c	Sl. No.	SSV Pair	p	n	a	b	c	Sl. No.	SSV Pair	p	n	a	b	c
1	I_4-V_6	C	A	C	A	C	1	I_4-V_1	C	A	C	A	A	1	I_4-V_2	C	A	C	C	A
2	I_5-V_6	C	B	C	B	C	2	I_5-V_1	C	B	C	B	B	2	I_5-V_2	C	B	C	C	B
3	I_5-V_1	C	B	C	B	B	3	I_5-V_2	C	B	C	C	B	3	I_5-V_3	C	B	B	C	B
4	I_4-V_1	C	A	C	A	A	4	I_4-V_2	C	A	C	C	A	4	I_4-V_3	C	A	A	C	A
5	I_o-V_o	C	C	C	C	C	5	I_o-V_o	C	C	C	C	C	5	I_o-V_o	C	C	C	C	C

SI = V & Sv = IV		(D)					SI = V & Sv = V		(E)					SI = V & Sv = VI		(F)				
		Switching Combinations							Switching Combinations							Switching Combinations				
Sl. No.	SSV Pair	p	n	a	b	c	Sl. No.	SSV Pair	p	n	a	b	c	Sl. No.	SSV Pair	p	n	a	b	c
1	I_4-V_3	C	A	A	C	A	1	I_4-V_4	C	A	A	C	C	1	I_4-V_5	C	A	A	A	C
2	I_5-V_3	C	B	B	C	B	2	I_5-V_4	C	B	B	C	C	2	I_5-V_5	C	B	B	B	C
3	I_5-V_4	C	B	B	C	C	3	I_5-V_5	C	B	B	B	C	3	I_5-V_6	C	B	C	B	C
4	I_4-V_4	C	A	A	C	C	4	I_4-V_5	C	A	A	A	C	4	I_4-V_6	C	A	C	A	C
5	I_o-V_o	C	C	C	C	C	5	I_o-V_o	C	C	C	C	C	5	I_o-V_o	C	C	C	C	C

(*Continued*)

TABLE 5.1 TO 5.6 (*Continued*)

TABLE 5.5

SI = III & Sv = I	(A)					SI = III & Sv = II	(B)					SI = III & Sv = III	(C)				
Sl. No.	SSV Pair	Switching Combinations					Sl. No.	SSV Pair	Switching Combinations				Sl. No.	SSV Pair	Switching Combinations			

Sl. No.	SSV Pair	p	n	a	b	c	Sl. No.	SSV Pair	p	n	a	b	c	Sl. No.	SSV Pair	p	n	a	b	c
1	I_2-V_6	B	C	B	C	B	1	I_2-V_1	B	C	B	C	C	1	I_2-V_2	B	C	B	B	C
2	I_3-V_6	B	A	B	A	B	2	I_3-V_1	B	A	B	A	A	2	I_3-V_2	B	A	B	B	A
3	I_3-V_1	B	A	B	A	A	3	I_3-V_2	B	A	B	B	A	3	I_3-V_3	B	A	A	B	A
4	I_2-V_1	B	C	B	C	C	4	I_2-V_2	B	C	B	B	C	4	I_2-V_3	B	C	C	B	C
5	I_0-V_0	B	B	B	B	B	5	I_0-V_0	B	B	B	B	B	5	I_0-V_0	B	B	B	B	B

SI = III & Sv = IV	(D)					SI = III & Sv = V	(E)					SI = III & Sv = VI	(F)				

Sl. No.	SSV Pair	p	n	a	b	c	Sl. No.	SSV Pair	p	n	a	b	c	Sl. No.	SSV Pair	p	n	a	b	c
1	I_2-V_3	B	C	C	B	C	1	I_2-V_4	B	C	C	B	B	1	I_2-V_5	B	C	C	C	B
2	I_3-V_3	B	A	A	B	A	2	I_3-V_4	B	A	A	B	B	2	I_3-V_5	B	A	A	A	B
3	I_3-V_4	B	A	A	B	B	3	I_3-V_5	B	A	A	A	B	3	I_3-V_6	B	A	B	A	B
4	I_2-V_4	B	C	C	B	B	4	I_2-V_5	B	C	C	C	B	4	I_2-V_6	B	C	B	C	B
5	I_0-V_0	B	B	B	B	B	5	I_0-V_0	B	B	B	B	B	5	I_0-V_0	B	B	B	B	B

(*Continued*)

TABLE 5.1 TO 5.6 (*Continued*)

TABLE 5.6

SI = VI& Sv = I		(A) Switching Combinations					SI = VI & Sv = II		(B) Switching Combinations					SI = VI & Sv = III		(C) Switching Combinations				
Sl. No.	SSV Pair	p	n	a	b	c	Sl. No.	SSV Pair	p	n	a	b	c	Sl. No.	SSV Pair	p	n	a	b	c
1	I_5–V_6	C	B	C	B	C	1	I_5–V_1	C	B	C	B	B	1	I_5–V_2	C	B	C	C	B
2	I_6–V_6	A	B	A	B	A	2	I_6–V_1	A	B	A	B	B	2	I_6–V_2	A	B	A	A	B
3	I_6–V_1	A	B	A	B	B	3	I_6–V_2	A	B	A	A	B	3	I_6–V_3	A	B	B	A	B
4	I_5–V_1	C	B	C	B	B	4	I_5–V_2	C	B	C	C	B	4	I_5–V_3	C	B	B	C	B
5	I_o–V_o	B	B	B	B	B	5	I_o–V_o	B	B	B	B	B	5	I_o–V_o	B	B	B	B	B

SI = VI & Sv = IV		(D) Switching Combinations					SI = VI & Sv = V		(E) Switching Combinations					SI = VI & Sv = VI		(F) Switching Combinations				
Sl. No.	SSV Pair	p	n	a	b	c	Sl. No.	SSV Pair	p	n	a	b	c	Sl. No.	SSV Pair	p	n	a	b	c
1	I_5–V_3	C	B	B	C	B	1	I_5–V_4	C	B	B	C	C	1	I_5–V_5	C	B	B	B	C
2	I_6–V_3	A	B	B	A	B	2	I_6–V_4	A	B	B	A	A	2	I_6–V_5	A	B	B	B	A
3	I_6–V_4	A	B	B	A	A	3	I_6–V_5	A	B	B	B	A	3	I_6–V_6	A	B	A	B	A
4	I_5–V_4	C	B	B	C	C	4	I_5–V_5	C	B	B	B	C	4	I_5–V_6	C	B	C	B	C
5	I_o–V_o	B	B	B	B	B	5	I_o–V_o	B	B	B	B	B	5	I_o–V_o	B	B	B	B	B

5.4 Model of Indirect Space-Vector-Modulated Three-Phase Matrix Converter

The model of the three-phase AC to three-phase AC indirect SVM MC is developed in SIMULINK [8,9]. This model is shown in Figure 5.9. The individual subsystems are explained below.

The power circuit of the model is shown in in Figure 5.9. This is developed using the Power Systems block set in SIMULINK [8]. This mainly consists of three-phase AC source, bidirectional switch matrix, output filter and R–L load. The arrangement of the bidirectional switch matrix using IGBTs is shown in Figure 2.2 of Chapter 2. The modulation algorithm is developed in several sub-units using Embedded MATLAB Function, MATLAB Function, Math Function, Logical and Bit Operator and using Sources block set in SIMULINK [8,9]. The various sub-units are explained below.

5.4.1 Duty-Cycle Sequence and Sector Switch Function Generator

This is developed using Embedded MATLAB Function block in SIMULINK, as shown in Figure 5.9. With peak line-to-neutral input voltage, desired peak output phase voltage, input frequency, output frequency, time, and angular frequency of reference frame ω_c as input parameters, the three-phase output voltage can be resolved into dq-axis component voltages [10] and the absolute value of the angle of the output phase voltage, vout_angle can be calculated using the function atan2(voq,vod). Using input voltage and input phase displacement angle, a similar procedure is used to calculate the absolute value of the input current angle, i_in_angle. The value of reference frame frequency ω_c is zero. This is illustrated in Program segment 4.1 in Chapter 4. These two values of the angles are used to calculate the output voltage sector SV, input current sector SI using MATLAB Function and theta_inv and theta_rect using the REM and SUBTRACT modules. Program Segment 5.1 illustrates part of the MATLAB code for duty-cycle sequencing and sector switch function generation. Duty-cycle dI, dII, dIII, dIV and d0 are calculated using equations 5.45–5.49, respectively, assuming unity input power factor. This is used to calculate the timings A, B, C and D and E by comparison of the cumulative sector timing with a triangle carrier as shown in Program segment 5.1. This triangle carrier has a period T_s, peak value $T_s/2$ V and minimum value zero, as shown in Figure 4.4 of Chapter 4, where T_s is the carrier switching period. The PWM signal timing is shown in Figure 5.10. Cumulative sector timing is obtained by appropriately adding the value of duty ratios and multiplying by $T_s/2$, as shown in Program segment 5.1. Sector switch functions ssf1 to ssf18 are determined using if-then-else statement. For example, sector switch function ssf1 is HIGH only when sv = si = 1 or sv = si = 4 and is LOW otherwise.

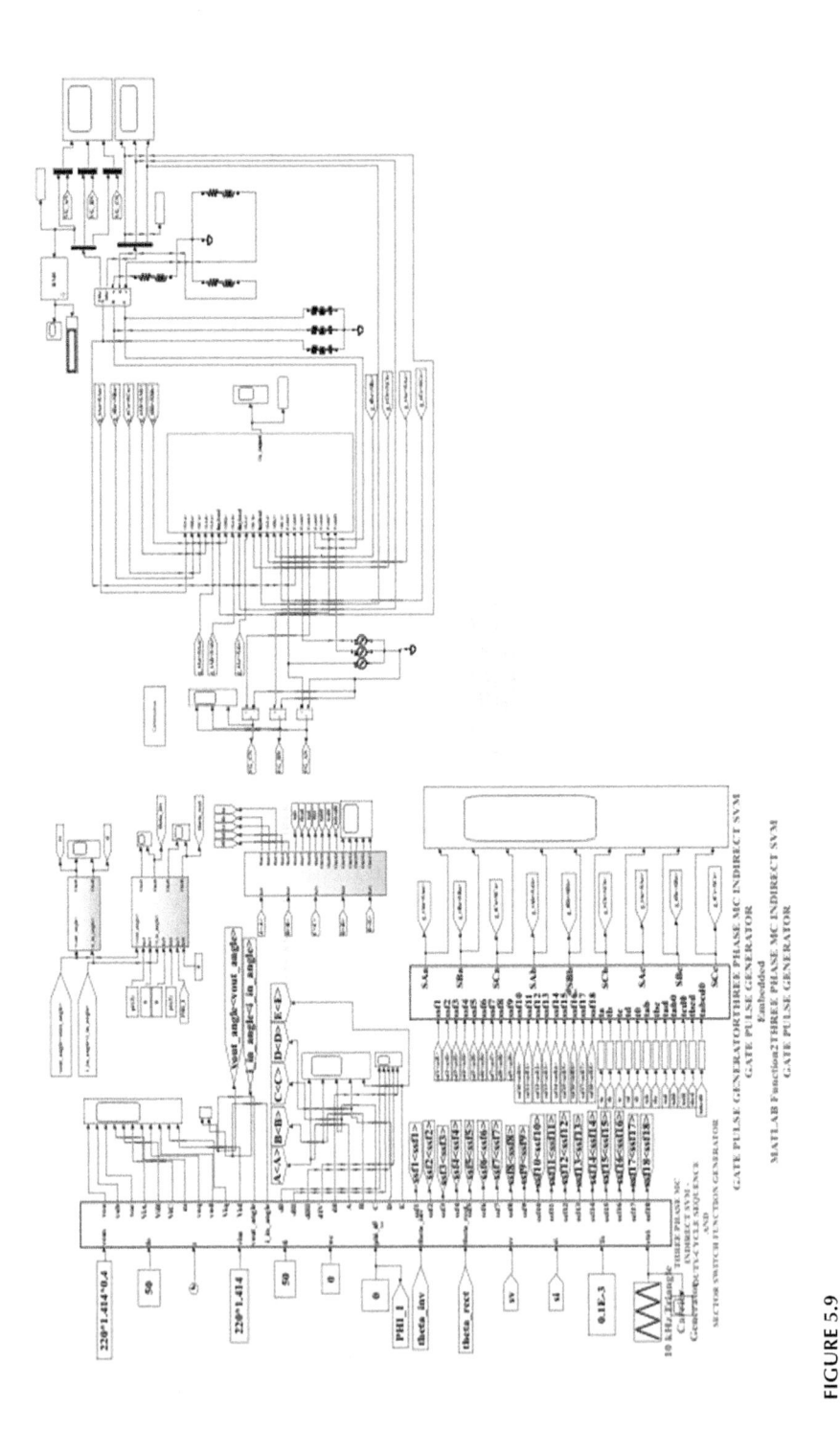

FIGURE 5.9
Three-phase AC to three-phase AC indirect space-vector-modulated matrix converter.

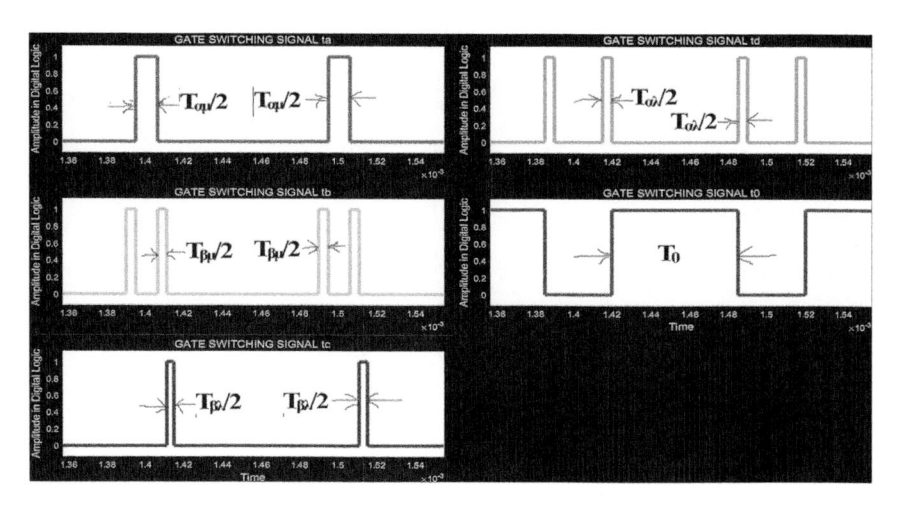

FIGURE 5.10
Three-phase ISVM MC – PWM gate signal timing.

Program Segment 5.1

```
%% Dr.Narayanaswamy. P.R. Iyer
Function [m,dI,dII,dIII,dIV,d0,A,B,C,D,E,ssf1,ssf2,ssf3,ssf4,
ssf5,ssf6,ssf7,ssf8,ssf9,ssf10,ssf11,ssf12,ssf13,ssf14,ssf15,
ssf16,ssf17, ssf18] = fcn(vom,fo,t,vim,fi,wc,phi_i,theta_inv,
theta_rect,sv,si,Ts,vtri)
m = vo/vim;
dI = (m*sin(pi/(3)-theta_rect)*sin(pi/(3)-theta_inv));
%%d_alfa_meu
dII = (m*sin(theta_rect)*sin(pi/(3)-theta_inv));%%d_beta_meu
dIII = (m*sin(theta_rect)*sin(theta_inv));%%d_beta_lamda
dIV = (m*sin(pi/(3)-theta_rect)*sin(theta_inv));%%d_alfa_lamda
d0 = (1 - dI - dII - dIII - dIV);
if (vtri <= dI*Ts/(2))
A = 1;  else
A = 0;
end
if (vtri <= (dII + dI)*Ts/(2))
B = 1;  else
B = 0;
end
if (vtri <= (dIII + dII + dI)*Ts/(2))
C = 1;  else
C = 0;
end
if (vtri <= (dIV + dIII + dII + dI)*Ts/(2))
D = 1; else
D = 0;
end
```

```
if (vtri <= (dIV + dIII + dII + dI + d0)*Ts/(2))
E = 1; else
E = 0;
end
if (sv == 1 && si == 1 || sv == 4 && si == 4)
  %%sector 1. sector switch function ssf1.
  ssf1 = 1;
else
  ssf1 = 0;
end
if (sv == 2 && si == 1 || sv == 5 && si == 4)
  %%sector 2. sector switch function ssf2.
  ssf2 = 1;
else
  ssf2 = 0;
end
%%Similar statement for ssf3 to ssf18.
.
.
.
if (sv == 6 && si == 3 || sv == 3 && si == 6)
  %%sector 18. sector switch function ssf18.
  ssf18 = 1;
else
  ssf18 - 0;
end
```

5.4.2 Output Voltage and Input Current Sector Calculator

This is developed using MATLAB Function, a mux and a demux as shown in Figure 4.5 of Chapter 4. The inputs to the mux are vout_angle and i_in_angle. The outputs from the demux are SV and SI. The source code to determine SV and SI is shown in Program segment 4.3 of Chapter 4.

5.4.3 Output Voltage and Input Current Reference Angle Calculator

This is shown in Figure 5.11. This uses two REM function blocks and SUBTRACT modules. The first REM block has the inputs v_out_angle and $\pi/(3)$. The output of the first REM block gives θ_{inv} defined in Figure 5.4. The second REM block has the inputs i_in_angle and $\pi/(3)$. The output of the second REM block is subtracted from input p.f. angle φ_i to obtain θ_{rect} defined in Figure 5.8.

5.4.4 Gate Pulse Timing Calculator

This is shown in Figure 5.12. The inputs are A, B, C, D and E, respectively. The individual gate pulse timings, t_a, t_b, t_c, t_d and t_0 are, respectively, $dI^*T_s/(2)$,

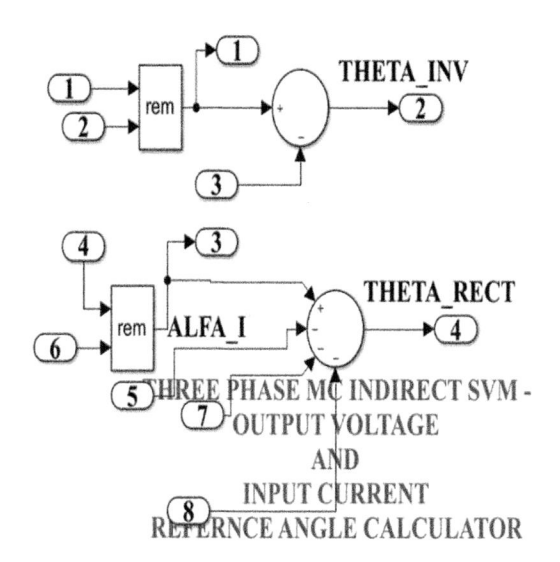

FIGURE 5.11
Output voltage and input current reference angle calculator.

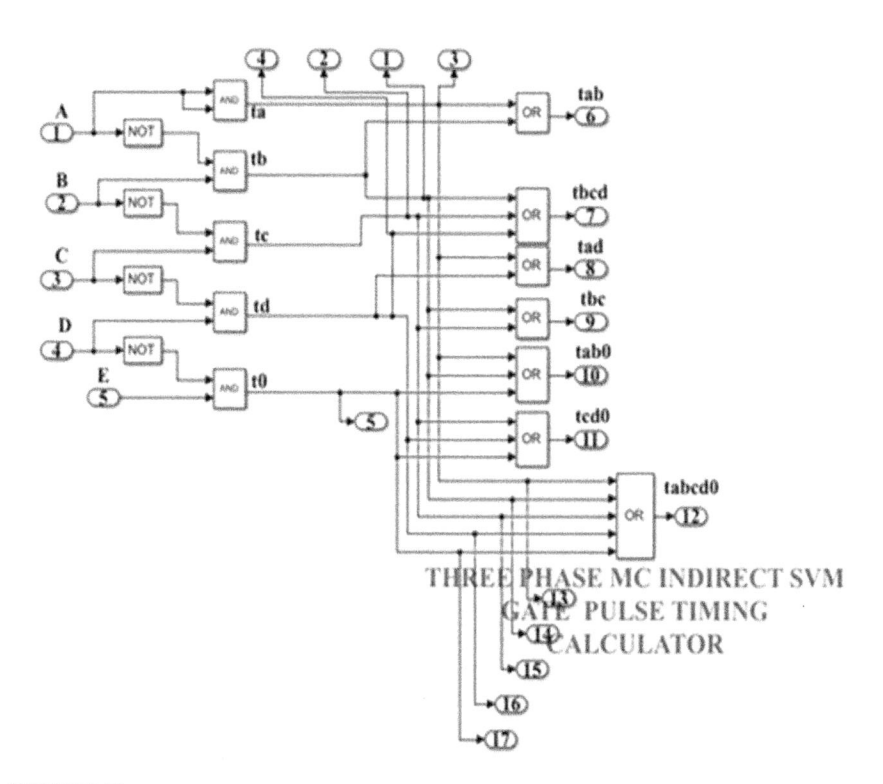

FIGURE 5.12
Gate pulse timing calculator.

dII*T_s/(2), dIII*T_s/(2), dIV*T_s/(2) and d0*T_s/(2) where each of these duty cycles dI, dII, dIII, dIV and d0, respectively, corresponds to $d_{\alpha\mu},\ d_{\beta\mu},\ d_{\beta\lambda},\ d_{\alpha\lambda}$ and d_0 as defined in equations 5.45, 5.47, 5.48, 5.46 and 5.49. This individual gate pulse duration is obtained as follows.

$$t_a = (A \ \& \ A)$$

$$t_b = (\sim A \ \& \ B)$$

$$t_c = (\sim B \ \& \ C) \tag{5.50}$$

$$t_d = (\sim C \ \& \ D)$$

$$t_0 = (\sim D \ \& \ E)$$

where symbol (&) represents logical AND and symbol (~) represents NOT operation. To easily generate the MC gate pulse, the timing pulses in equation 5.50 are given to OR gates to obtain the following additional timing pulse:

$$t_{ab} = (t_a \ \| \ t_b)$$

$$t_{bcd} = (t_b \ \| \ t_c \ \| \ t_d)$$

$$t_{ad} = (t_a \ \| \ t_d)$$

$$t_{bc} = (t_b \ \| \ t_c) \tag{5.51}$$

$$t_{ab0} = (t_a \ \| \ t_b \ \| \ t_0)$$

$$t_{cd0} = (t_c \ \| \ t_d \ \| \ t_0)$$

$$t_{abcd0} = (t_a \ \| \ t_b \ \| \ t_c \ \| \ t_d \ \| \ t_0)$$

where symbol (‖) represents logical OR operation. The complete gate timing pattern and the timing for the bidirectional switch S_{Aa} using indirect SVM algorithm are given in Table 5.7.

5.4.5 Gate Pulse Generator

The gate pulse generator for the MC is developed using Embedded MATLAB as shown in Figure 5.9. The sector switch functions ssf1 to ssf18, gate pulse timing calculator outputs defined in equations 5.50 and 5.51 form the input modules and the outputs are the gate pulses for the nine bidirectional switches of the MC. In Table 5.7, the gate timing for the S_{Aa} bidirectional switch is shown. This gate timing for switch S_{Aa} can be combined by logically

TABLE 5.7

Three-Phase ISVM MC Gate Pulse Pattern and Timing

Sl. No.	SI, SV	ssf_x High	Bidirectional Switch Gate Pulse Pattern					S_{Aa} Timing
			t_a	t_b	t_c	t_d	t_0	
1	I,I or IV,IV	ssf1	ABA	ACA	ACC	ABB	AAA	t_{abcd0}
2	I,II or IV,V	ssf2	ABB	ACC	AAC	AAB	AAA	t_{abcd0}
3	I,III or IV,VI	ssf3	AAB	AAC	CAC	BAB	AAA	t_{ab0}
4	I,IV or IV,I	ssf4	BAB	CAC	CAA	BAA	AAA	t_0
5	I,V or IV,II	ssf5	BAA	CAA	CCA	BBA	AAA	t_0
6	I,VI or IV,III	ssf6	BBA	CCA	ACA	ABA	AAA	t_{cd0}
7	II,I or V,IV	ssf7	ACA	BCB	BCC	ACC	CCC	t_{ad}
8	II,II or V,V	ssf8	ACC	BCC	BBC	AAC	CCC	t_{ad}
9	II,III or V,VI	ssf9	AAC	BBC	CBC	CAC	CCC	t_a
10	II,IV or V,I	ssf10	CAC	CBC	CBB	CAA	CCC	0
11	II,V or V,II	ssf11	CAA	CBB	CCB	CCA	CCC	0
12	II,VI or V,III	ssf12	CCA	CCB	BCB	ACA	CCC	t_d
13	III,I or VI,IV	ssf13	BCB	BAB	BAA	BCC	BBB	0
14	III,II or VI,V	ssf14	BCC	BAA	BBA	BBC	BBB	0
15	III,III or VI,VI	ssf15	BBC	BBA	ABA	CBC	BBB	t_c
16	III,IV or VI,I	ssf16	CBC	ABA	ABB	CBB	BBB	t_{bc}
17	III,V or VI,II	ssf17	CBB	ABB	AAB	CCB	BBB	t_{bc}
18	III,VI or VI,III	ssf18	CCB	AAB	BAB	BCB	BBB	t_b

ANDing the respective value with sector switch function `ssf_x` and ORing the value for each sector. Thus, the gate timing pulse for switch S_{Aa} can be expressed as

$$S_{Aa} = \text{ssf}1 \,\&\, \left(t_{abcd0}\right) \,\|\, \text{ssf}2 \,\&\, \left(t_{abcd0}\right) \,\|\, \text{ssf}3 \,\&\, \left(t_{ab0}\right) \,\|\, \cdots \,\|\, \text{ssf}18 \,\&\, \left(t_b\right) \qquad (5.52)$$

Program segment 5.2 gives the method of generating the gate timing pulses for the nine bidirectional switches using the gate pulse pattern given in Table 5.7.

Program Segment 5.2

```
function [SAa,SBa,SCa,SAb,SBb,SCb,SAc,SBc,SCc] =
fcn(ssf1,ssf2, ssf3,ssf4,ssf5,ssf6,ssf7,ssf8,ssf9,ssf10,ssf11,
ssf12, ssf13,ssf14,ssf15,ssf16,ssf17,ssf18,ta,tb,tc,td,t0,tab,
tbc,tad,tab0,tcd0,tbcd,tabcd0)
%% Dr.Narayanaswamy. P.R.Iyer
SAa = ssf1&(tabcd0)||ssf2&(tabcd0)||ssf3&(tab0)||ssf4&(t0)||
ssf5&(t0)|| ssf6&(tcd0)||ssf7&(tad)||ssf8&(tad)||ssf9&(ta)||
```

```
ssf10&(0)||ssf11&(0)|| ssf12&(td)||ssf13&(0)||ssf14&(0)||
ssf15& (tc)||ssf16&(tbc)|| ssf17& (tbc)||ssf18&(tb);
SBa = ssf1&(0)||ssf2&(0)||ssf3&(td)||ssf4&(tad)||ssf5&(tad)||
ssf6&(ta)|| ssf7&(tbc)||ssf8&(tbc)||ssf9&(tb)||ssf10&(0)||
ssf11&(0)||ssf12&(tc)||ssf13&(tabcd0)||ssf14&(tabcd0)||ssf15&
(tab0)||ssf16&(t0)||ssf17& (t0)||ssf18&(tcd0);
SCa = ssf1&(0)||ssf2&(0)||ssf3&(tc)||ssf4&(tbc)||ssf5&(tbc)||
ssf6&(tb)|| ssf7&(t0)||ssf8&(t0)||ssf9&(tcd0)||ssf10&(tabcd0)||
ssf11&(tabcd0)|| ssf12&(tab0)||ssf13&(0)||ssf14&(0)||ssf15&(td)
||ssf16&(tad)|| ssf17& (tad)||ssf18&(ta);
SAb = ssf1&(t0)||ssf2&(tcd0)||ssf3&(tabcd0)||ssf4&(tabcd0)||
ssf5&(tab0)|| ssf6&(t0)||ssf7&(0)||ssf8&(td)||ssf9&(tad)||
ssf10&(tad)||ssf11&(ta) ||ssf12&(0)||ssf13&(tbc)||ssf14&(tb)||
ssf15&(0)|| ssf16&(0)|| ssf17& (tc)||ssf18&(tbc);
SBb = ssf1&(tad)||ssf2&(ta)||ssf3&(0)||ssf4&(0)||ssf5&(td)||
ssf6&(tad)|| ssf7&(0)||ssf8&(tc)||ssf9&(tbc)||ssf10&(tbc)||
ssf11&(tb)||ssf12&(0)||ssf13&(t0)||ssf14&(tcd0)||ssf15&(tabcd0)
||ssf16&(tabcd0)|| ssf17& (tab0)||ssf18&(t0);
SCb = ssf1&(tbc)||ssf2&(tb)||ssf3&(0)||ssf4&(0)||ssf5&(tc)||
ssf6&(tbc)|| ssf7&(tabcd0)||ssf8&(tab0)||ssf9&(t0)||ssf10&(t0)
||ssf11&(tcd0)|| ssf12&(tabcd0)||ssf13&(tad)||ssf14&(ta)||
ssf15&(0)||ssf16&(0)|| ssf17&(td)||ssf18&(tad);
SAc = sf1&(tab0)||ssf2&(t0)||ssf3&(t0)||ssf4&(tcd0)||ssf5&
(tabcd0)||ssf6&(tabcd0)||ssf7&(ta)||ssf8&(0)||ssf9&(0)||
ssf10&(td)|| ssf11&(tad)|| ssf12&(tad)||ssf13&(tc)||ssf14&(tbc)
||ssf15&(tbc)||ssf16&(tb)|| ssf17&(0)||ssf18&(0);
SBc = ssf1&(td)||ssf2&(tad)||ssf3&(tad)||ssf4&(ta)||ssf5&(0)||
ssf6&(0)|| ssf7&(tb)||ssf8&(0)||ssf9&(0)||ssf10&(tc)||ssf11&
(tbc)||ssf12&(tbc) ||ssf13&(tab0)||ssf14&(t0)||ssf15&(t0)||
ssf16&(tcd0)|| ssf17&(tabcd0)||ssf18&(tabcd0);
SCc = ssf1&(tc)||ssf2&(tbc)||ssf3&(tbc)||ssf4&(tb)||ssf5&(0)||
ssf6&(0)|| ssf7&(tcd0)||ssf8&(tabcd0)||ssf9&(tabcd0)||ssf10&
(tab0)||ssf11&(t0)||ssf12&(t0)||ssf13&(td)||ssf14&(tad)||
ssf15&(tad)||ssf16&(ta)|| ssf17&(0)||ssf18&(0);
```

Given here is the method to calculate the dwell time for the bidirectional switch S_{Aa}. In Table 5.7, for serial number 1 corresponding to ssf1 HIGH, the input phase 'A' occupies the position of output phase 'a' in all columns corresponding to t_a, t_b, t_c, t_d and t_0. Thus, when ssf1 is HIGH, dwell time for S_{Aa} is $(t_a \parallel t_b \parallel t_c \parallel t_d \parallel t_0)$ which is by definition t_{abcd0}. Similarly, when ssf2 is HIGH this dwell time for S_{Aa} is t_{abcd0}. When ssf3 is HIGH, 'A' is seen in the position of output phase 'a' in timing columns t_a, t_b and t_0 and S_{Aa} dwell time is t_{ab0}. Output phase a, b and c must be considered from left to right in order in the timing columns from t_a to t_0. In this way, the dwell time for all the nine bidirectional switches of the MC can be calculated for all 18 sector switch functions. This is shown in Program segment 5.2 [9].

5.5 Simulation Results

The simulation of the three-phase indirect SVM MC was carried out in SIMULINK [8]. The parameters used for simulation are shown in Table 5.8. The ode15S (Stiff/NDF) solver is used. Simulation is carried out for two different output frequencies with all other parameters constant. The simulation results for the harmonic spectrum of line-to-neutral output voltage, input current, load current and line-to-line output voltage for 50-Hz output frequency are shown in Figure 5.13a–d and their respective oscilloscope waveforms are shown in Figure 5.14a–d. The same results in the above order for 20-Hz output frequency are shown in Figures 5.15a–d and 5.16a–d, respectively. The simulation results are tabulated in Table 5.9.

5.6 Discussion of Results

Indirect SVM for three-phase MC assumes the three-phase MC to be considered equivalent to a three-phase rectifier-inverter converter and the SVM technique to be applied separately to the three-phase rectifier and inverter and the combined duty-cycle timing for the MC is the product of the corresponding individual duty-cycle timing for the three-phase inverter and individual duty-cycle timing for the three-phase rectifier. It is also seen from Table 5.9 that as the output frequency is reduced, the THD of the load current is reduced whereas the THD of line-to-neutral output voltage, line-to-line output voltages and input current are increased for this modulation index. The indirect SVM has eight commutations per carrier switching period. Also, it is seen from the simulation results that the harmonic

TABLE 5.8

Three-Phase ISVM MC Model Parameters

Sl. No.	Parameter	Value	Units
1	RMS line-to-neutral input voltage	220	V
2	Input frequency f_i	50	Hz
3	Modulation index q	0.4	-
4	Output frequency f_o	50, 20	Hz
5	Carrier sampling frequency f_{sw}	10	kHz
6	R–L–C output filter	10, 0.01e–3, 25.356e–6	Ω, H, F
7	R–L load	100, 500e–3	Ω, H
8	Input phase displacement φ_i	0	rad

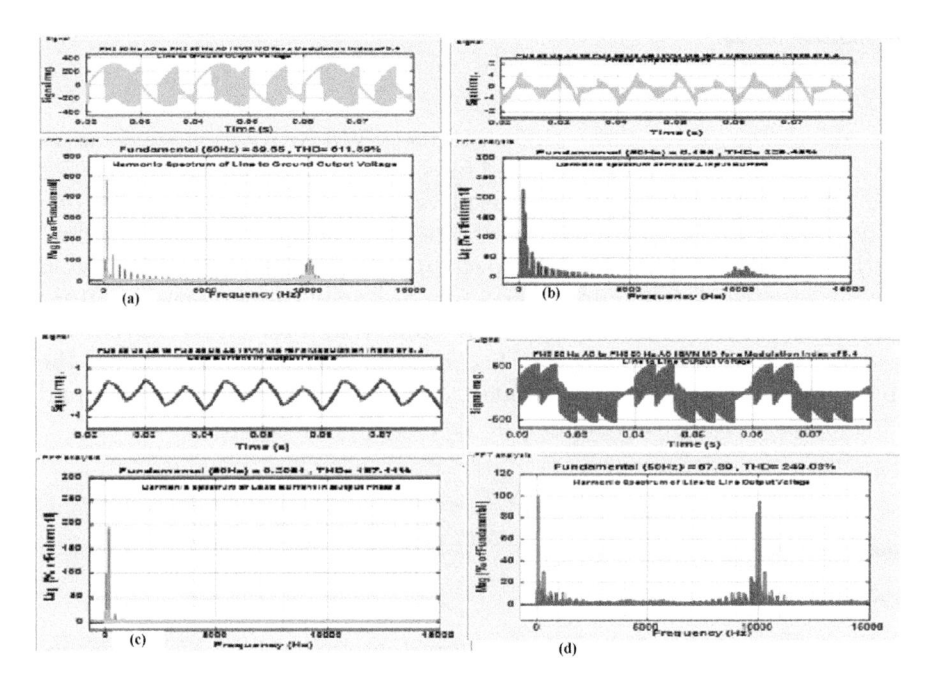

FIGURE 5.13
Three-phase 50-Hz AC to three-phase 50-Hz AC ISVM MC – harmonic spectrum: (a) line-to-ground output voltage, (b) phase A input current, (c) load current in output phase A and (d) line-to-line output voltage.

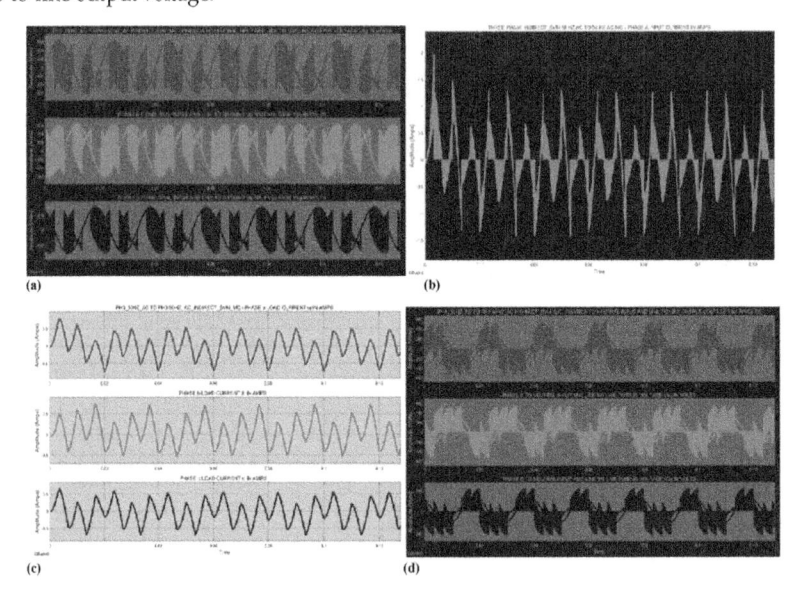

FIGURE 5.14
Three-phase 50-Hz AC to three-phase 50-Hz AC ISVM MC: (a) line-to-ground output voltage, (b) phase A input current, (c) load current and (d) line-to-line output voltage.

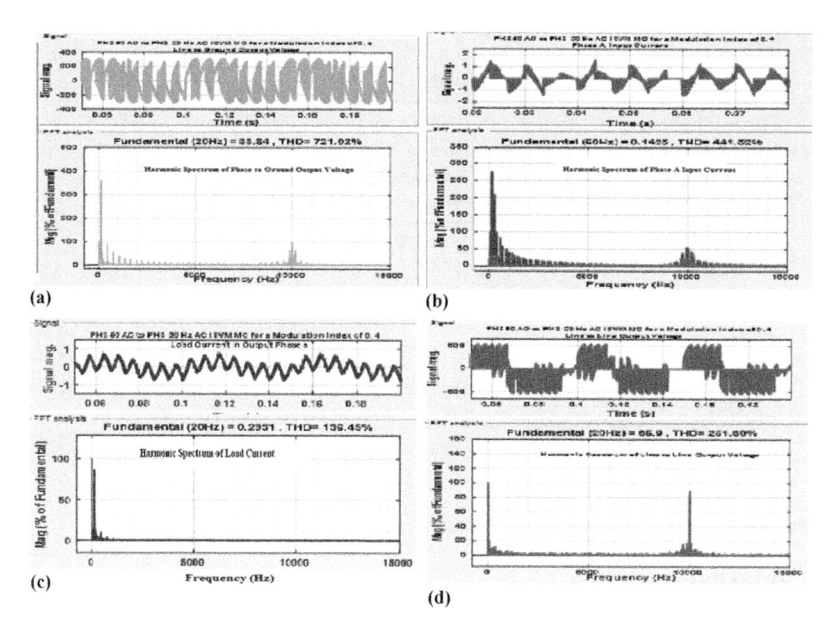

(a)　(b)　(c)　(d)

FIGURE 5.15
Three-phase 50-Hz AC to three-phase 20-Hz AC ISVM MC – harmonic spectrum: (a) line-to-ground output voltage, (b) phase A input current, (c) load current and (d) line-to-line output voltage.

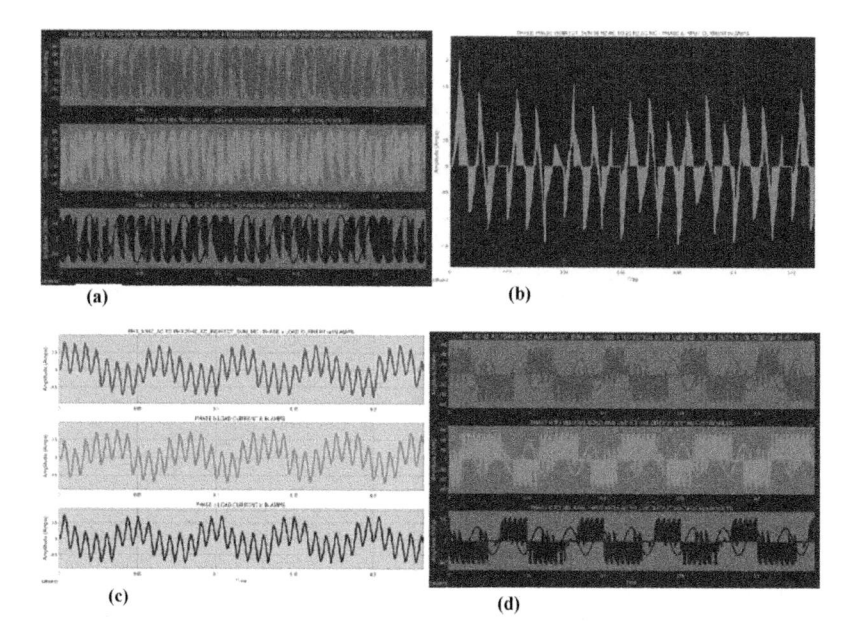

(a)　(b)　(c)　(d)

FIGURE 5.16
Three-phase 50-Hz AC to three-phase 20-Hz AC ISVM MC: (a) line-to-ground output voltage, (b) phase A input current, (c) load current and (d) line-to-line output voltage.

TABLE 5.9

Three-Phase ISVM MC Simulation Results

Sl. No.	Three-Phase Indirect SVM MC Input–Output Frequency (Hz)	Modulation Index q	THD of Line-to-Line Output Voltage (p.u.)	THD of Line-to-Neutral Output Voltage (p.u.)	THD of Input Current (p.u.)	THD of Load Current (p.u.)
1	50–50	0.4	2.49	6.11	3.29	1.97
2	50–20	0.4	2.51	7.21	4.41	1.36

components of line-to-neutral output voltage, line-to-line output voltage, input current and load current tend to concentrate at integral multiples of carrier switching frequency.

5.7 Conclusions

This chapter presents an easy method of modelling a three-phase indirect SVM MC using Embedded MATLAB Function, MATLAB Function, Math block set, Logic and Bit Operations toolbox and Sources block set. The input current, line-to-neutral and line-to-line output voltage harmonic distortions are increased with reduction in the output frequency, whereas for the load current, this harmonic distortion is reduced for this given modulation index. The three-phase AC to three-phase AC output voltage–input current SVM algorithm is systematically reviewed. A simple geometric presentation of the three-phase AC to three-phase AC MC SVM procedure is given. The indirect SVM algorithm uses eight commutations per carrier switching period. Harmonic components of line-to-neutral output voltage, line-to-line output voltage, input current and load current tend to concentrate at integral multiples of switching frequency. The model can be easily adapted for hardware implementation.

References

1. L. Huber and D. Borojevic: Space vector modulated three-phase to three-phase matrix converter with input power factor correction, *IEEE Transactions on Industry Applications*, Vol. 31, No. 6, 1995, pp. 1234–1245.
2. L. Huber, D. Borojevic, X.F. Zhuang and F.C. Lee: Design and implementation of a three phase to three phase matrix converter with input power factor correction, *IEEE Applied Power Electronics Conference and Exposition*, San Diego, CA, 1993, pp. 860–865.

3. L. Huber and D. Borojevic: Space vector modulation with unity input power factor for forced commutated cycloconverters, *IEEE Industry Application Society Annual Meeting*, Dearborn, MI, 1991, pp. 1032–1041.

4. L. Huber and D. Borojevic: Space vector modulator for forced commutated cycloconverters, *IEEE Industry Application Society Annual Meeting*, San Diego, CA, 1989, pp. 871–876.

5. L. Huber, D. Borojevic and N. Burany: Voltage space vector based PWM control of forced commutated cycloconverters, *IEEE - IECON*, Philadelphia, PA, 1989, pp. 106–111.

6. L. Huber, D. Borojevic and N. Burany: Analysis, design and implementation of the space-vector modulator for forced-commutated cycloconverters, *IEE-Proceedings, Part B*, Vol. 139, No. 2, 1992, pp. 103–113.

7. H.J. Cha and P. Enjetti: Matrix converter–fed ASDs, *IEEE Industry Applications Magazine*, Vol. 10, No. 4, 2004, pp. 33–39.

8. The Mathworks Inc.: www.mathworks.com, MATLAB/SIMULINK user manual, MATLAB R2017b, 2017.

9. N.P.R. Iyer: Modelling, simulation and real time implementation of a three phase AC to AC matrix converter, Ph.D. thesis, Ch. 9, Department of ECE, Curtin University, Perth, WA, Australia, February 2012.

10. P.C. Krause, O. Wasynczuk and S.D. Sudhoff: *Analysis of Electrical Machinery*, IEEE Press, New York, 1995.

6

Programmable AC to DC Rectifier Using Matrix Converter Topology

6.1 Introduction

Three-phase AC to three-phase AC matrix converter (MC) topology-based single and dual fixed supply AC to DC rectifier is reported in the literature [1–3]. But these reports only indicate that dual DC output voltages with fixed value are possible by setting the desired AC output voltage phase angle leading by 30° and the frequency of output voltage zero in the model [1,2]. Detailed modelling studies using SIMULINK reveals that with the frequency of desired AC output voltage set to zero, as the AC output voltage phase angle is varied from 0 to $+\pi$ and 0 to $-\pi$, dual DC output voltages in multitude of combinations are possible such as (a) both voltages positive and unequal, (b) both voltages positive and equal, (c) any one voltage zero and the other positive, (d) any one voltage positive and the other negative with unequal modulus value, (e) any one voltage positive and the other negative with equal modulus value, (f) any one voltage zero and the other negative, (g) both voltages negative and unequal, and (h) both voltages negative and equal. In this chapter, a detailed insight into this finding is made with a mathematical derivation for the dual DC output voltage magnitude. Theoretical derivation for output voltage magnitude is presented which is confirmed by model simulation [9,10]. The model is extended to verify the speed control, acceleration and brake by plugging of separately excited DC (SEDC) motors [9,10].

Similarly, in the single-ended bidirectional AC to DC MC topology also it is possible to obtain a single variable DC output voltage by changing the output voltage phase angle from 0 to $+\pi$ and 0 to $-\pi$ [1,2]. Theoretical derivation for output voltage magnitude is presented and the result is confirmed by model simulation. The model is used to verify the speed control and brake plugging of an SEDC motor.

Real-time applications of this method for SEDC motor are highlighted. Additional application of this single and dual programmable rectifier topology as a variable output voltage, variable-frequency pure sine-wave AC power supply is also presented.

6.2 Output Voltage Amplitude Limit of Direct AC to AC Converters

The three-phase AC to three-phase AC MC is shown in Figure 2.1 of Chapter 2. The model of the three-phase AC to three-phase AC MC in terms of modulation duty cycle has been derived in equations 2.1–2.8 in Section 2.2 of Chapter 2 [4,5]. Venturini modulation algorithm which gives the solution of the modulation duty cycle for the MC is also presented in equations 2.9–2.17 in Section 2.3 of Chapter 2 [4,5]. Now consider the three-phase AC input voltage and desired three-phase AC output voltage of the MC as defined in equations 2.9 and 2.11 in Section 2.3 of Chapter 2 [6,7]. For the three-phase output voltage waveform synthesised or reconstructed from the three-phase input voltage waveform without distortion, it is necessary to switch the bidirectional switches of the MC at a high sampling rate or carrier frequency much higher than the input frequency and the desired output frequency. It is also a requirement that the output voltage waveform so synthesised or reconstructed from the input voltage waveform must lie within the upper and lower bounds of the input voltage waveform at every instant of time. This requirement can be expressed as follows [6,7]:

$$v_{i\_LB}(t) \leq v_o(t) \leq v_{i\_UB}(t) \tag{6.1}$$

where $v_{i\_LB}(t)$ and $v_{i\_UB}(t)$ represent the lower and upper bounds of the input voltage $v_i(t)$.

Now considering the general case of the three-phase input and three-phase output voltage waveform defined in equations 2.9 and 2.11 in section 2.3 of Chapter 2, where the input and output frequencies ω_i and ω_o are not correlated, equation 6.1 can be expressed as follows [6,7]:

$$\underset{\substack{\min \\ 0 \leq \omega_i t \leq 2\pi}}{v_{i\_UB}(t)} \quad \geq \quad \underset{\substack{\max \\ 0 \leq \omega_o t \leq 2\pi}}{v_{o\_UB}(t)} \tag{6.2}$$

where

$\underset{\max}{v_{o\_UB}}$ is the maximum value of output voltage upper bound max

$\underset{\min}{v_{i\_UB}}$ is the minimum value of input voltage upper bound min.

The condition given by equation 6.2 is shown in Figure 6.1. Thus, for a distortionless synthesis or reconstruction of the output voltage waveform from the three-phase input voltage, the maximum value of the output voltage at any time should not exceed the input voltage value where $v_{iA} = v_{iB} = v_{iC}$ or at the intersecting point of the three-phase input voltages. This can be expressed as follows [6,7]:

$$V_{im} * \cos(\omega_i t) = V_{im} * \cos\left(\omega_i t - \frac{2\pi}{3}\right) = V_{im} * \cos\left(\omega_i t - \frac{4\pi}{3}\right) \tag{6.3}$$

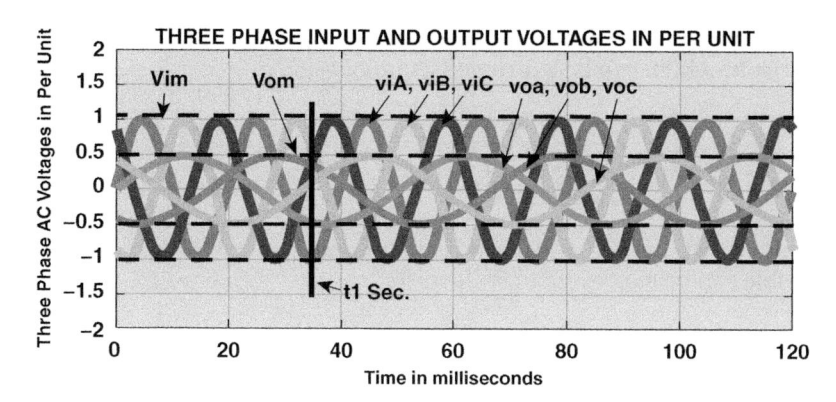

FIGURE 6.1
Three-phase AC input and output voltages.

Equation 6.3 is valid at intervals of $\pi/3$ rad and the magnitude at the intersecting point of the three-phase input voltages is $(0.5*V_{im})$ [6,7]. This is illustrated in Figure 6.1. The condition given in equation 6.2 requires the following equation to be valid [6,7]:

$$v_o(t) \le 0.5 * v_i(t)$$
$$V_{om} = 0.5 * V_{im}$$

(6.4)

Equation 6.4 is independent of the conversion algorithm used. The maximum value of the output voltage amplitude is obtained when the maximum value of the difference between the output voltage upper bound and output voltage lower bound coincides with minimum value of the difference between the input voltage upper bound and the input voltage lower bound. Referring to Figure 6.1, the maximum value of the upper bound of the output voltage v_{oa} varies from 0 to $0.5*V_{im}$ with respect to time while for the other two output phase v_{ob} and v_{oc} the lower bound of the output voltage varies from $-0.5*V_{im}$ to $+0.5*V_{im}$ and $+0.5*V_{im}$ to $-0.5*V_{im}$, respectively. At any instant of time during the positive half cycle of the output voltage v_{oa}, this maximum value of the difference between the upper and lower bounds of the output voltage is therefore the phasor difference between v_{oa} and v_{ob} or v_{oa} and v_{oc} which can be expressed as follows [6,7]:

$$\left(\underset{\substack{max \\ 0 \le \omega_o t \le 2\pi}}{v_o\_UB(t) - v_o\_LB(t)} \right) = \left(\vec{V}_{oa} - \vec{V}_{ob} \right) = \left(\vec{V}_{oa} - \vec{V}_{oc} \right) = \sqrt{3} * V_{om}$$

(6.5)

Equation 6.5 gives the maximum value of the line-to-line output voltage. From Figure 6.1, at any time interval t_1 seconds, for the three-phase input voltage, the minimum value of the lower bound for v_{iA} is $-V_{im}$ and the minimum values of the upper bound for the input voltages V_{iB} and V_{iC} are $+0.5*V_{im}$, respectively, which can be expressed as follows [6,7]:

$$\left(v_{i\_UB} - v_{i\_LB} \right)_{\substack{min \\ 0 \leq \omega_i t \leq 2\pi}} = \left[+0.5*V_{im} - (-V_{im}) \right] = \frac{3V_{im}}{2} \qquad (6.6)$$

Equating equations 6.5 and 6.6, the maximum output voltage amplitude obtainable can be expressed as follows [6,7]:

$$V_{om} = \frac{\sqrt{3}}{2} V_{im} = 0.866 V_{im} \qquad (6.7)$$

$$V_o = 0.866 V_i$$

Equation 6.7 gives the maximum output voltage amplitude obtainable for a three-phase AC to three-phase AC MC assuming no carrier modulation.

6.3 Principle of Dual Programmable AC to DC Rectifier

For the three-phase AC input voltage and the output voltage as defined in equations 2.9 and 2.11, the Venturini modulation algorithm for modulation duty cycle which satisfies equations 2.11 and 2.12 assuming unity input phase displacement factor is given in equation 2.17 in Section 2.3 of Chapter 2. For the above three-phase MC to work as an AC to DC rectifier, the output frequency f_o is to be set to zero [1,2]. This makes equations 2.10 and 2.11 given in Section 2.3 of Chapter 2 as follows:

$$i_o = \begin{bmatrix} i_a \\ i_b \\ i_c \end{bmatrix} = I_{om} * \begin{bmatrix} \cos(\varphi_o) \\ \cos\left(\varphi_o - \dfrac{2\pi}{3}\right) \\ \cos\left(\varphi_o + \dfrac{2\pi}{3}\right) \end{bmatrix} \qquad (6.8)$$

$$v_o = \begin{bmatrix} v_a \\ v_b \\ v_c \end{bmatrix} = q * V_{im} * \begin{bmatrix} \cos(\varphi_o) \\ \cos\left(\varphi_o - \dfrac{2\pi}{3}\right) \\ \cos\left(\varphi_o + \dfrac{2\pi}{3}\right) \end{bmatrix} \qquad (6.9)$$

In equation 6.9, if the output voltage phase angle φ_o is set to +30°, it is seen that the output voltage v_b becomes zero, whereas v_a and v_c have equal magnitude [1,2]. In any three-phase AC to three-phase AC MC, the intrinsic output voltage limit irrespective of control algorithm used is given by equation 6.7 [6,7]. In the linear region of the modulation index q, the input-output voltage relation in an MC can be expressed as follows:

$$V_o = q * v_i \tag{6.10}$$

Combining equations 6.7 and 6.10, we have the following:

$$v_{om} = 0.866 * q * V_{im} \tag{6.11}$$

Three-phase output voltage in equation 2.11 of Chapter 2 is shown in Figure 6.2a. With b as virtual ground or virtual reference point, for any arbitrary angular frequency ω_o of the output voltage, v_{ab} and v_{cb} which correspond to v_{o1} and $v_{o2,}$ leads v_a by $\pi/6$ and $\pi/2$ rad, respectively. Now noting that for the three-phase MC to work as a dual AC to DC rectifier the output frequency ω_o is zero, combining equation 6.11 and noting the phase lead of v_{ab} and v_{cb} with respect to $v_a,$ equations 6.12 and 6.13 follow:

$$V_{ab} = V_{o1} = 0.866 * q * V_{im} * \cos\left(\varphi_o + \frac{\pi}{6}\right) \tag{6.12}$$

$$V_{cb} = -V_{bc} = V_{o2} = 0.866 * q * V_{im} * \cos\left(\varphi_o + \frac{\pi}{2}\right) \tag{6.13}$$

The three-phase AC voltage generated by the power grid is a sine wave of the form $V_{iK} = V_{im}*\sin(\omega_i t - \gamma)$, where $\gamma = 0, 2\pi/3, 4\pi/3$ for $K = A, B, C$, respectively. Considering this factor, a phase lag of $\pi/2$ rad is given to the three-phase AC input and output voltage defined in equations 2.9 and 2.11 of Chapter 2. This permits three-phase sine-wave AC input voltage to be used for practical

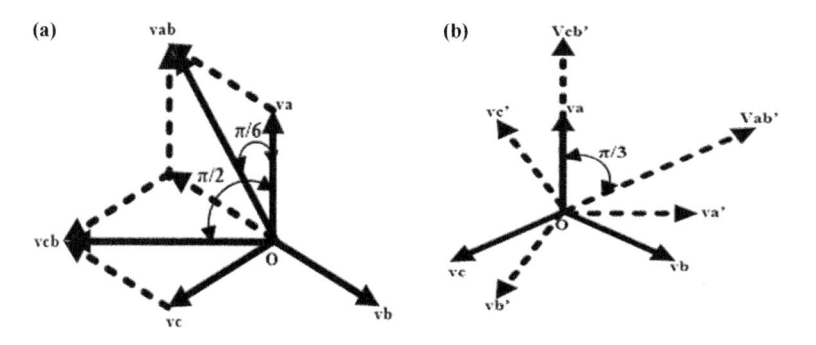

FIGURE 6.2
Phasor diagram: (a) three-phase cosine-wave output voltage and (b) three-phase sine-wave output voltage.

implementation. With a phase lag of $\pi/2\,\text{rad}$ given to the three-phase output voltage defined in equation 2.11 of Chapter 2, the new phasor diagram is shown in Figure 6.2b. The new position of v_a, v_b and v_c are marked v'_a, v'_b and, v'_c, respectively. The new position of v_{ab} and v_{cb} are marked v'_{ab} and v'_{cb} which lag the original position of v_{ab} and v_{cb} in Figure 6.2a by $\pi/2\,\text{rad}$. Thus, the new position of v'_{ab} and v'_{cb} in Figure 6.2b, lag the original position of v_a in Figure 6.2a by $\pi/3$ and $0\,\text{rad}$, respectively. Now noting that for the three-phase MC to work as a dual AC to DC rectifier the output frequency ω_o is zero, combining equation 6.11 and noting the phase lag of v'_{ab} and v'_{cb} with respect to v_a, equations 6.14 and 6.15 follows:

$$V'_{ab} = V_{o1} = 0.866 * q * V_{im} * \cos\left(\varphi_o - \frac{\pi}{3} \right) \tag{6.14}$$

$$V'_{cb} = -V'_{bc} = V_{o2} = 0.866 * q * V_{im} * \cos(\varphi_o) \tag{6.15}$$

As can be seen from equations 6.14 and 6.15, when φ_o is varied from 0 to $+\pi$ and 0 to $-\pi$, multitude of combination of dual DC output voltages are possible. Thus by suitably changing the value of φ_o in the model of three-phase MC, dual programmable DC output voltages can be obtained.

6.4 Model of Dual Programmable AC to DC Rectifier

The model of the dual programmable AC to DC rectifier is shown in Figure 6.3. The model is developed in SIMULINK [8]. Here Venturini algorithm for unity input phase displacement factor is assumed in generating the gate pulses for the bidirectional switches. Also three-phase sine-wave input and output voltages are used for developing the model. The model consists of (a) bidirectional switch gate pulse generator, (b) output voltage phase-angle-varying device, (c) three-phase sine-wave AC voltage source, (d) MC bidirectional switch matrix, (e) R–L–C output filter and (f) R–L load. The parameters used for the model are given in Table 6.1.

Bidirectional switch gate pulse generator is shown in Figure 6.4a. This consists of two embedded MATLAB functions. The first MATLAB function generates the nine modulation functions according to equations 2.17 of Chapter 2, using inputs q, V_{im}, f_i, f_o, phi_o and the time module. The source code is shown in Program segment 6.1. The second MATLAB function generates the gate timing pulse for the nine bidirectional switches, by comparison of the modulation functions with a saw-tooth carrier V_{saw}. The code for generating the timing pulse for the three bidirectional switches connected between input phase A, B, C and output phase a, b, c is given in Program segment 2.2 of Chapter 2. Output voltage phase-angle-varying device is shown in Figure 6.3. This is using slider gain block. The constant block with value $+\pi/12$ is given

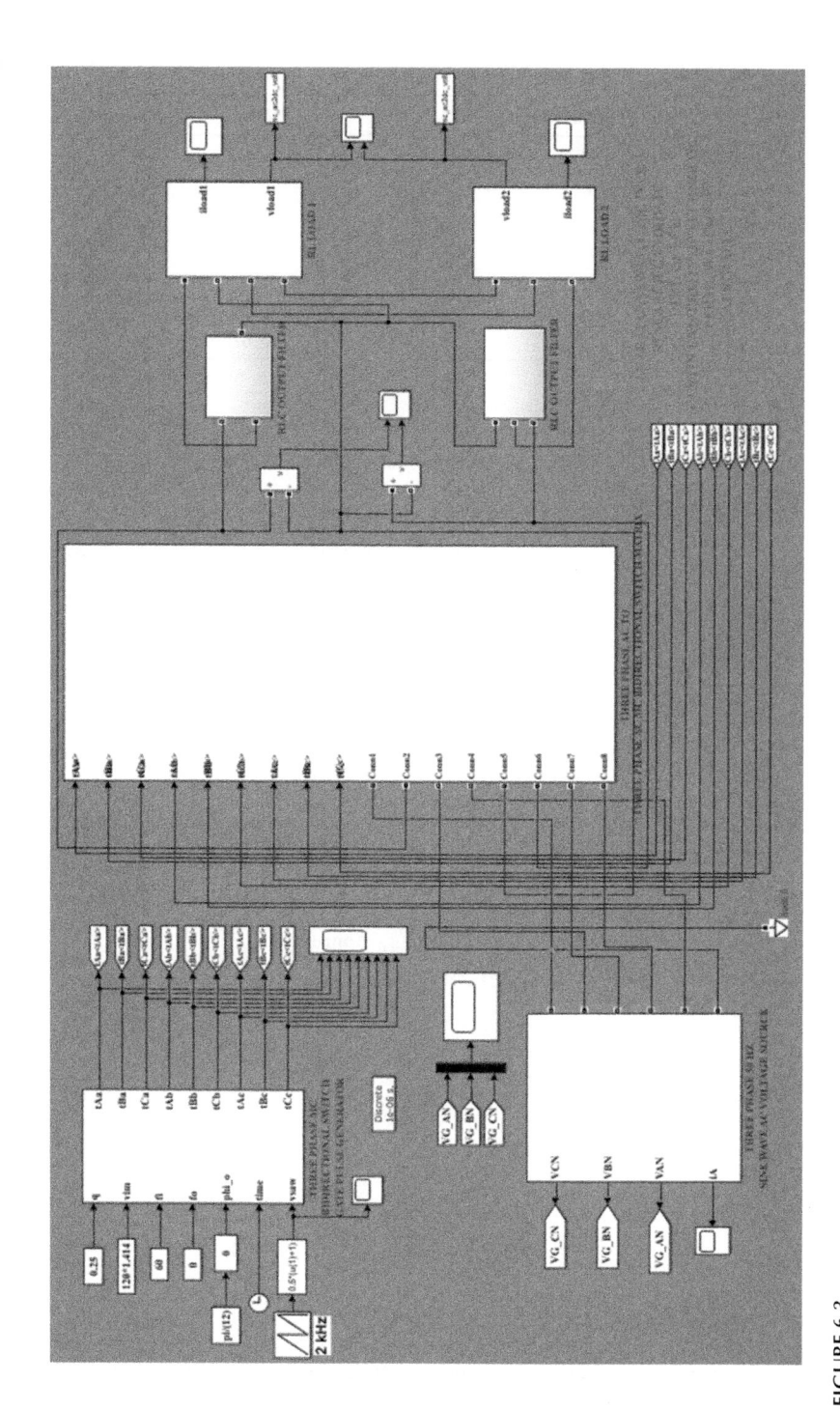

FIGURE 6.3
Dual programmable AC to DC rectifier model using three-phase matrix converter topology.

TABLE 6.1

Dual Programmable Rectifier Model Parameters

Sl. No.	Parameter	Value	Units
1	RMS line-to-neutral input voltage V_{im}	120	V
2	Modulation index q	0.25	-
3	Input frequency	60	Hz
4	Output frequency	0	Hz
5	Saw-tooth carrier switching frequency	2	kHz
6	Output R–L–C filter	50, 10e–6, 633.9e–6	Ω, H, F
7	R–L load	50, 0.5	Ω, H

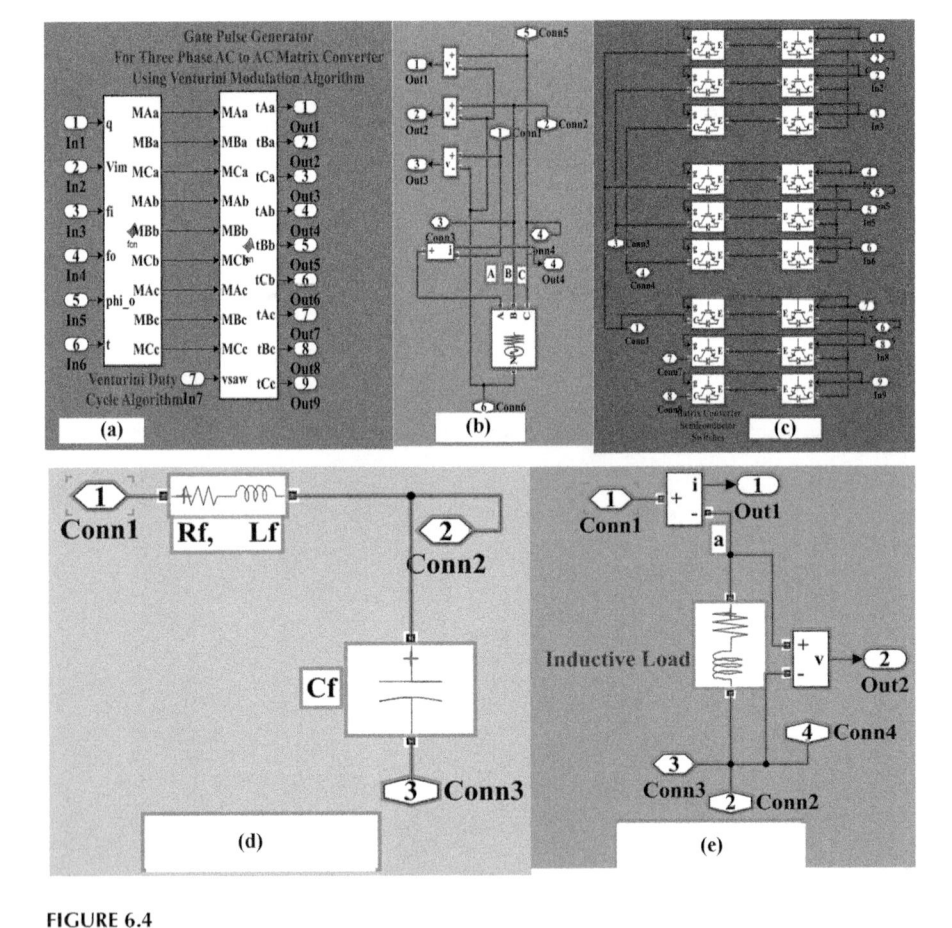

FIGURE 6.4

Dual programmable AC to DC rectifier model subsystems (a) Venturini duty-cycle algorithm and gate pulse generator, (b) three-phase sine-wave AC generator, (c) bidirectional switch matrix, (d) R–L–C filter and (e) R–L load.

as input to the slider gain block whose multiplication constant can be varied from −12 to +12. Appropriate gain multiplier value is entered in the current value field of slider gain block. The output of slider gain provides the variable output voltage phase angle φ_o in radians. Three-phase AC sine-wave voltage source from Power systems block set is shown in Figure 6.4b. This three-phase sine-wave AC source generates three-phase voltage with a line-to-line RMS value of 207.84 V and a frequency of 60 Hz. The nine MC bidirectional switch matrix using IGBTs and diodes is shown in Figure 6.4c. R–L–C output filter is shown in Figure 6.4d. The L and C values are so chosen to resonate at the carrier switching frequency of 2 kHz. The output voltage is tapped across the filter capacitor. R–L load is shown in Figure 6.4e. The theoretically computed values of the dual DC output voltages V_{o1} and V_{o2} using equations 6.14 and 6.15 for the parameters shown in Table 6.1 are given in Table 6.2 for various values of output voltage phase angle φ_o in radians.

Program Segment 6.1

```
function [MAa,MBa,MCa,MAb,MBb,MCb,MAc,MBc,MCc] =
fcn(q,Vim,fi,fo,phi_o,t)
%% Dr.Narayanaswamy.P.R.Iyer.
ViA = Vim*cos(2*pi*fi*t - pi/(2));
ViB = Vim*cos(2*pi*fi*t - pi/(2) - 2*pi/(3));
ViC = Vim*cos(2*pi*fi*t - pi/(2) + 2*pi/(3));
voa = q*Vim*(cos(2*pi*fo*t - pi/(2) + phi_o));
vob = q*Vim*(cos(2*pi*fo*t - pi/(2) + phi_o - 2*pi/(3)));
voc = q*Vim*(cos(2*pi*fo*t - pi/(2) + phi_o + 2*pi/(3)));
%%switch modulation
MAa = (1/(3) + (2*ViA*voa)/(3*Vim*Vim));
MBa = (1/(3) + (2*ViB*voa)/(3*Vim*Vim));
```

TABLE 6.2

Dual Rectifier Computed Values

Sl. No.	φ_o (rad)	V_{o1} (V)	V_{o2} (V)	Sl. No.	φ_o (rad)	V_{o1} (V)	V_{o2} (V)
1	0	18.3679	36.7357	13	$\pm\pi$	−18.3679	−36.7357
2	$+\pi/12$	25.9761	35.4840	14	$-11\pi/12$	−25.9761	−35.4840
3	$+\pi/6$	31.8141	31.8141	15	$-5\pi/6$	−31.8141	−31.8141
4	$+\pi/4$	35.4840	25.9761	16	$-3\pi/4$	−35.4840	−25.9761
5	$+\pi/3$	36.7357	18.3679	17	$-2\pi/3$	−36.7357	−18.3679
6	$+5\pi/12$	35.4840	9.5079	18	$-7\pi/12$	−35.4840	−9.5079
7	$+\pi/2$	31.8141	0	19	$-\pi/2$	−31.8141	0
8	$+7\pi/12$	25.9761	−9.5079	20	$-5\pi/12$	−25.9761	9.5079
9	$+2\pi/3$	18.3679	−18.3679	21	$-\pi/3$	−18.3679	18.3679
10	$+3\pi/4$	9.5079	−25.9761	22	$-\pi/4$	−9.5079	25.9761
11	$+5\pi/6$	0	−31.8141	23	$-\pi/6$	0	31.8141
12	$+11\pi/12$	−9.5079	−35.4840	24	$-\pi/12$	9.5079	35.4840

TABLE 6.3

Dual Rectifier Model Simulation Results

Sl. No.	φ_o (rad)	V_{o1} (V)	V_{o2} (V)	Sl. No.	φ_o (rad)	V_{o1} (V)	V_{o2} (V)
1	0	18.5	36.75	13	$\pm\pi$	−18.4	−36.75
2	$+\pi/12$	26	35.5	14	$-11\pi/12$	−26	−35.5
3	$+\pi/6$	32	32	15	$-5\pi/6$	−31.75	−31.75
4	$+\pi/4$	35.5	26	16	$-3\pi/4$	−36.5	−26
5	$+\pi/3$	36.75	18.5	17	$-2\pi/3$	−36.75	−18.5
6	$+5\pi/12$	35.75	9.5	18	$-7\pi/12$	−36.5	−9.5
7	$+\pi/2$	31.75	0	19	$-\pi/2$	−31.75	0
8	$+7\pi/12$	26	−9.6	20	$-5\pi/12$	−26	9.5
9	$+2\pi/3$	18.5	−18.5	21	$-\pi/3$	−18.5	18.5
10	$+3\pi/4$	9.5	−26	22	$-\pi/4$	−9.5	26
11	$+5\pi/6$	0	−32	23	$-\pi/6$	0	31.75
12	$+11\pi/12$	−9.6	−35.75	24	$-\pi/12$	9.5	35.5

```
MCa = (1/(3) + (2*ViC*voa)/(3*Vim*Vim));
MAb = (1/(3) + (2*ViA*vob)/(3*Vim*Vim));
MBb = (1/(3) + (2*ViB*vob)/(3*Vim*Vim));
MCb = (1/(3) + (2*ViC*vob)/(3*Vim*Vim));
MAc = (1/(3) + (2*ViA*voc)/(3*Vim*Vim));
MBc = (1/(3) + (2*ViB*voc)/(3*Vim*Vim));
MCc = (1/(3) + (2*ViC*voc)/(3*Vim*Vim));
```

6.4.1 Simulation Results

The simulation of the dual programmable AC to DC rectifier was carried out in SIMULINK [8]. The ode15s (Stiff/NDF) solver is used. The simulation results for the value of φ_o shown in the order in Table 6.2 are shown in Figure 6.5a–x and the three-phase input voltage in Figure 6.6a. Graph of v_{o1} and v_{o2} is shown in Figure 6.6b. The simulation results are shown in Table 6.3.

The model simulation results indicate that the magnitude and sign of v_{o1} and v_{o2} closely agree with the theoretically computed value shown in Table 6.2.

6.5 Principle of Single Programmable AC to DC Rectifier

The single-ended bidirectional MC topology can be used to develop a single programmable AC to DC rectifier. Referring to Figure 2.1 in Chapter 2, for a single-ended bidirectional topology, remove all the three bidirectional switches S_{Ab}, S_{Bb} and S_{Cb} connected between input phase A, B, C and output phase b, remove the R–L load connected between output phase a, b, c

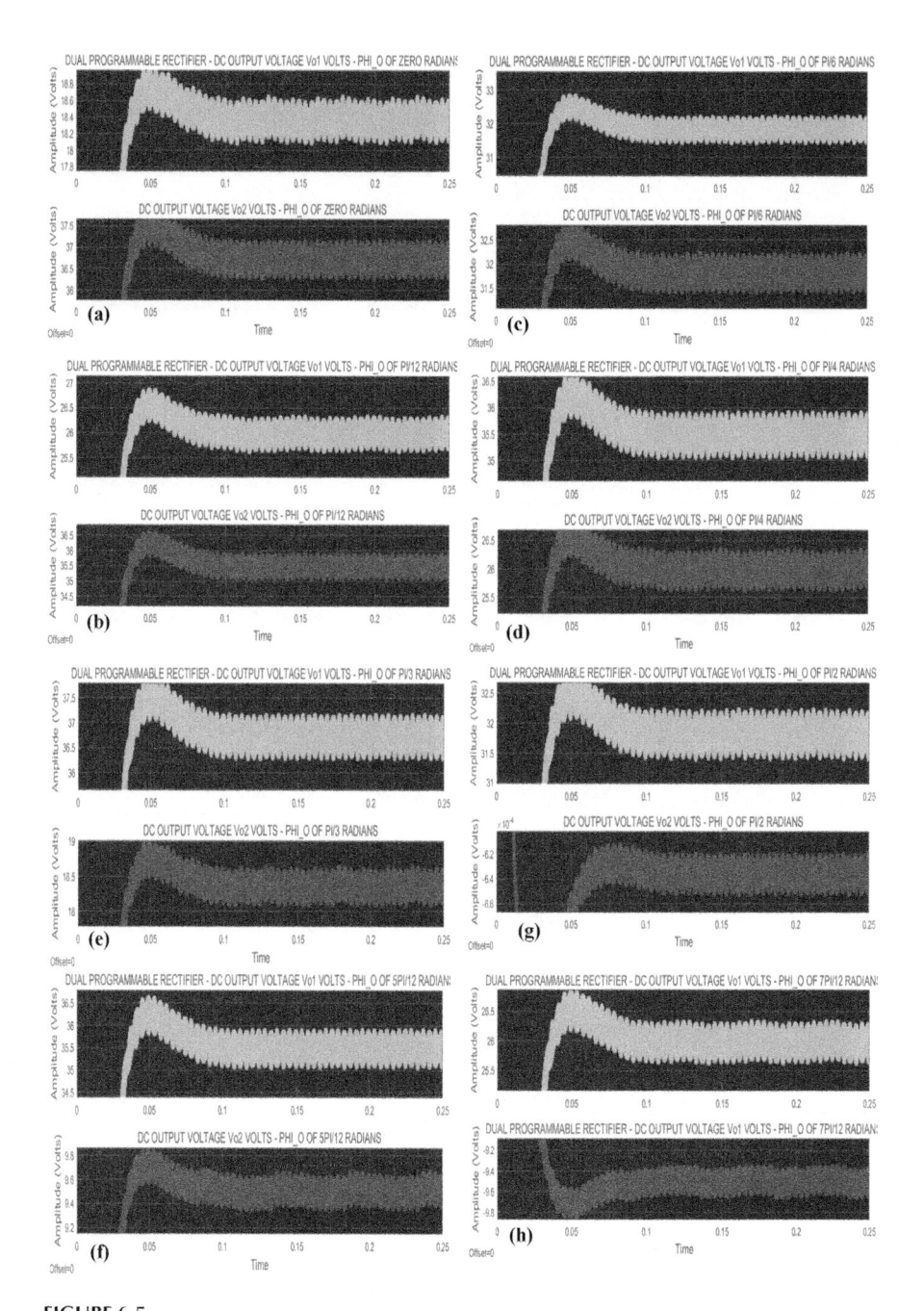

FIGURE 6.5
Dual programmable rectifier simulation results. (a)–(y) V_{o1} (top) and V_{o2} (bottom) for different phi_o values.

(*Continued*)

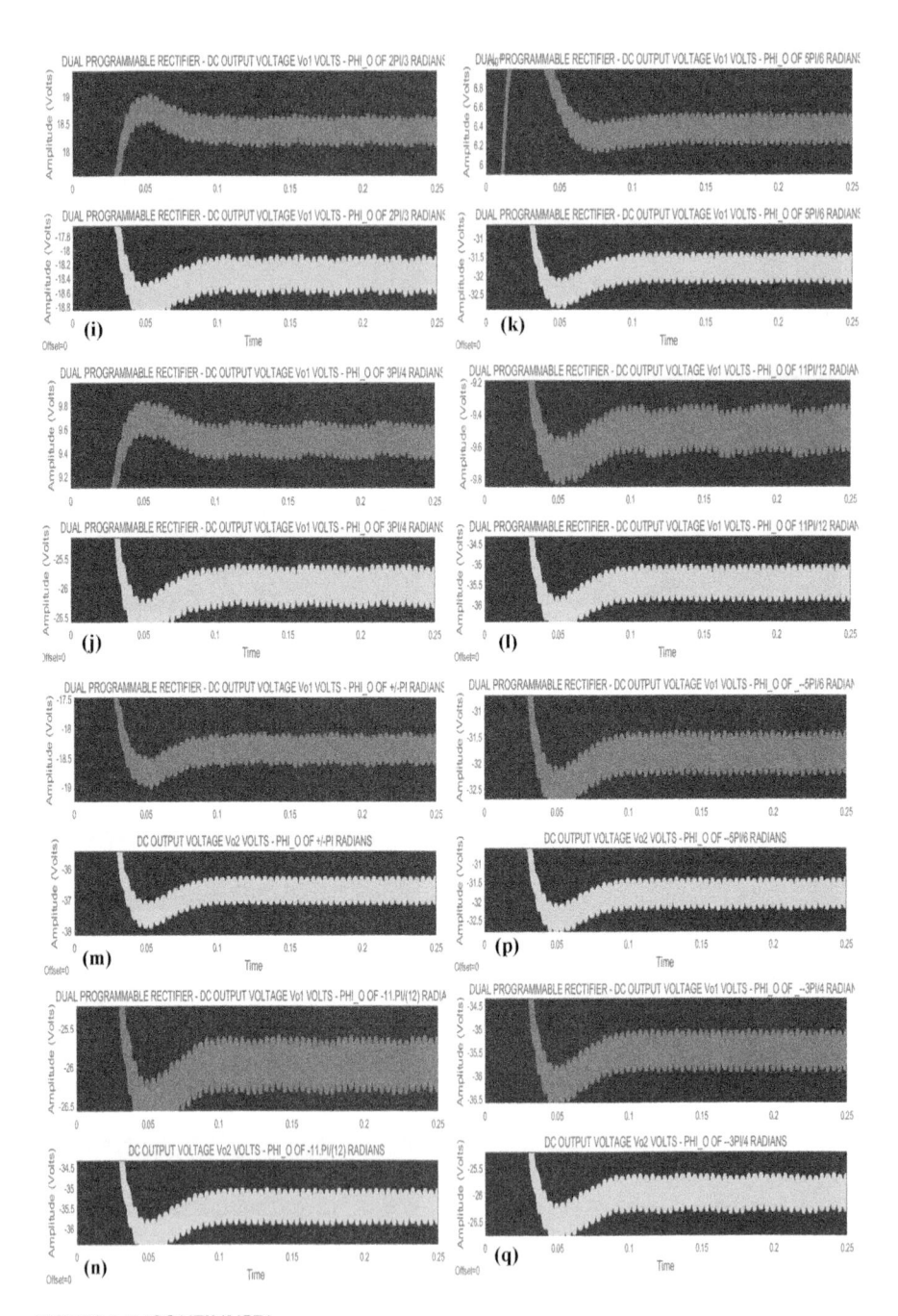

FIGURE 6.5 (CONTINUED)
Dual programmable rectifier simulation results. (a)–(y) V_{o1} (top) and V_{o2} (bottom) for different phi_o values.

(Continued)

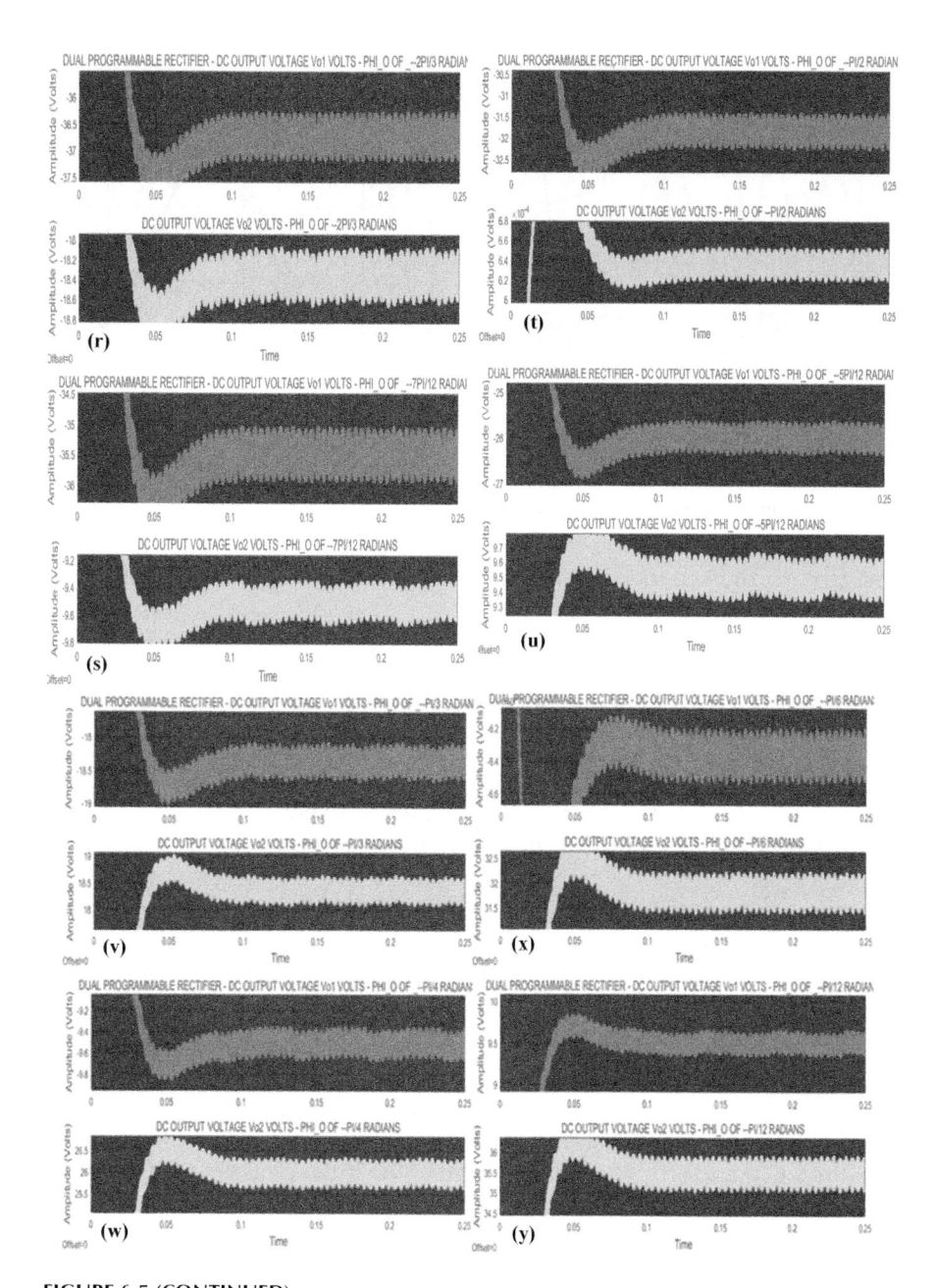

FIGURE 6.5 (CONTINUED)
Dual programmable rectifier simulation results. (a)–(y) V_{o1} (top) and V_{o2} (bottom) for different phi_o values.

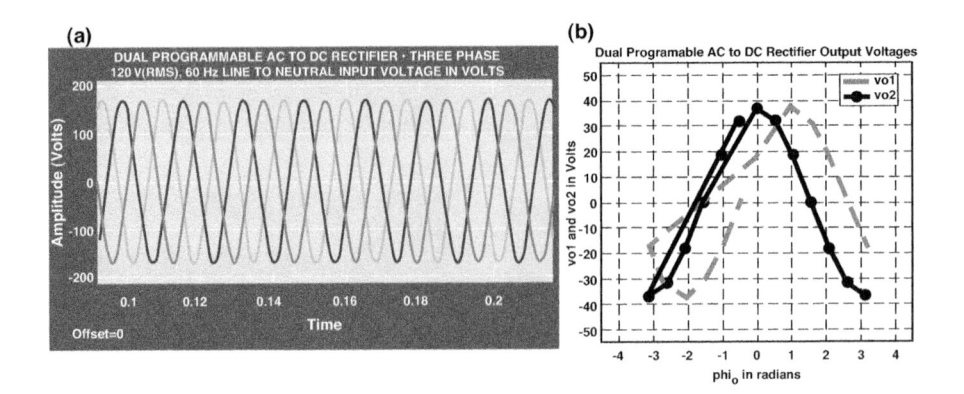

FIGURE 6.6
Dual programmable AC to DC rectifier: (a) three-phase 120V (RMS), 60Hz input voltage and (b) plot of V_{o1} and V_{o2} voltages with phi_o.

and ground, then connect a suitable low pass filter and a single-phase load between output phase a and c.

The modulation functions assuming unity input displacement factor for this MC can be expressed as in equation 2.17 in Chapter 2. Three-phase sine-wave input voltage applied to the MC is defined below:

$$v_{iK} = V_{im} * \sin\left(2\pi * f_i * t - \gamma_K\right) \qquad (6.16)$$

where γ_K is 0, $2\pi/3$, $4\pi/3$ for $K = A$, B and C, respectively.

The output voltage of the MC is defined below:

$$v_{on} = q * V_{im} * \sin\left(2\pi * f_o * t - \beta_n + \varphi_o\right) \qquad (6.17)$$

where β_n is 0, $4\pi/3$ for $n = a$ and c, respectively, φ_o is the output voltage phase angle and $V_{om} = q*V_{im}$. The above MC, with its output phase b disconnected, to work as a single programmable rectifier, the output frequency is set to zero [1,2]. Varying φ_o gives variable DC output voltage positive negative or zero. The derivation is given below.

In the linear region, for a given modulation index q, the input-output voltage relation and the maximum output voltage obtainable are given by equations 6.10 and 6.11, respectively. Two of the three-phase output voltages defined in equation 6.17 are shown in Figure 6.7. As the output voltage is tapped across the terminals a and c, the phasor v_{ac} represents this value. This phasor diagram is drawn for any arbitrary frequency f_o for the output voltage. Now noting that for the three-phase MC to work as a single programmable rectifier, the output frequency f_o is zero and the phase lag is $\pi/6$ rad. From Figure 6.7, the output voltage v_{ac} can be expressed as follows [9,10]:

$$V_{ac} = 0.866 * q * V_{im} * \sin\left(\phi_o - \frac{\pi}{6}\right) \qquad (6.18)$$

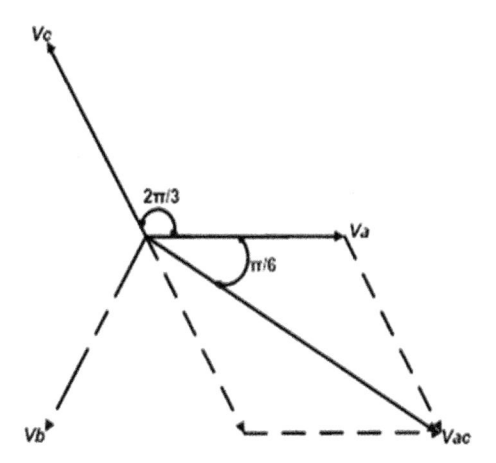

FIGURE 6.7
Three-phase output voltage phasor diagram.

From equation 6.18, it is clear that as φ_o is varied from 0 to $+\pi$ and 0 to $-\pi$, multitude of combination of output voltage is possible such as positive increasing, zero and negative decreasing.

6.6 Model of Single Programmable AC to DC Rectifier

The model of single programmable AC to DC Rectifier is shown in Figure 6.8 [9,10]. The model subsystems are shown in Figure 6.9a–e. Here modulation functions defined in equation 2.17 are generated using Embedded MATLAB Function shown in Figure 6.9a, as per source code shown in Program segment 6.2. To generate the gate pulse for output phase a, the modulation functions M_{Aa} is compared with a saw-tooth carrier V_{saw} using comparator. The comparator output is HIGH if M_{Aa} is greater than or equal to $V_{saw,}$ else this output is LOW. This gives gate pulse t_{Aa}. Similar comparison is made for $(M_{Aa} + M_{Ba})$ to generate t_{ABa}. These two gate pulses are applied to an EXCLUSIVE-OR gate to obtain t_{Ba}. The gate pulse t_{ABa} is inverted using a NOT gate to obtain t_{Ca}. Similarly M_{Ac} and $(M_{Ac} + M_{Bc})$ are compared with V_{saw} in two separate comparators and the gate pulse t_{Ac}, t_{Bc} and t_{Cc} are derived as given above (Figure 6.9a). Comparators output must have logic level compatibility. AND gates are optional. These form six gate pulses for the six bidirectional switches between the three input phase A, B and C and the two output phase a and c. The output voltage phase angle φ_o is varied from 0 to $+\pi$ and 0 to $-\pi$ using a slider gain block. The six bidirectional switch matrix, three-phase AC voltage source, R–L–C output filter and R–L load are shown in Figure 6.9b–e, respectively. The L and C values for the output filter are chosen to resonate at a carrier switching frequency of 2 kHz.

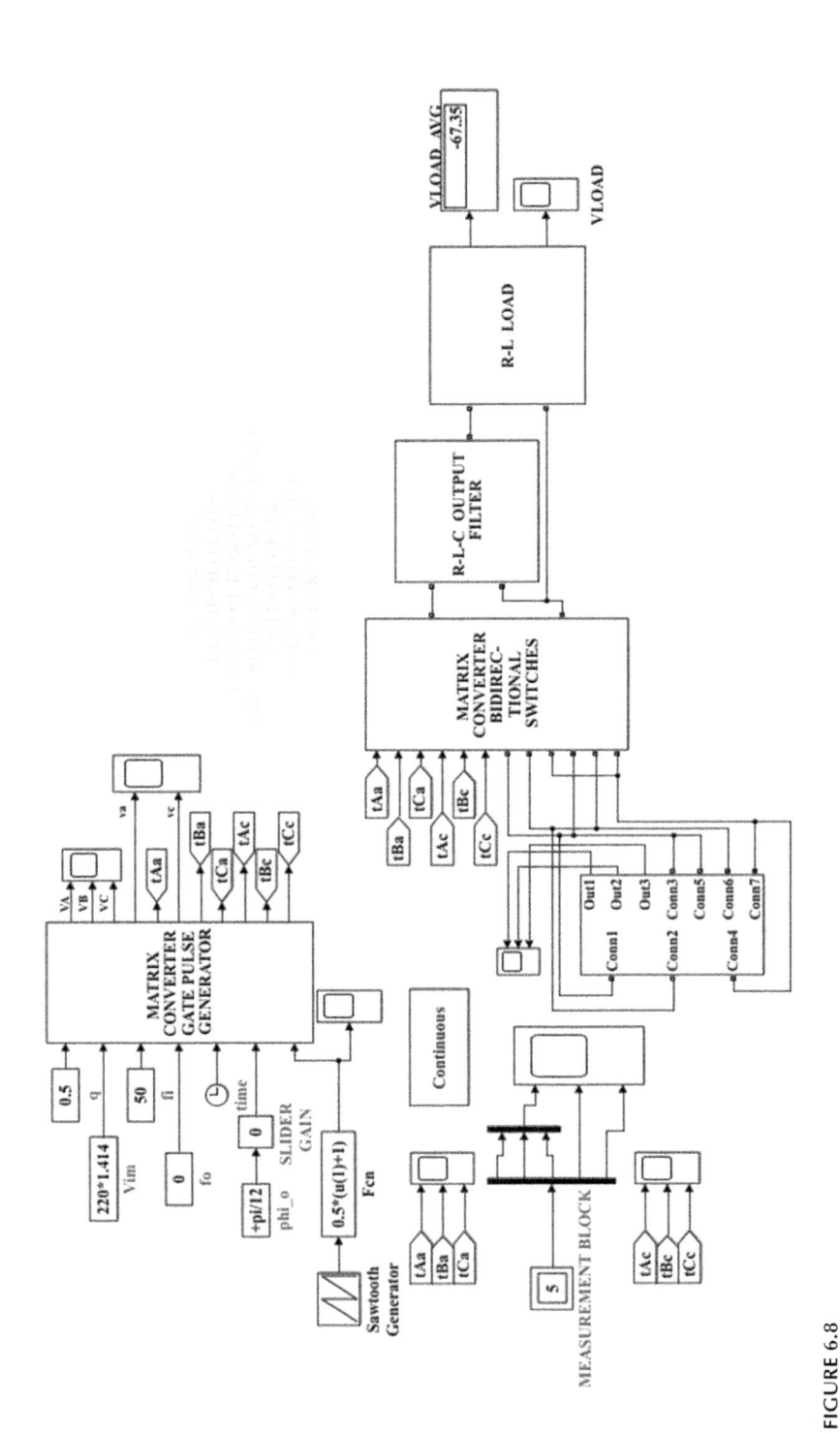

FIGURE 6.8

Single programmable AC to DC rectifier model using three-phase matrix converter topology.

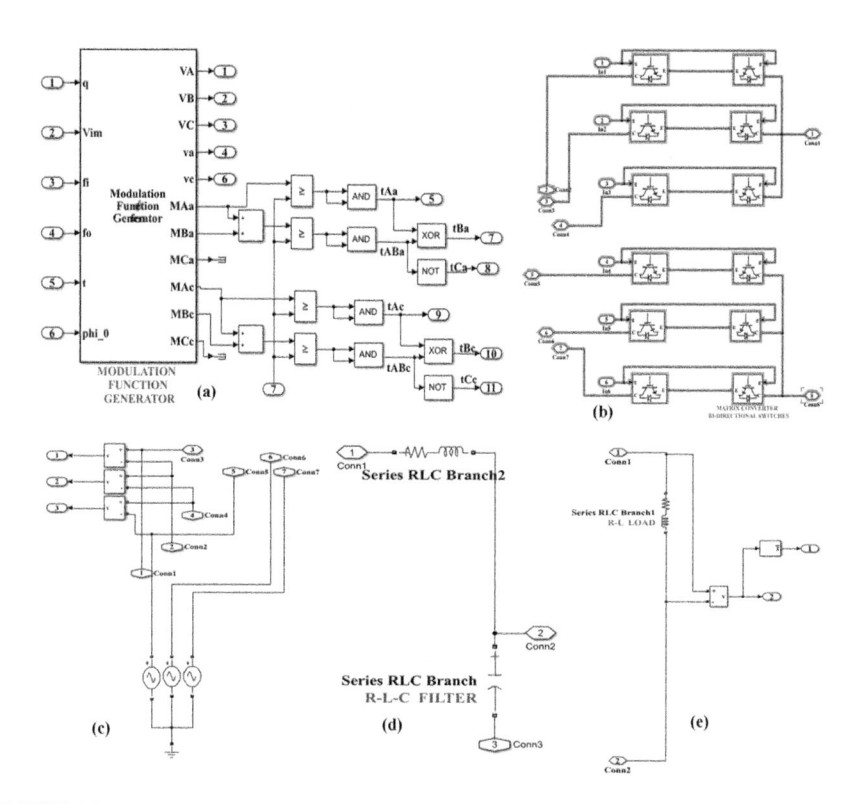

FIGURE 6.9
Single Programmable AC to DC rectifier model subsystems (a) Modulation function generator,
(b) bidirectional switch matrix, (c) three-phase AC voltage source, (d) R–L–C output filter and
(e) R–L load.

Program Segment 6.2

```
function [VA,VB,VC,va,vc,MAa,MBa,MCa,MAc,MBc,MCc] =
fcn(q,Vim,fi,fo,t,phi_0)
%% Dr. Narayanaswamy P R Iyer.
%% Three Phase Input Voltages.
VA = Vim*sin(2*pi*fi*t);
VB = Vim*sin(2*pi*fi*t - 2*pi/(3));
VC = Vim*sin(2*pi*fi*t - 4*pi/(3));
%% Phase Voltages for Output Phase a and c.
va = q*Vim*sin(2*pi*fo*t + phi_0);
vc = q*Vim*sin(2*pi*fo*t + phi_0 - 4*pi/(3));
%% Modulation Functions.
MAa = 1/(3) + (2/(3))*VA*va/(Vim*Vim);
MBa = 1/(3) + (2/(3))*VB*va/(Vim*Vim);
MCa = 1/(3) + (2/(3))*VC*va/(Vim*Vim);
MAc = 1/(3) + (2/(3))*VA*vc/(Vim*Vim);
MBc = 1/(3) + (2/(3))*VB*vc/(Vim*Vim);
MCc = 1/(3) + (2/(3))*VC*vc/(Vim*Vim);
```

6.6.1 Simulation Results

The simulation of the single programmable rectifier model is carried out in SIMULINK [8]. The ode23tb (stiff/TR-BDF2) solver is used. Selected simulation results are shown in Figure 6.10a–n. In this simulation, data used are shown in Table 6.4. Further using the data in Table 6.4 and equation 6.18, the values of output voltage are calculated and these values along with that obtained by model simulation are tabulated in Table 6.5. Three-phase input voltage is shown in Figure 6.11. Plot of output voltage magnitude versus output voltage phase angle φ_o is shown in Figure 6.12.

6.7 Case Study: Speed Control and Brake by Plugging of Separately Excited DC Motor Using Single Programmable AC to DC Rectifier

The model of the single programmable AC to DC rectifier-fed SEDC motor is shown in Figure 6.13. The model for gate pulse generation is the same as explained in Section 6.6. The differences in this model are that the load used is an SEDC motor shown in Figure 6.14. The DC motor model parameters are shown in Table 6.6 [8]. Also the output voltage phase angle (φ_o) varying device is a multiport switch. The Repeating Sequence1 forms the control port. With eight different values for starting times, ports 0–7 are entered in the Repeating Sequence1 block. Eight inputs with ports 0–7 are connected to a constant block marked $\pi/12$ with appropriate gain multipliers. For example, at any time, if the output value is port 3 in the Repeating Sequence1 block, then the output φ_o of Multiport Switch will be $\pi/2$ rad from Figure 6.13.

6.7.1 Simulation Results

The simulation of the single programmable AC to DC rectifier-fed SEDC motor is carried out using SIMULINK [8]. The ode23tb (Stiff/TR-BDF2) solver is used. At intervals of 0.3 s, the port numbers from 0 to 7, respectively, are given as input from control port of multiport switch. The respective outputs of φ_o during this time intervals are $+\pi/6$, $+\pi/3$, $+\pi/3$, $+\pi/2$, $+2\pi/3$, $+2\pi/3$, $-\pi/6$ and 0 rad. The output voltage V_o of output filter which is also the armature applied voltage for DC machine 1 can be obtained from Table 6.5. These values from Table 6.5 indicate that DC machine 1 is accelerating, coasting, accelerating, coasting, decelerating and reversing speed which corresponds to braking. The simulation results for the motor speed, armature current, electromagnetic torque and applied armature voltage of DC machines 1 are shown in Figure 6.15a–d, which confirm the above finding using Table 6.5. The armature voltage in Figure 6.15d will not be able to attain the value in Table 6.5 due to given time interval.

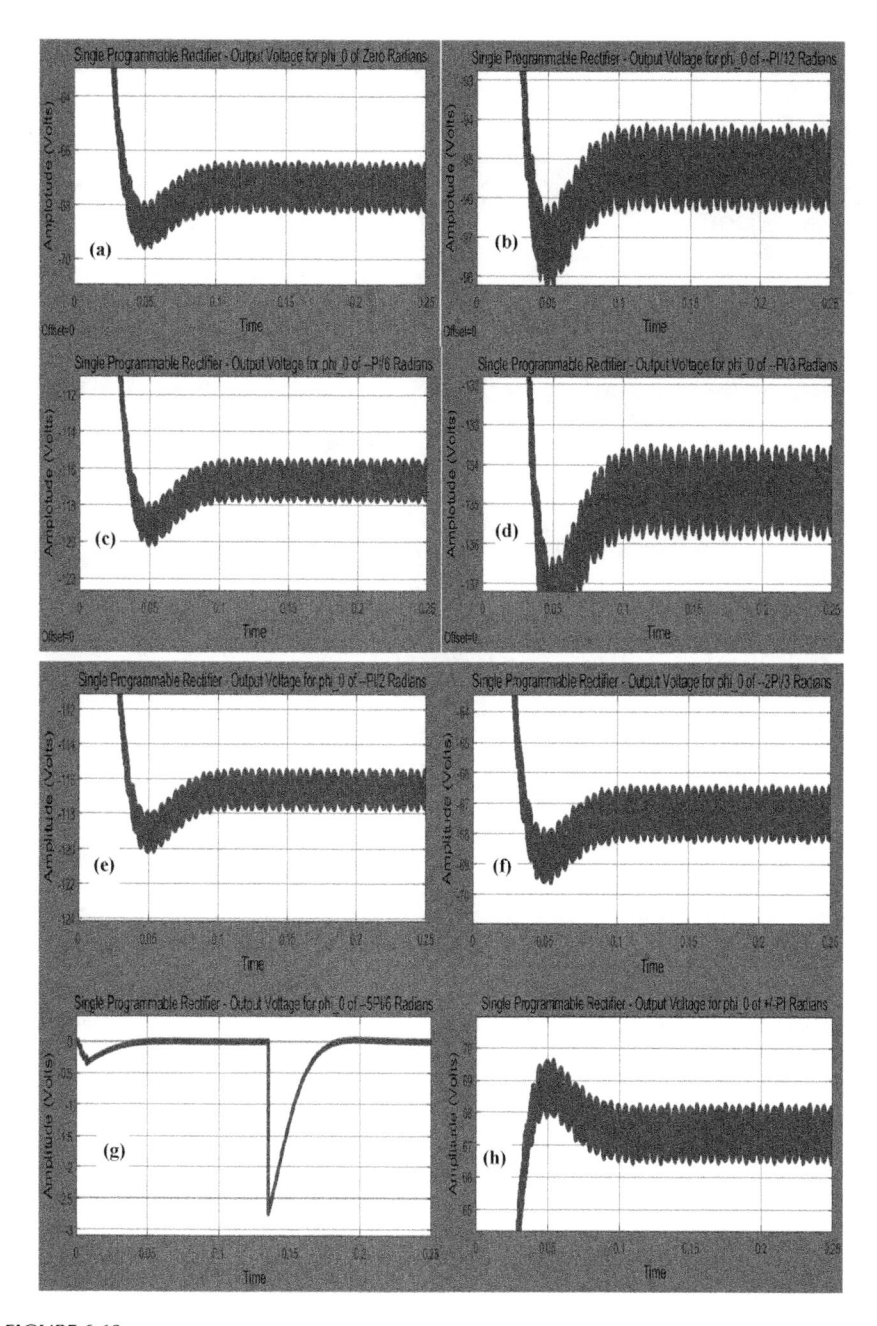

FIGURE 6.10
(a)–(n) Single programmable AC to DC rectifier – simulation results for output voltage phase angle phi_0 values: 0, $-\pi/12$, $-\pi/6$, $-\pi/3$, $-\pi/2$, $-2\pi/3$, $-5\pi/6$, $\pm\pi$, $+\pi/12$, $+\pi/6$, $+\pi/3$, $+\pi/2$, $+2\pi/3$ and $+5\pi/6$ rad.

(Continued)

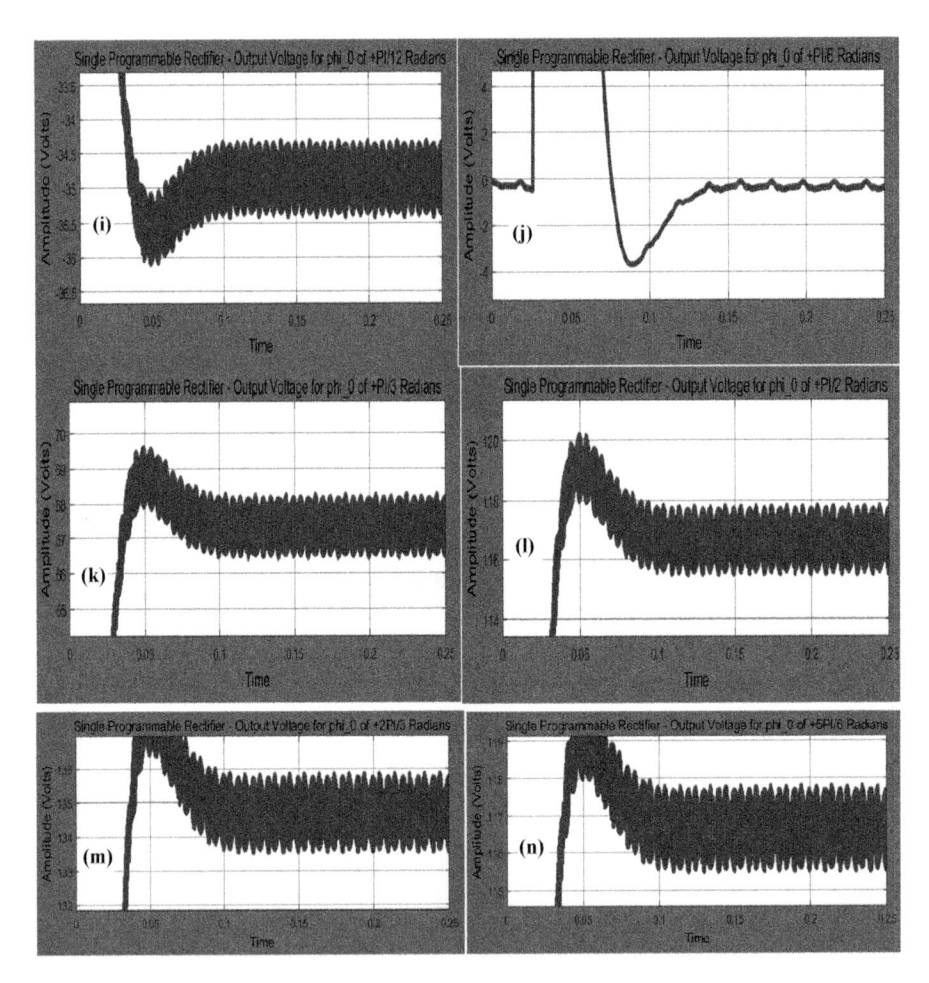

FIGURE 6.10 (CONTINUED)
(a)–(n) Single programmable AC to DC rectifier – simulation results for output voltage phase angle phi_0 values: 0, $-\pi/12$, $-\pi/6$, $-\pi/3$, $-\pi/2$, $-2\pi/3$, $-5\pi/6$, $\pm\pi$, $+\pi/12$, $+\pi/6$, $+\pi/3$, $+\pi/2$, $+2\pi/3$ and $+5\pi/6$ rad.

TABLE 6.4

Single Programmable Rectifier – Model Parameters

Sl. No.	Three-Phase Input Voltage Line-to-Neutral RMS Value (V)	Input Frequency (Hz)	Output Frequency (Hz)	Carrier Frequency (kHz)	Modulation Index q
1	220	50	0	2	0.5

TABLE 6.5

Single Programmable Rectifier – Theoretical and Model Simulation Results

Sl. No.	Output Voltage Phase Angle φ_o (rad)	Theoretical Output Voltage (V)	Model Simulation Output Voltage (V)	Sl. No.	Output Voltage Phase Angle φ_o (rad)	Theoretical Output Voltage (V)	Model Simulation Output Voltage (V)
1	0	−67.3488	−67.35	8	±π	+67.3488	+67.35
2	−π/12	−95.2456	−95.24	9	+π/12	−34.8623	−34.86
3	−π/6	−116.6516	−116.6	10	+π/6	0	0
4	−π/3	−134.6976	−134.7	11	+π/3	+67.3488	+67.35
5	−π/2	−116.6516	−116.6	12	+π/2	+116.6516	+116.6
6	−2π/3	−67.3488	−67.35	13	+2π/3	+134.6976	+134.7
7	−5π/6	0	0	14	+5π/6	+116.6516	+116.6

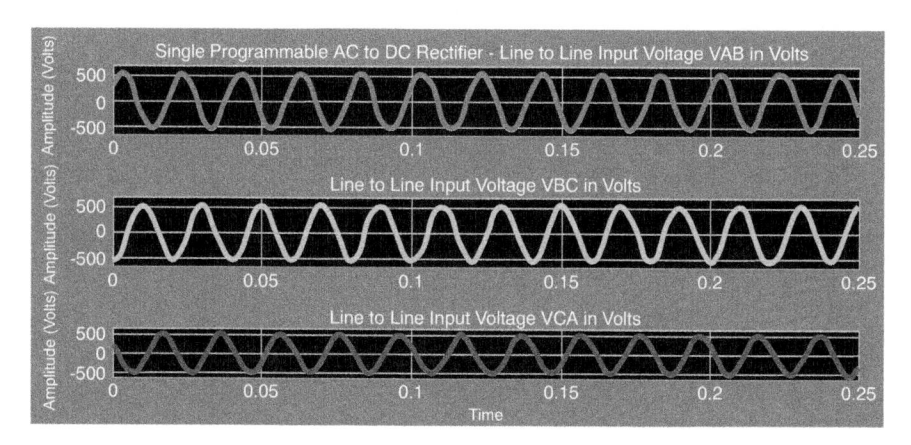

FIGURE 6.11
Single programmable AC to DC rectifier – three-phase AC input voltage.

6.8 Case Study: Variable-Frequency Variable-Voltage Pure Sine-Wave AC Power Supply

The single and dual programmable AC to DC rectifier topology can also be used as a single and dual variable-frequency variable-voltage pure sine-wave AC power supply. The model is the same as given in Figures 6.3 and 6.8. The magnitude and frequency of the output voltage can be varied by varying the modulation index q and output frequency f_o Hz, respectively, in the model. Changing the output voltage phase angle φ_o shifts the phase

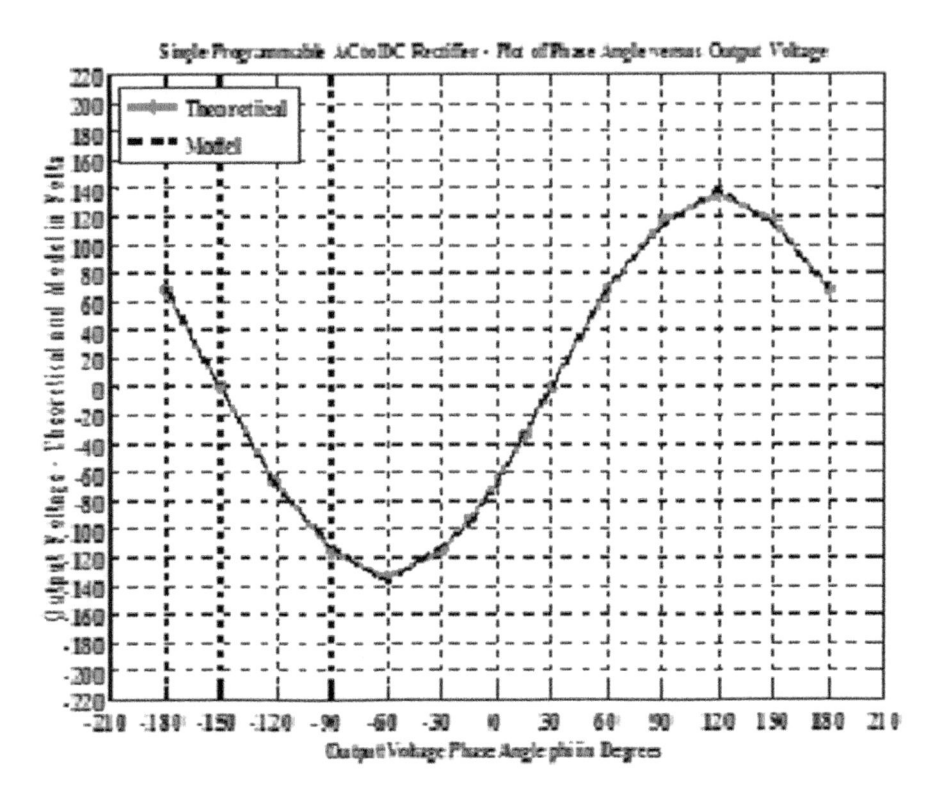

FIGURE 6.12
Single programmable AC to DC rectifier – plot or output voltage phase angle phi_0 versus output voltage.

displacement of the output voltage and has no effect on its magnitude and frequency. The coefficients for the cosine-wave and sine-wave expressions given by equations 6.14, 6.15 and 6.18 give the peak value for the output voltage and the value obtained could be less than this value due to voltage drop in the inductor and resistor or attenuation in the output filter.

6.8.1 Simulation Results

The simulation of the single and dual sine-wave AC power supply is carried out using SIMULINK [8]. The input data used are given in Tables 6.1 and 6.4, respectively. The output frequencies used are 20 and 60 Hz for the former power supply and 20 and 100 Hz for the latter power supply. In both cases, modulation indices of 0.25 and 0.5 are used. The single sine-wave AC power supply output voltage simulation results for frequencies of 20 and 60 Hz are shown in Figures 6.16 and 6.17, respectively. The dual sine-wave AC power supply output voltage simulation results for frequencies of 20 and 100 Hz are shown in Figures 6.18 and 6.19, respectively.

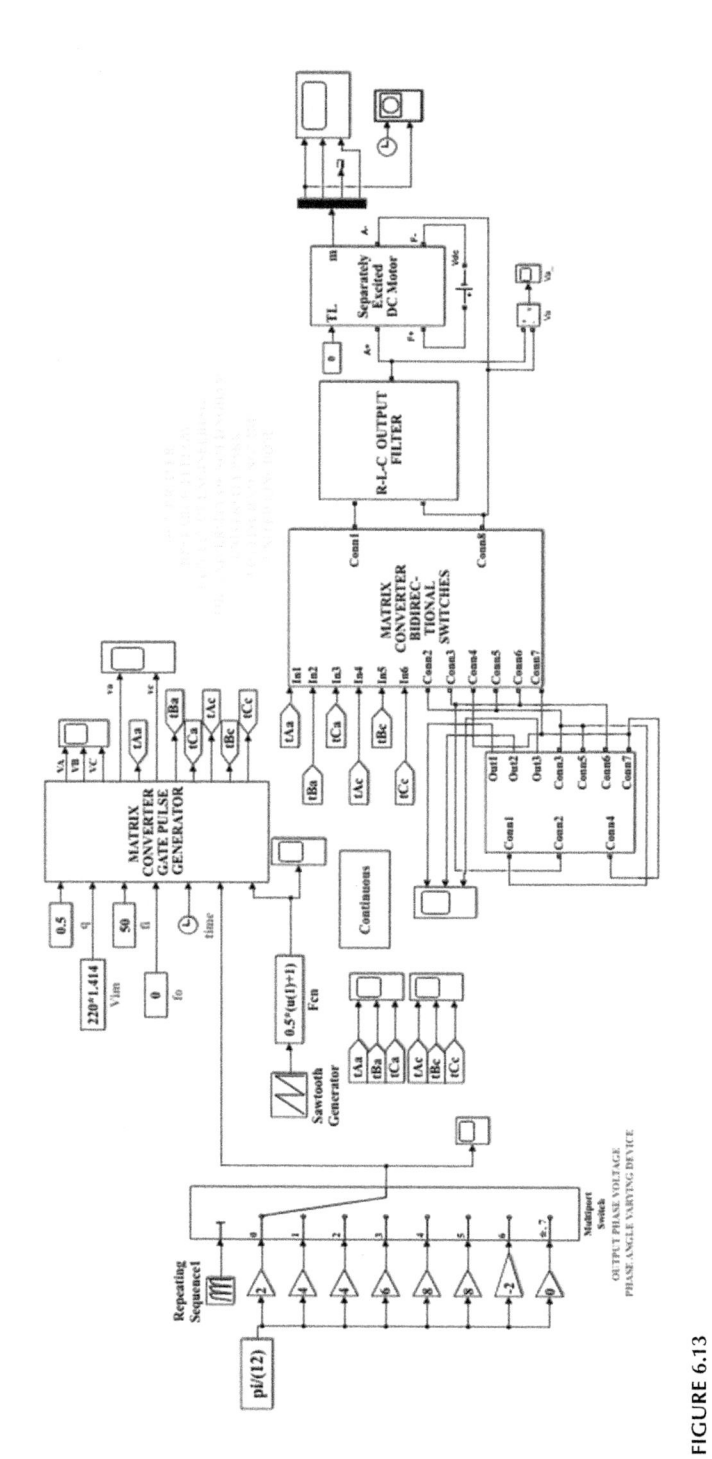

FIGURE 6.13
Model of single programmable AC to DC rectifier-fed separately excite DC motor drive.

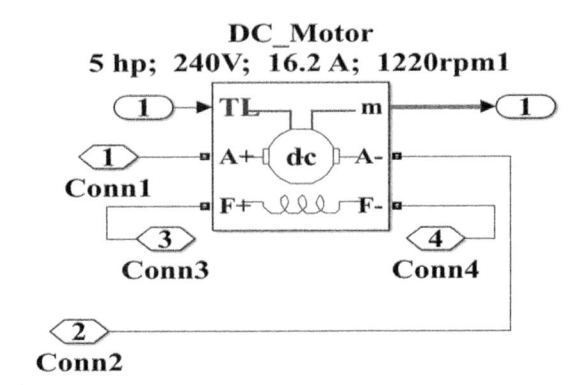

DC_Motor
5 hp; 240V; 16.2 A; 1220rpm1

FIGURE 6.14
DC motor model subsystem.

TABLE 6.6

DC Motor Model

Sl. No.	Parameters	DC Machine 1 Value	Units
1	Power output	5	hp
2	Rated terminal voltage	240	V
3	Speed	1,220	rpm
4	Field voltage	240	V
5	Armature resistance	0.6	Ω
6	Armature inductance	0.012	H
7	Field resistance	240	Ω
8	Field inductance	120	H
9	Field-armature mutual inductance	1.8	H
10	Total inertia	1.0	kg m^2
11	Damping coefficient	0	N m s

6.9 Case Study: Speed Control and Brake by Plugging of Two Separately Excited DC Motors Using Dual Programmable AC to DC Rectifier

The model of the dual programmable rectifier-fed DC motors is shown in Figure 6.20. The model and gate pulse generation are the same as explained in Section 6.4 except that the R–L loads are replaced by two SEDC motors. The model subsystem for the DC motor is the same as explained in Figure 6.14. The two SEDC motor ratings are as shown in Table 6.7 [8]. A multiport

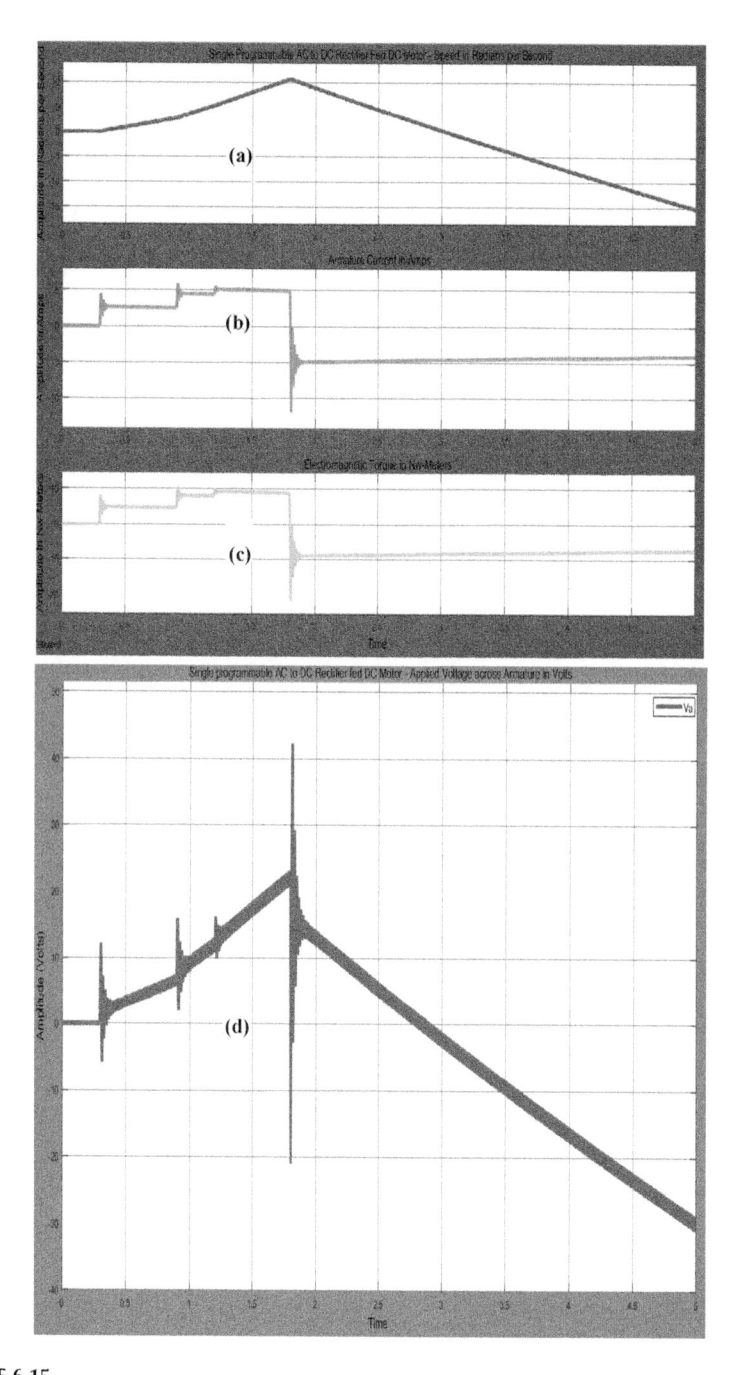

FIGURE 6.15
Single programmable AC to DC rectifier-fed DC motor drive: (a) Speed in mechanical radians per second, (b) armature current in amperes, (c) electromagnetic torque in Newton-metres and (d) applied voltage across armature in volts.

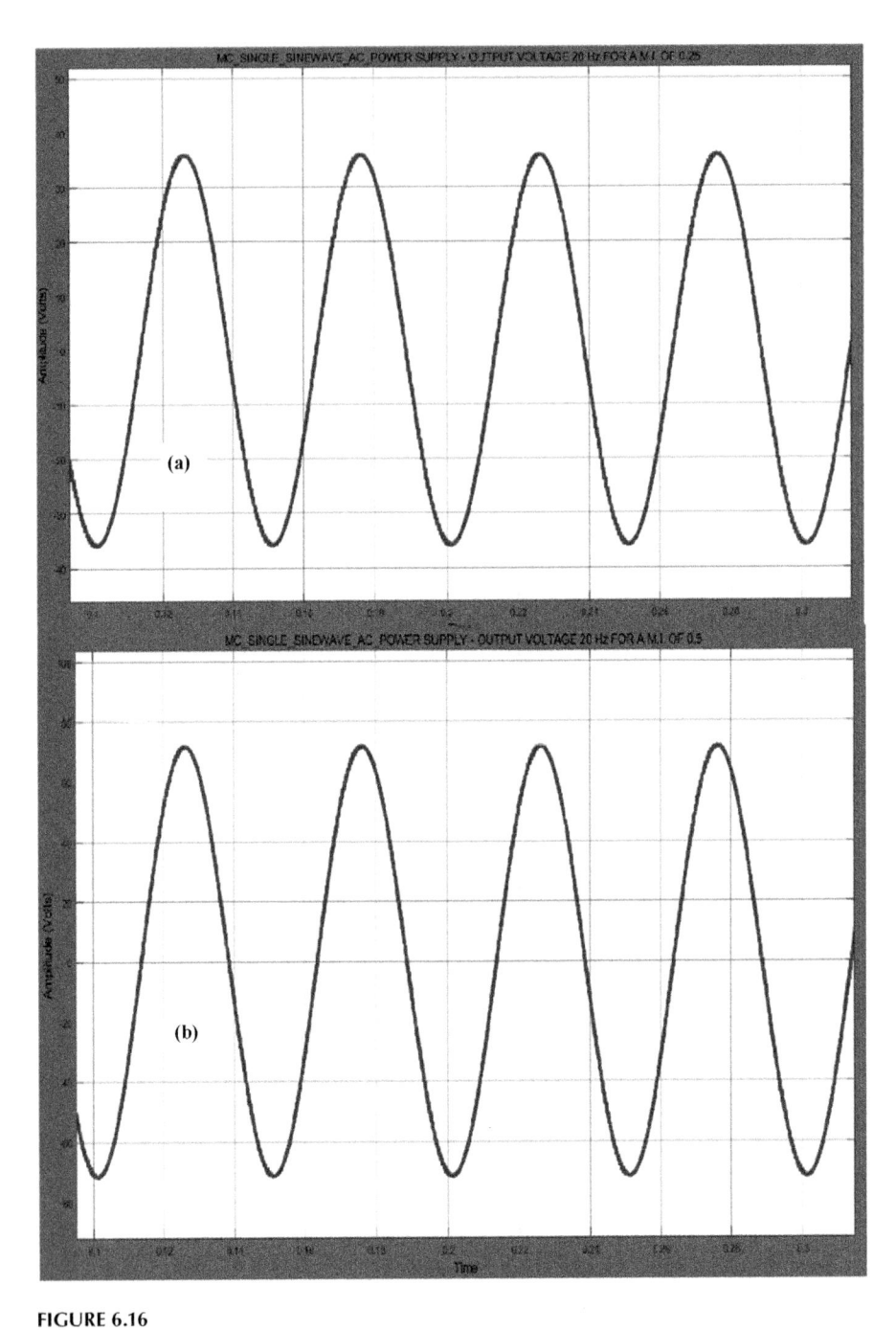

FIGURE 6.16
MC single sine-wave AC power supply: (a) output voltage frequency 20 Hz for a MI of 0.25 and (b) output voltage frequency 20 Hz for a MI of 0.5.

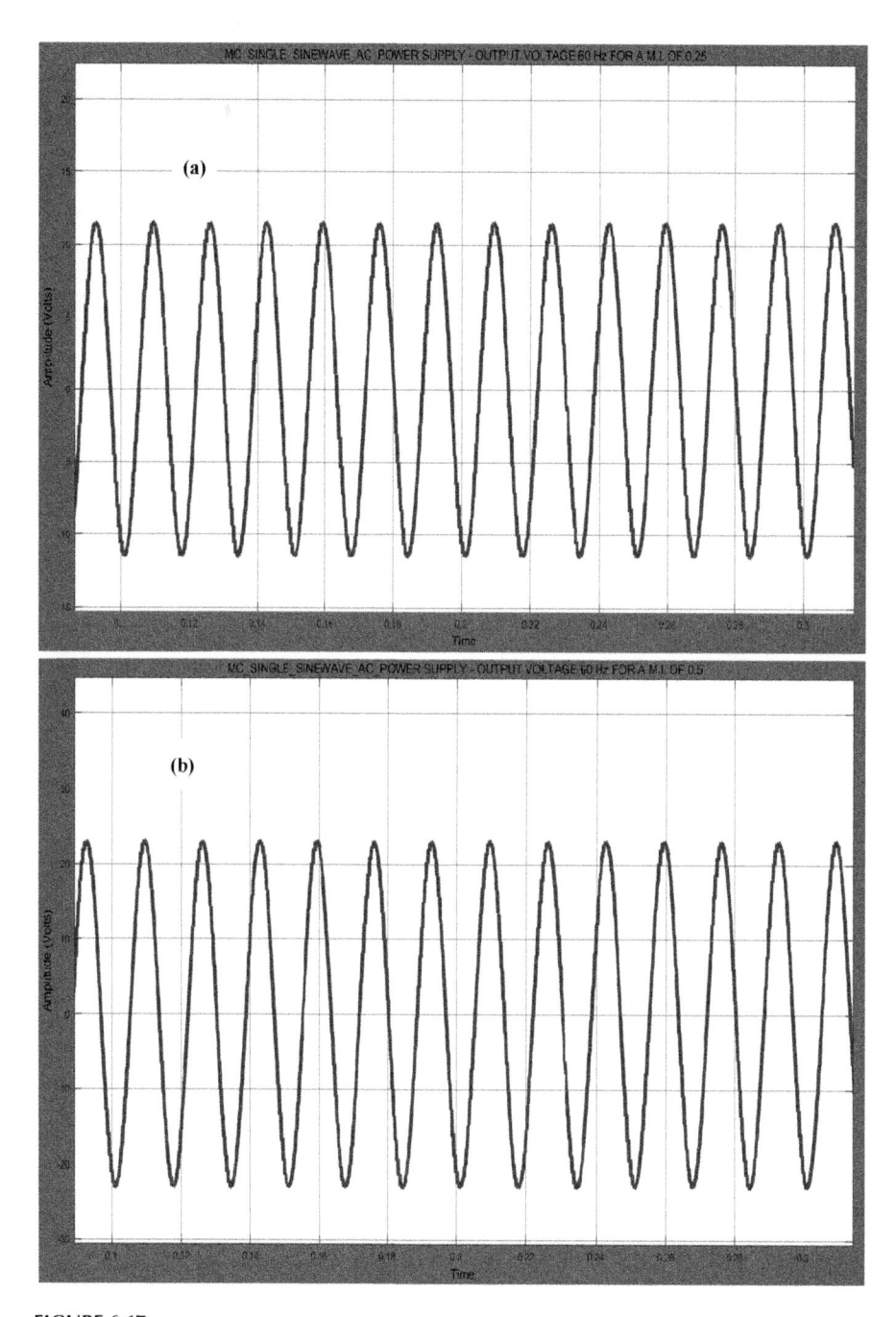

FIGURE 6.17
Single sine-wave AC power supply: (a) output voltage frequency 60 Hz for an MI of 0.25 and (b) output voltage frequency 60 Hz for an MI of 0.5.

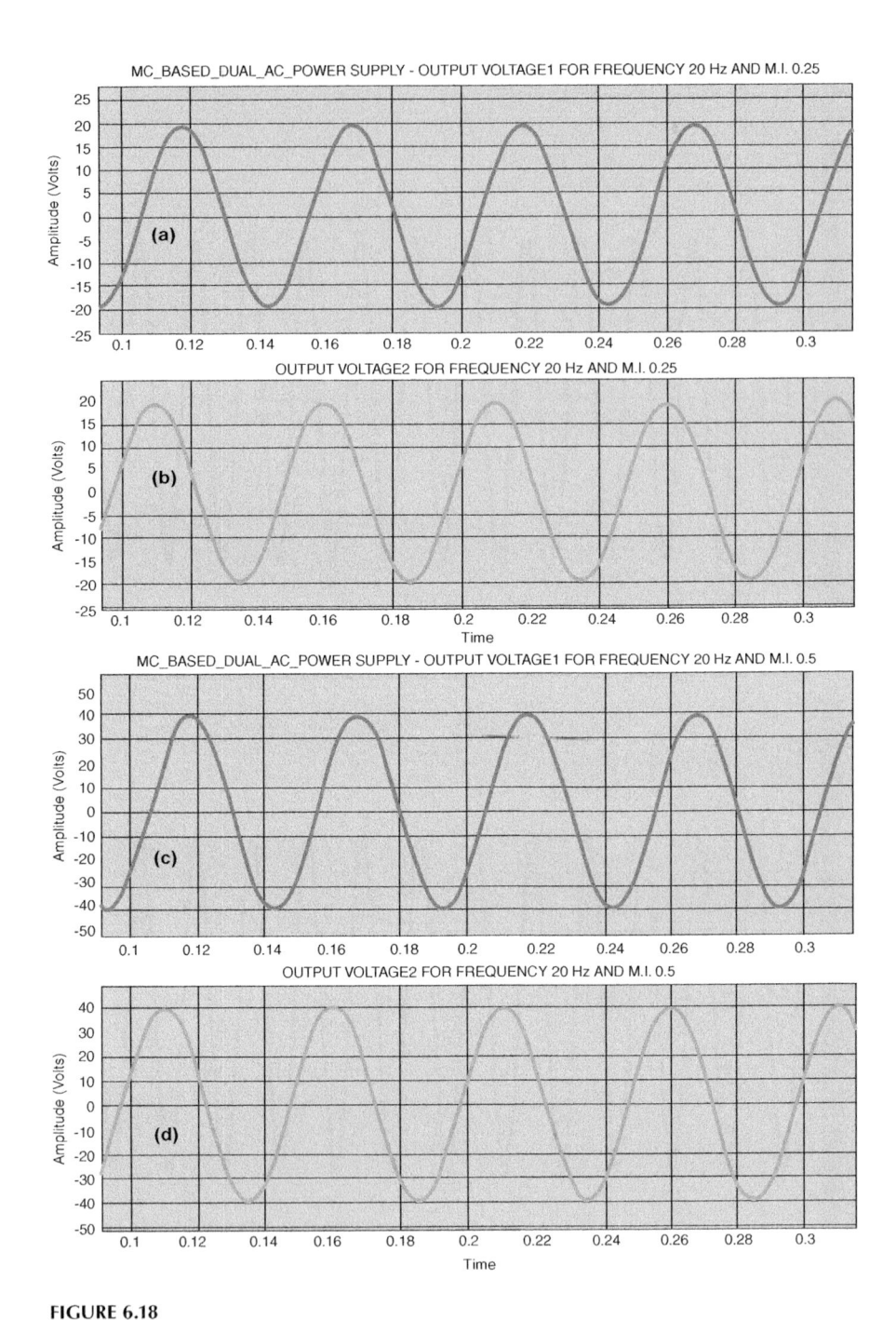

FIGURE 6.18
Dual sine-wave AC power supply: (a) V_{o1} and (b) V_{o2} output voltages for a frequency of 20 Hz and MI 0.25, (c) V_{o1} and (d) V_{o2} output voltage for a frequency of 20 Hz and MI 0.5.

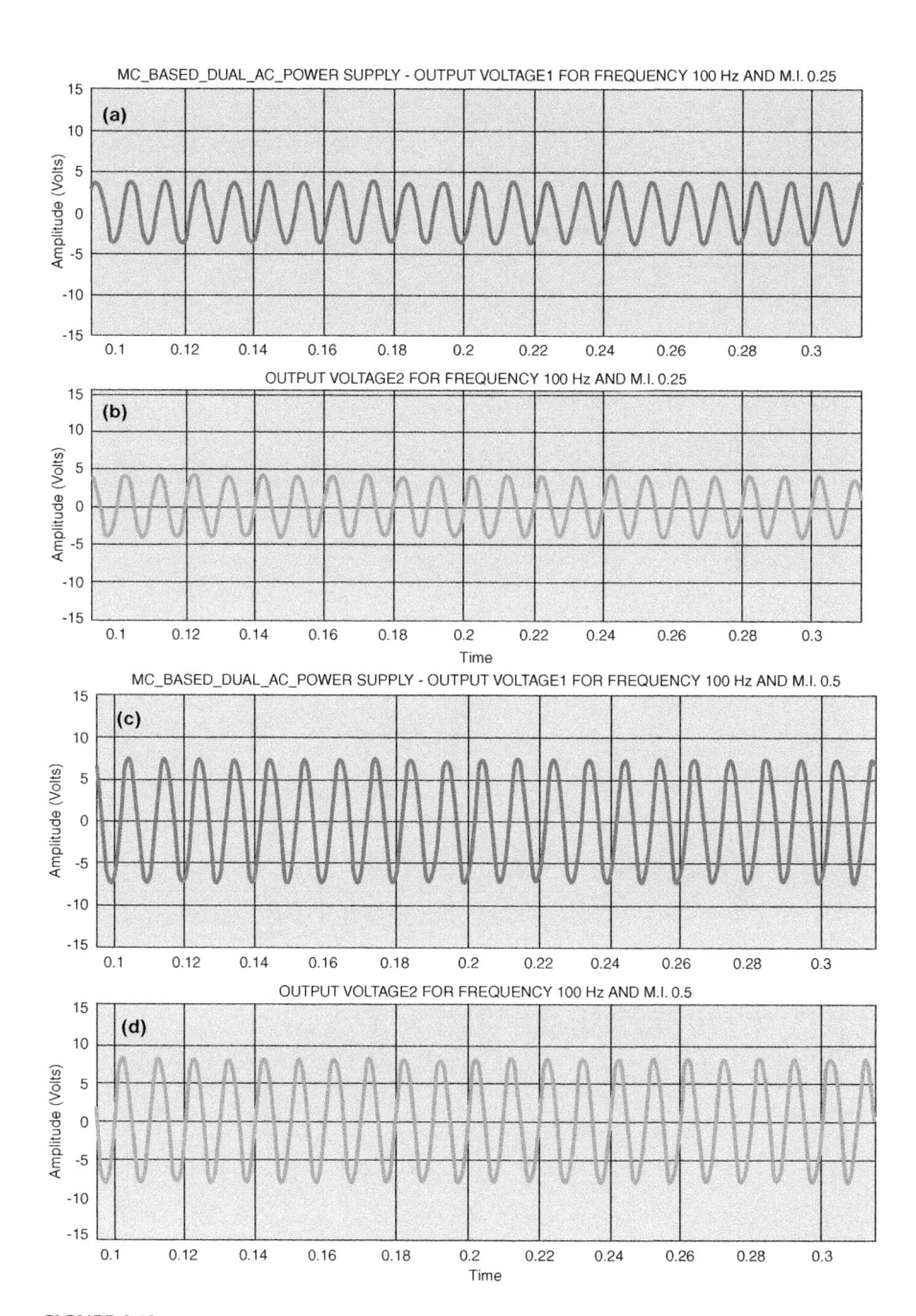

FIGURE 6.19
Dual sine-wave AC power supply: (a) V_{o1} and (b) V_{o2} output voltages for a frequency of 100 Hz and MI 0.25, (c) V_{o1} and (d) V_{o2} output voltages for a frequency of 100 Hz and MI 0.5.

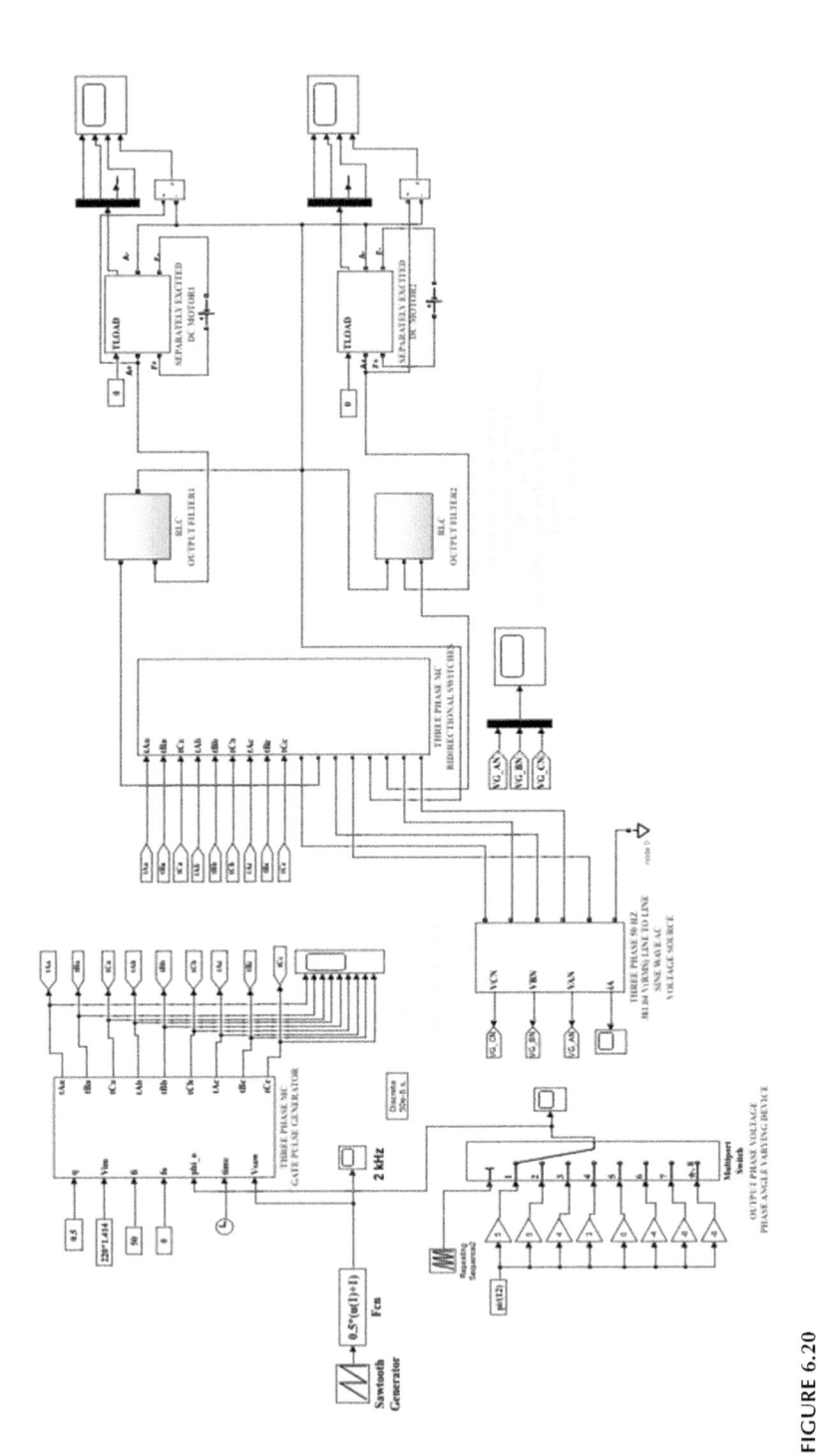

FIGURE 6.20
Model of dual programmable AC to DC rectifier-fed separately excited DC motors.

TABLE 6.7

SEDC Model Parameters

Sl. No.	Parameters	DC Machine 1		DC Machine 2	
		Value	Units	Value	Units
1	Power output	10	hp	5	hp
2	Rated terminal voltage	500	V	240	V
3	Speed	1,750	rpm	1,750	rpm
4	Field voltage	300	V	300	V
5	Armature resistance	4.712	Ω	2.581	Ω
6	Armature inductance	0.0527	H	0.028	H
7	Field resistance	180	Ω	281.3	Ω
8	Field inductance	71.4	H	156	H
9	Armature mutual inductance	1.345	H	0.9483	H
10	Total inertia	0.0425	kg m²	0.0221	kg m²
11	Damping coefficient	0.0034	N m s	0.0029	N m s

switch is used to vary the output voltage phase angle φ_o. This functions the same way as explained in Section 6.7. At intervals of 0.2 s from 0 to 1.8 s, the output φ_o of the multiport switch is $+5\pi/12$, $+5\pi/12$, $+\pi/3$, $+\pi/4$, 0, $-\pi/3$, $-2\pi/3$ and $-2\pi/3$ rad. The values of v_{o1} and v_{o2} referring to Table 6.3 indicate that the first SEDC motor accelerates, coasts, accelerates, decelerates and reverses speed which is braked by plugging whereas the second SEDC accelerates, coasts, accelerates, coasts, decelerates and the cycle repeats.

6.9.1 Simulation Results

The simulation of the dual programmable rectifier-fed DC motor drive was carried out using SIMULINK [8]. The fixed-step discrete solver is used. The simulation results for speed, armature current, electromagnetic torque and the applied armature voltage for SEDC motors 1 and 2 are shown in Figures 6.21a–d and 6.22a–d, respectively. Simulation results for DC motors 1 and 2 confirm the speed prediction in Section 6.9.

6.10 Real-Time Implementation

Results in Sections 6.7 and 6.9 indicate that this single and dual programmable AC to DC rectifier can be used for the speed control of single and two SEDCM drives thus rendering it suitable for hybrid electric vehicle [11] and electric traction applications. Some of these applications are highlighted in Figure 6.23a–c. The dSPACE implementation of the saw-tooth carrier and

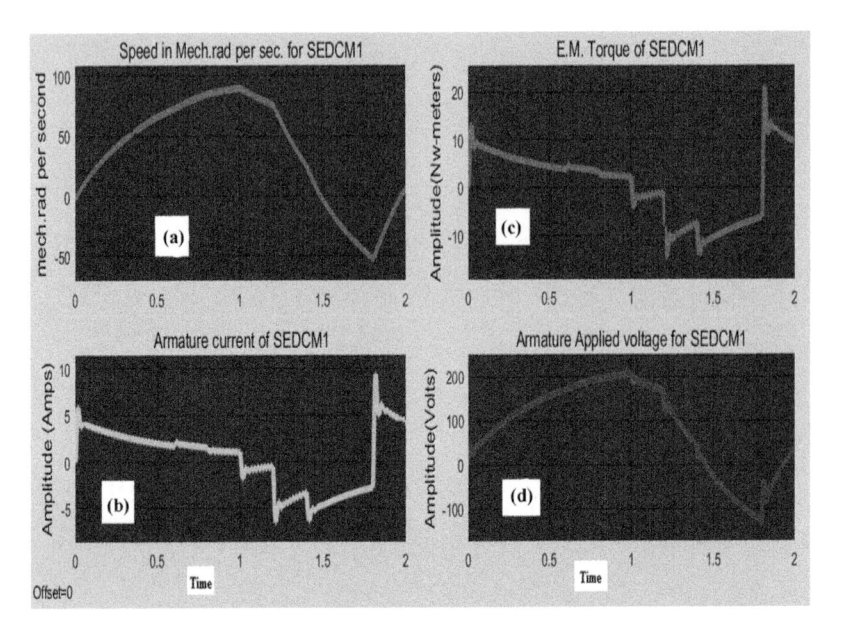

FIGURE 6.21
Dual programmable rectifier-fed SEDC motor – simulation results for SEDC motor 1: (a) speed in mechanical radians per seconds, (b) armature current in amperes, (c) electromagnetic torque in Newton-metres and (d) armature applied voltage in volts.

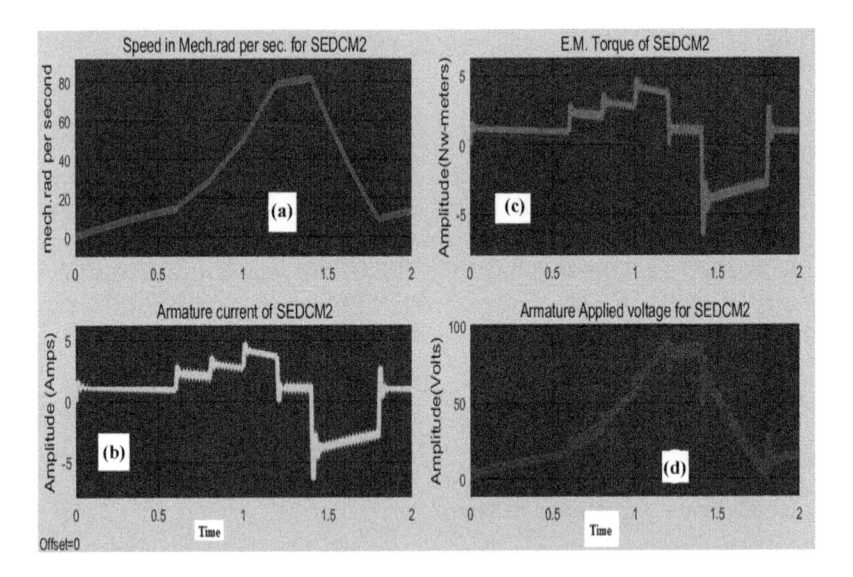

FIGURE 6.22
Dual programmable rectifier-fed DC motor – simulation results for SEDC motor 2: (a) speed in mechanical radians per seconds, (b) armature current in amperes, (c) electromagnetic torque in Newton-metres and (d) armature applied voltage in volts.

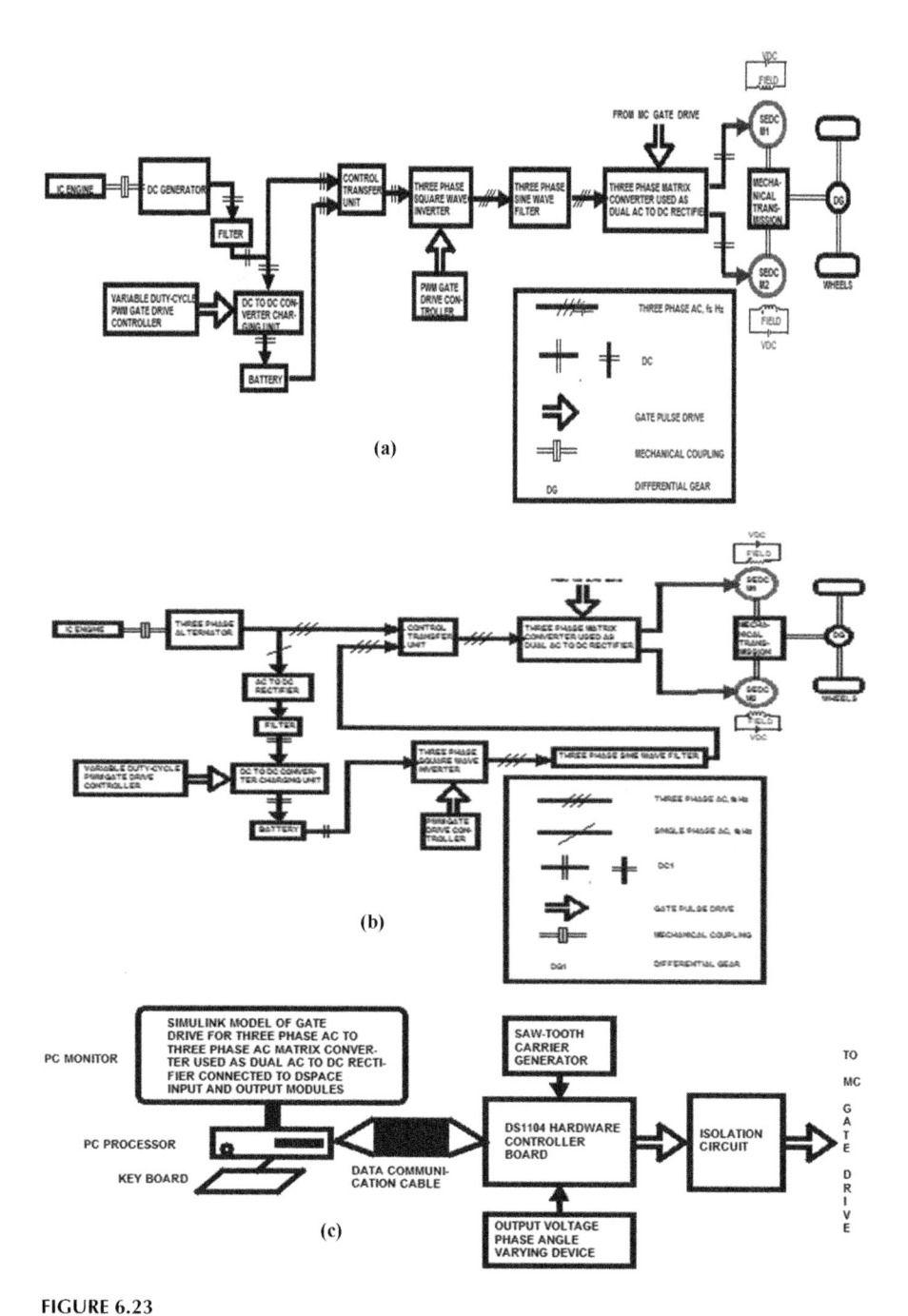

FIGURE 6.23
Dual programmable AC to DC rectifier-fed SEDC motor drives scheme of implementation for real-time HIL simulation: (a) Driven by DC generator, (b) driven by three-phase alternator and (c) scheme of implementation using dSPACE DS1104 hardware controller board.

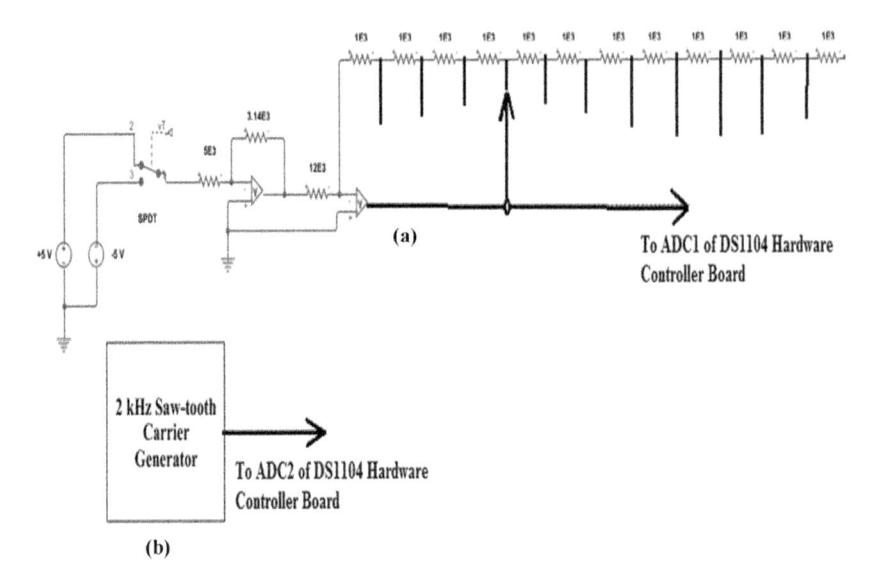

(a)

To ADC1 of DS1104 Hardware
Controller Board

2 kHz Saw-tooth
Carrier
Generator

To ADC2 of DS1104 Hardware
Controller Board

(b)

FIGURE 6.24
dSPACE implementation: (a) output voltage phase angle varying device and (b) saw-tooth carrier signal.

output voltage phase-angle-varying device for the three-phase MC used as single/dual programmable AC to DC rectifier is shown in Figure 6.24.

6.11 Discussion of Results

The theoretically calculated results using formula and model simulation results for dual programmable rectifier shown in Tables 6.2 and 6.3 well agree with zero percentage error [9,10]. Similarly, for the single programmable rectifier the theoretical and model simulation results shown in Table 6.5 closely well agree and the error is negligible. Both single and dual programmable rectifiers can be used for speed control and brake by plugging of SEDC motors and hence find applications in hybrid electric vehicles [9,11]. This is also illustrated by model simulation in Figures 6.15, 6.21 and 6.22 [9,10]. The application of the single and dual rectifier topology as variable-voltage and variable-frequency single and dual pure sine-wave AC power supply is also illustrated successfully by model simulation as shown in Figures 6.16–6.19. Neglecting voltage drop across armature resistance, the speed and applied armature voltage curves in Figures 6.21 and 6.22 agree well. A real-time implementation scheme for dual programmable rectifier using dSPACE DS1104 hardware controller board is provided which is also applicable to single programmable rectifier.

6.12 Conclusions

A novel concept of single and dual programmable AC to DC rectifier using three-phase AC to three-phase AC MC topology is presented in this chapter. Simulation results and theoretically computed values closely well agree. The Alesina–Venturini algorithm for the maximum output voltage amplitude limit forms the foundation for the operation of the single and dual programmable AC to DC rectifier. The proposed schemes find application in the speed control, acceleration and brake by plugging of SEDC motor. Also the application of single and dual rectifier topology as a variable-voltage variable-frequency pure sine-wave AC power supply is verified successfully by model simulation.

References

1. D.G. Holmes and T.A. Lipo: Implementation of a controlled rectifier using AC–AC matrix converter theory, *IEEE-PESC*, Milwaukee, WI, 1989, pp. 353–359.
2. D.G. Holmes and T.A. Lipo: Implementation of a controlled rectifier using AC–AC matrix converter theory, *IEEE Transactions on Power Electronics*, Vol. 7, No. 1, 1992, pp. 240–250.
3. S. Huseinbegovic and O. Tanovic: Matrix converter based AC/DC rectifier. *IEEE-SIBIRCON II*, Irkutsk, Russia, 2010, pp. 653–658.
4. P. Wheeler, J. Rodriguez, J.C. Clare and L. Empringham: Matrix converters: a technology review, *IEEE Transactions on Industrial Electronics*, Vol. 49, No. 2, 2002, pp. 276–288.
5. P. Wheeler, J. Clare, L. Empringham, M. Apap and M. Bland: Matrix converters, *Power Engineering Journal*, Vol. 16, 2002, pp. 273–282.
6. A. Alesina and M. Venturini: Intrinsic amplitude limits and optimum design of 9-switches direct PWM AC–AC converters, *IEEE-PESC*, Kyoto, Japan, 1988, pp. 1284–1291.
7. A. Alesina and M. Venturini: Analysis and design of optimum amplitude nine-switch direct AC–AC converters, *IEEE Transactions on Power Electronics*, Vol. 4, No. 1, 1989, pp. 101–112.
8. Mathworks Inc.: www.mathworks.com, MATLAB/SIMULINK users' manual, MATLAB R2017b, 2017.
9. N.P.R. Iyer: Modelling, simulation and real time implementation of a three phase AC to AC matrix converter, Ph.D. thesis, Ch.10 and Ch.14, Department of ECE, Curtin University, Perth, WA, Australia, February 2012.
10. N.P.R. Iyer: A dual programmable AC to DC rectifier using three-phase matrix converter topology—analysis aspects, *Electrical Engineering*, Vol. 100, No. 2, 2018, pp. 1183–1194. DOI: 10.1007/s00202-017-0572-9.
11. M. Ehsani, Y. Gao and A. Emadi: *Modern Electric, Hybrid Electric and Fuel Cell Vehicles*, CRC Press and Taylor & Francis, Raton, FL, 2010, pp. 105–168.

7

Delta-Sigma Modulation of Three-Phase Matrix Converters

7.1 Introduction

Matrix converter (MC) directly converts the AC input voltage at any given frequency to AC output voltage with arbitrary amplitude at any unrestricted frequency without the need for a DC-link capacitor storage element at the input side. Since inception, several carrier-based modulation techniques have been proposed for MC. All these techniques use either a saw-tooth carrier [1–8] or a triangle carrier [9–12] to be compared with modulation voltage signals to generate gate switching pulses for the bidirectional switches of MC. These modulation voltage signals are generated using the Venturini algorithm [1–5] or using other recently proposed algorithms [9–12]. This chapter examines the recently proposed delta-sigma modulation technique for MC [13–15]. A model of the three-phase AC to three-phase AC MC using delta-sigma modulation technique is developed in SIMULINK [16]. The Venturini [1–5] modulation algorithm is used for generating the modulation voltage signals. Simulation is carried out for a given sampling frequency. It is seen from simulation results that voltage harmonic peaks are reduced at integral multiples of sampling frequency [13–15]. Also simulation of three-phase induction motor (IM) fed by delta-sigma-modulated MC is presented. The delta-sigma modulation technique can reduce noise peaks in the output voltage as compared to conventional pulse-width modulation (PWM) technique and has the advantage of maintaining noise regulation [13–15].

7.2 Review of Matrix Converter Gate Pulse Generation

The three-phase AC to three-phase AC MC is shown in Figure 2.1 of Chapter 2. Gate pulses for three-phase MC bidirectional switches have to be generated to comply with some specific requirement. Referring to Figure 2.1 of Chapter 2, it is seen that if two or more bidirectional switches connected to same output phase such as S_{Aa}, S_{Ba} and S_{Ca} are turned ON simultaneously, there will be a dead short circuit. These gate pulses have to be on in sequence, one after the other. Also, it is required that any one bidirectional switch in each phase should remain closed. This is achieved by carrier PWM technique [1–5] and also by delta-sigma modulation technique [13–15].

7.2.1 Delta-Sigma PWM Technique

This is a recently proposed technique for three-phase MCs [13–15]. In delta-sigma modulation, high-frequency noise peaks in the output voltages due to switching operation do not occur and this becomes an advantage for clearing noise regulations [13–15]. Figure 7.1a,b shows the first-order delta-sigma modulator [17]. Analysis of Figure 7.1a gives the following:

$$X(z) - z^{-1} * Y(z) + z^{-1} * E(z) = E(z) \tag{7.1}$$

Analysis of Figure 7.1b gives the following:

$$X(z) - z^{-1} * Y(z) + z^{-1} * E(z) = E(z) \tag{7.2}$$

In sample time, equations 7.1 and 7.2 can be expressed as follows:

$$x(k) - y(k-1) + e(k-1) = e(k) \tag{7.3}$$

Equations 7.1 and 7.2 agree well and are identical for Figure 7.1a,b.
   Quantizer characteristics can be expressed as follows:

$$y(k) = sgn\big[e(k)\big] \tag{7.4}$$

The quantizer output is +1 if its input exceeds the threshold and –1 if its input is below threshold value. The quantizer output is used to control the bidirectional switches.

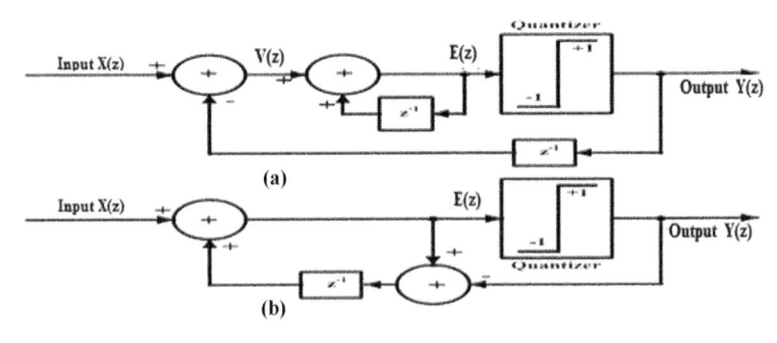

FIGURE 7.1
(a) and (b) First-order delta-sigma modulator.

7.3 Delta-Sigma Modulator Interface

This section discusses the implementation aspects of a delta-sigma modulator for a three-phase MC. The delta-sigma modulator for the output phase a of three-phase MC is shown in Figure 7.2. Similar circuit applies to output phase b and c of the MC. In Figure 7.2, the quantizer output is given to comparator. The comparator output is HIGH when its input is greater than or equal to zero, else the output is LOW. The comparator outputs A, B and C form the input to the combinational logic circuit which generates the gate pulse for the bidirectional switches. The combinational logic circuit must be so designed to comply with the gate drive requirement specified under Section 7.2. Table 7.1 shows one method of generating the gate pulse for the bidirectional switches connected to output phase a. Gate pulse t_{Aa}, t_{Ba} and t_{Ca} can be expressed as follows:

$$t_{Aa} = (A \cap B) \cup (\overline{B} \cap \overline{C}) \tag{7.5}$$

$$t_{Ba} = (\overline{A} \cap B) \tag{7.6}$$

$$t_{Ca} = (\overline{B} \cap C) \tag{7.7}$$

Gate pulse expressions for bidirectional switches connected to output phase b and c can be expressed as follows:

$$t_{Ab} = (\overline{E} \cap F) \tag{7.8}$$

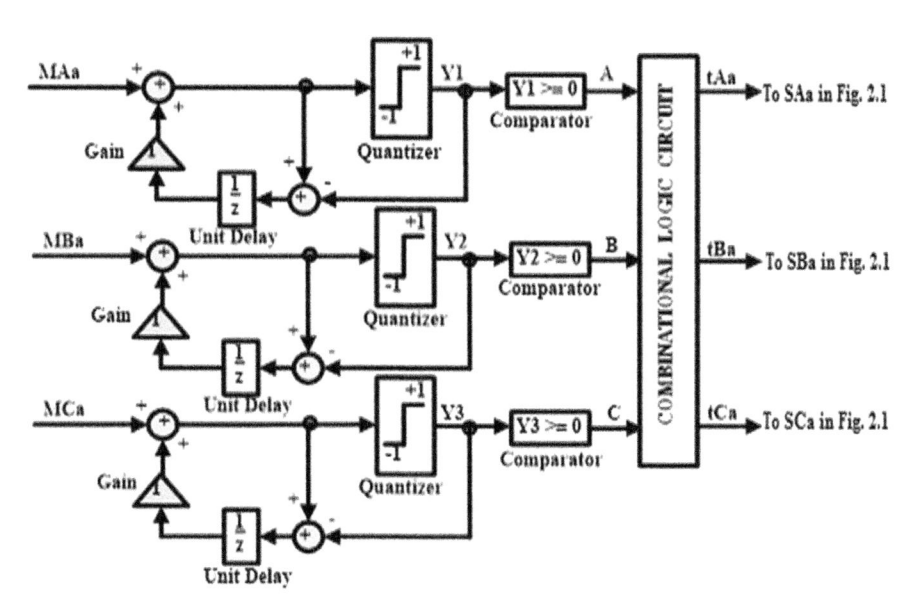

FIGURE 7.2
Delta-sigma modulator for output phase matrix converter.

TABLE 7.1

Truth Table for Delta-Sigma Modulator

Sl. No.	Inputs			Outputs		
	A	B	C	t_{Aa}	t_{Ba}	t_{Ca}
1	0	0	0	1	0	0
2	0	0	1	0	0	1
3	0	1	0	0	1	0
4	0	1	1	0	1	0
5	1	0	0	1	0	0
6	1	0	1	0	0	1
7	1	1	0	1	0	0
8	1	1	1	1	0	0

$$t_{Bb} = \left(\bar{E} \cap \bar{F}\right) \cup (D \cap E) \tag{7.9}$$

$$t_{Cb} = \left(\bar{D} \cap E\right) \tag{7.10}$$

$$t_{Ac} = \left(\bar{G} \cap H\right) \tag{7.11}$$

$$t_{Bc} = \left(\bar{H} \cap I\right) \tag{7.12}$$

$$t_{Cc} = (G \cap H) \cup \left(\bar{H} \cap \bar{I}\right) \tag{7.13}$$

In equations 7.5–7.13, the symbols $\cap, \cup, -$ represent logical AND, OR and NOT operations, respectively. For the output phase b and c, inputs to delta-sigma modulator in Figure 7.2, M_{Aa}, M_{Ba} and M_{Ca} are replaced by M_{Ab}, M_{Bb}, M_{Cb} and M_{Ac}, M_{Bc}, M_{Cc}, respectively. Similarly, the inputs A, B and C to the combinational logic circuit in Figure 7.2 are replaced by D, E and F for output phase b and G, H, I for output phase c, respectively [16].

7.4 Venturini Model of Three-Phase Matrix Converter Using Delta-Sigma Modulation

The Venturini model of the three-phase AC to three-phase AC MC using delta-sigma modulation is shown in Figure 7.3 [16]. The Embedded MATLAB Function generates the nine modulation duty cycles as defined in equation 2.17 of Chapter 2. The source code for generating the nine modulation functions is shown in Program segment 2.3 in Chapter 2. These nine modulation functions are applied as inputs to first-order delta-sigma modulators as shown in Figure 7.2. The combinational logic interface is designed as described in Section 7.3. The principle of operation of the delta-sigma modulator is explained in Section 7.3. The nine gate pulses from the output

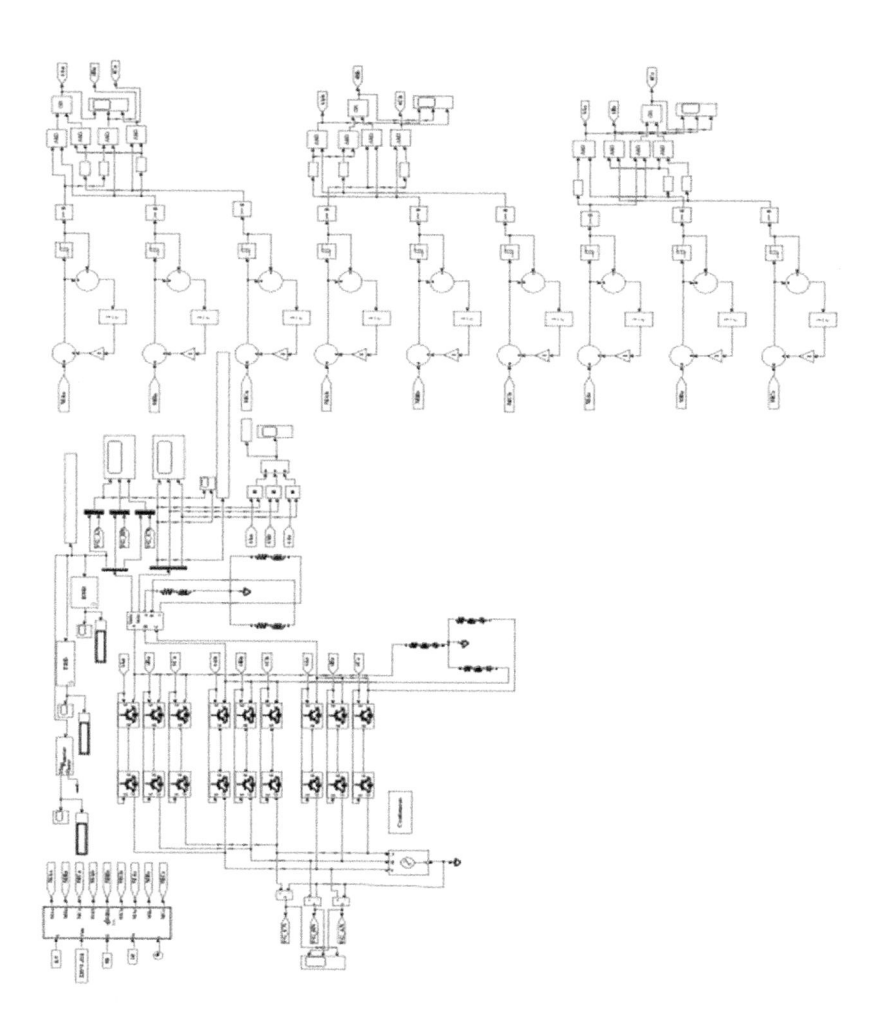

FIGURE 7.3
Venturini model of three-phase AC to three-phase AC matrix converter using delta-sigma modulation.

TABLE 7.2

Model Parameters for Delta-Sigma Modulator

Sl. No.	Parameter	Value	Units
1	RMS line-to-neutral input voltage	220	V
2	Input frequency	50	Hz
3	Output frequency	50	Hz
4	Sampling frequency	5	kHz
5	Modulation index	0.5	-
6	RL load	50, 0.5	Ω, H
7	Output RLC filter	10, 1e–3, 1.0132e–006	Ω, H, F

of combinational logic circuit are applied to the respective gates of the bidirectional switches.

7.4.1 Simulation Results

The simulation of the delta-sigma-modulated MC shown in Figure 7.3 was carried out using SIMULINK [18]. The simulation parameters are shown in Table 7.2. The ode23tb (Stiff/TR-BDF2) solver is used. The simulation results for the harmonic spectrum of line-to-line output voltage, input current, load current and line-to-neutral output voltage and their respective oscilloscope waveforms for an output frequency of 50 Hz are shown in Figures 7.4a–d and 7.5a–d, respectively. The simulation results are tabulated in Table 7.3.

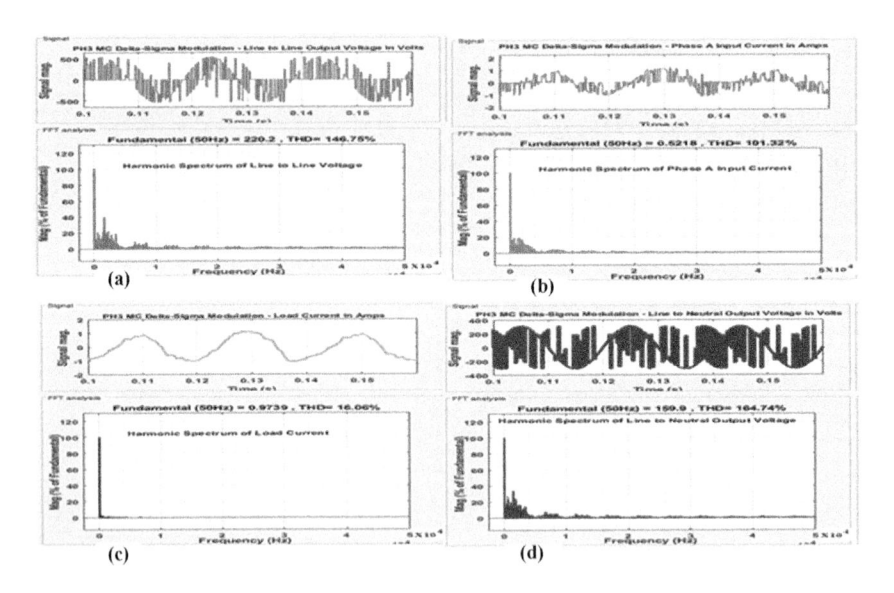

FIGURE 7.4

Delta-sigma-modulated three-phase 50-Hz MC harmonic spectrum. (a) Line-to-line output voltage, (b) phase A input current, (c) load current and (d) line-to-neutral output voltage.

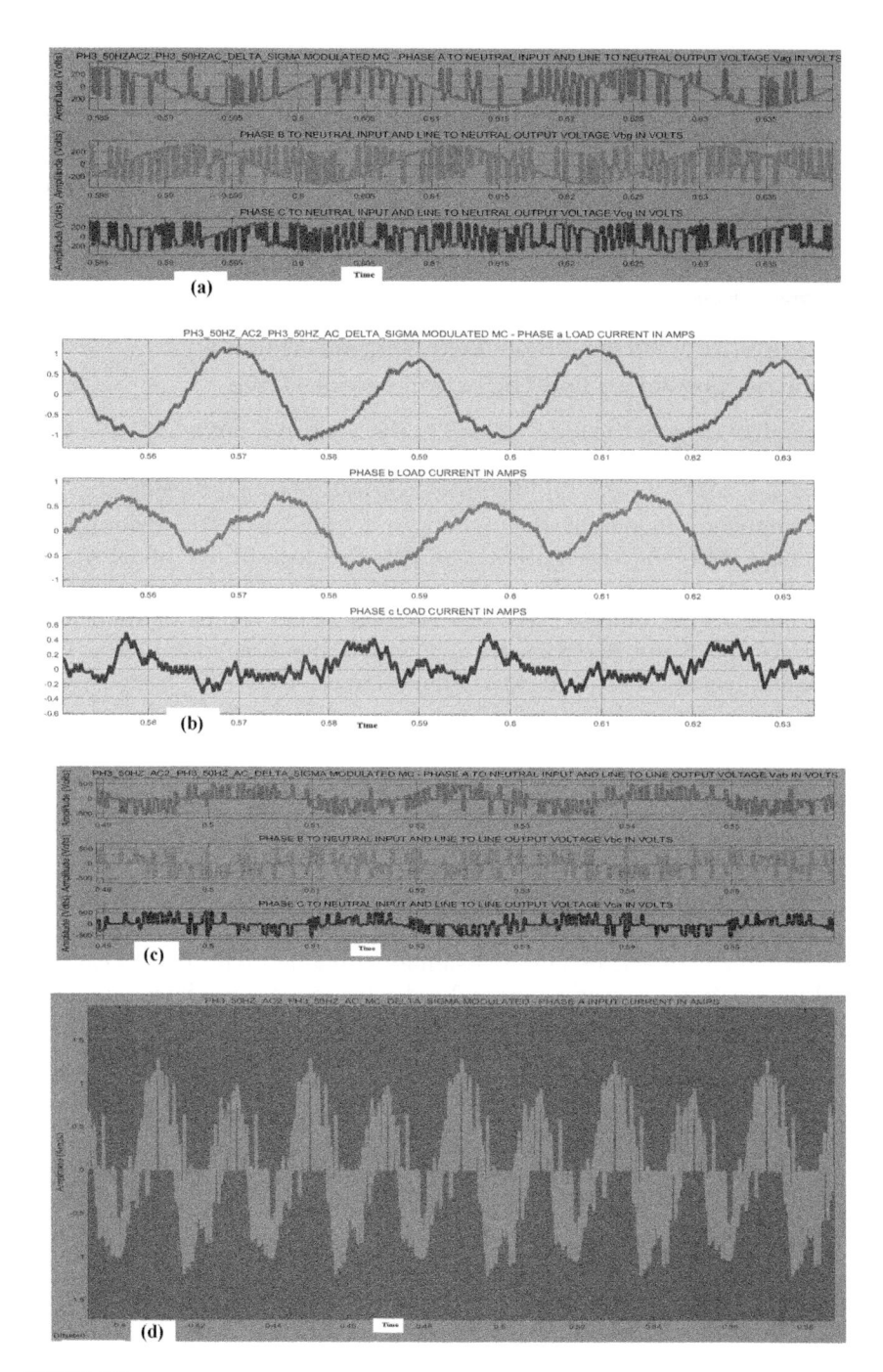

FIGURE 7.5
Three-phase 50 Hz AC delta-sigma-modulated MC: (a) line-to-neutral output voltage; (b) load current in phase a, b, and c; (c) line-to-line output voltage; and (d) phase A input current.

TABLE 7.3

Delta-Sigma-Modulated Three-Phase MC – Simulation Results

Sl. No.	Frequency Input– Output (Hz)	Line-to-Neutral Output Voltage THD (p.u.)	Line-to-Line Output Voltage THD (p.u.)	Input Current THD (p.u.)	Load Current THD (p.u.)
1	50 – 50	1.647	1.467	1.013	0.16

7.5 Case Study: Three-Phase Delta-Sigma-Modulated Matrix Converter Fed Induction Motor Drive

A three-phase cage IM is used as load to the Venturini model of MC using delta-sigma modulation. The parameters of the cage IM rated 4 kW, 400 V, three-phase, 50 Hz, 4 poles and 1,430 rpm are given in Table 7.4 [18].

The Venturini model of the delta-sigma-modulated MC driving the three-phase IM is shown Figure 7.6 [16]. The R–L load in the model shown in Figure 7.3 is disconnected and a three-phase cage IM whose parameters are in Table 7.4 is connected as load. The simulation results of stator currents, rotor speed and electromagnetic torque of the three-phase IM as load is shown in Figure 7.7 for zero external mechanical load.

7.6 Discussion of Results

Comparison of the simulation results for the harmonic spectrum of line-to-neutral and line output voltages as well as for input and load currents for the Venturini model of MC using delta-sigma modulation with that of carrier-based modulation given in Chapter 2 indicates the absence harmonic voltage components at integral multiples of sampling frequency. The proposed

TABLE 7.4

Three-Phase Induction Motor Model Parameters

Sl. No.	Parameter	Value	Units
1	Stator resistance R_s	1.405	Ω
2	Rotor resistance $R_{r'}$	1.395	Ω
3	Stator leakage inductance L_{ls}	0.005839	H
4	Rotor leakage inductance $L_{lr'}$	0.005839	H
5	Mutual inductance L_{lm}	0.1722	H
6	Rotor inertia	0.0131	kg m^2
7	Damping constant	0.002985	N m s
8	No. of poles	4	-

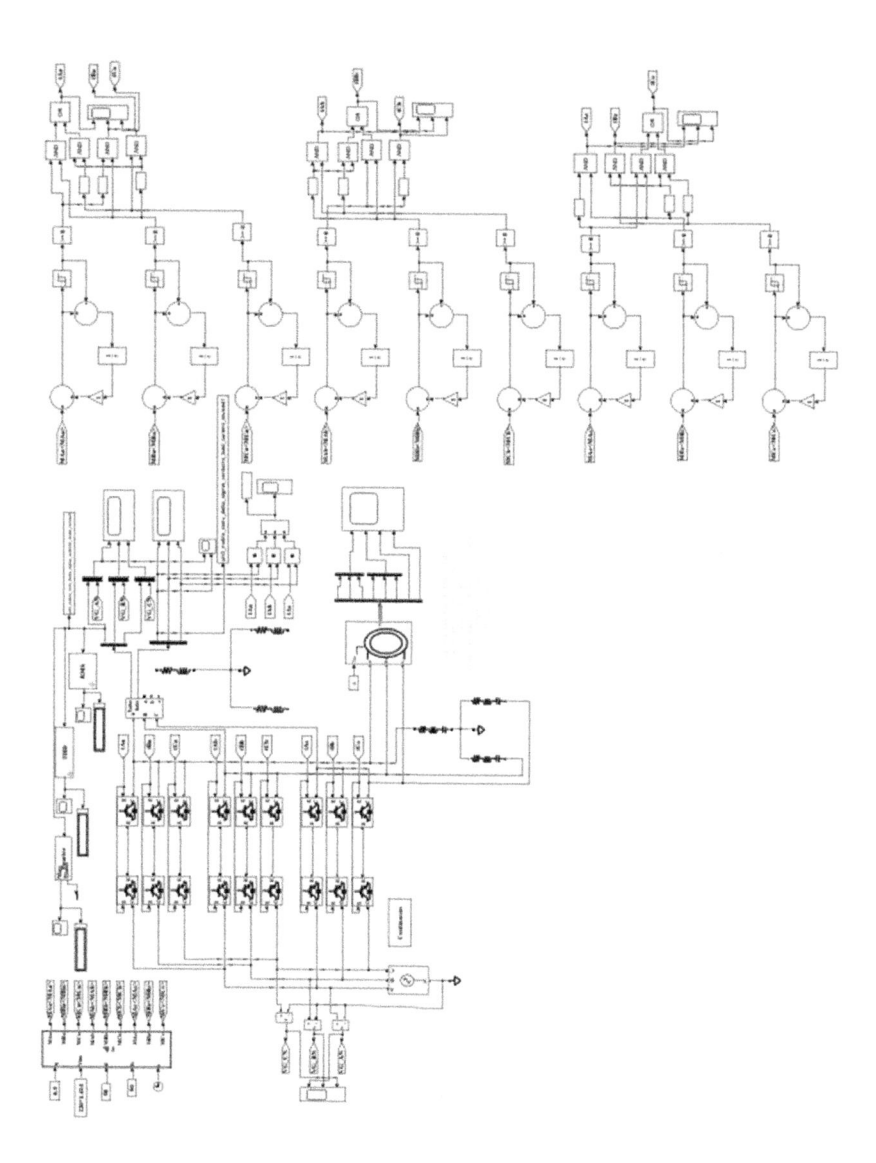

FIGURE 7.6

Delta-sigma-modulated three-phase matrix converter fed induction motor drive.

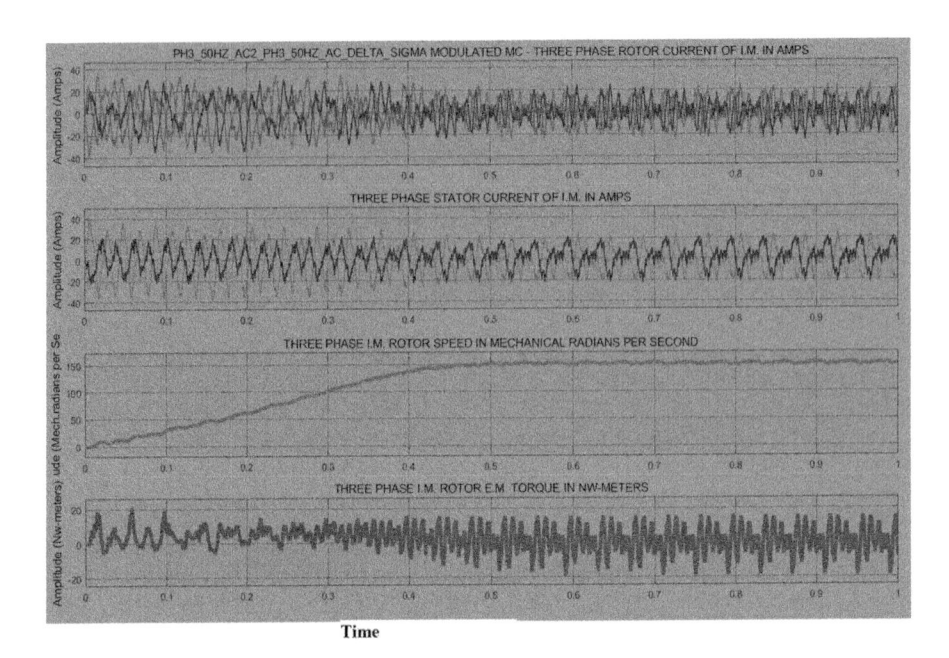

FIGURE 7.7

Three-phase delta-sigma-modulated MC fed I.M. drive simulation results.

method suppresses noise peaks in the output voltage and is found to be suitable for clearing noise regulation.

As for the simulation results of three-phase IM, it is seen that in both cases the stator starting current is high and finally settles to the no-load value. Similarly, the electromagnetic torque in both cases is initially high and finally settles to zero corresponding to the externally applied mechanical load. As for the rotor speed, it is seen that in both cases the rotor reaches close to the no-load speed of 1,430 rpm, which is 149.7 mech. rad/s.

7.7 Conclusions

A method of modelling MC using delta-sigma modulation technique is presented. Comparison of the THD of line-to-neutral and line-to-line output voltages for a 50-Hz output frequency indicates the absence of harmonic components of voltages at integral multiples of sampling frequency is worth noting. The proposed method suppresses noise peaks in the output voltage and is found to be suitable for clearing noise regulation. As for three-phase IM performance, it is seen that with delta-sigma modulation using the

Venturini model, the rotor reaches rated no-load speed of 149.7 mech. rad/s with zero external load.

References

1. A. Alesina and M.G.B. Venturini: Solid state power conversion: a Fourier analysis approach to generalized transformer synthesis, *IEEE Transactions on Circuits and Systems*, Vol. CAS-28, No. 4, 1981, pp. 319–330.
2. A. Alesina and M. Venturini: Intrinsic amplitude limits and optimum design of 9-switches direct PWM AC-AC converters, *IEEE-PESC*, Kyoto, Japan, 1988, pp. 1284–1291.
3. A. Alesina and M. Venturini: Analysis and design of optimum amplitude nine-switch direct AC-AC converters, *IEEE Transactions on Power Electronics*, Vol. 4, 1989, pp. 101–112.
4. P. Wheeler, J. Rodriguez, J.C. Clare, L. Empringham and A. Weinstein: Matrix converters – a technology review, *IEEE Transactions on Industrial Electronics*, Vol. 49, No. 2, 2002, pp. 276–288.
5. P. Wheeler, J. Clare, L. Empringham, M. Apap and M. Bland: Matrix converters, *Power Engineering Journal*, 2002, Vol. 16, No. 6, pp. 273–282.
6. S. Sunter and J.C. Clare: Development of a matrix converter induction motor drive, *IEEE MELECON*, Antalya, 1994, pp. 833–836.
7. S. Sunter and J.C. Clare: A true four quadrant matrix converter induction motor drive with servo performance, *IEEE PESC*, Italy, 1996, pp. 146–151.
8. S. Sunter and J.C. Clare: Feedforward indirect vector control of a matrix converter-fed induction motor drive, *The International Journal for Computation and Mathematics in Electrical and Electronic Engineering*, Vol. 19, No. 4, 2000, pp. 974–986.
9. K.K. Mohapatra, P. Jose, A. Drolia, G. Aggarwal, S. Thuta and N. Mohan: A novel carrier-based PWM scheme for matrix converters that is easy to implement, *IEEE-PESC*, Recife, Brazil, 2005, pp. 2410–2414.
10. K.K. Mohapatra, T. Satish and N. Mohan: A speed-sensorless direct torque control scheme for matrix converter driven induction motor, *IEEE-IECON*, Paris, France, 2006, pp. 1435–1440.
11. T. Sathish, K.K. Mohapatra and N. Mohan: Steady state over-modulation of matrix converter using simplified carrier based control, *IEEE-IECON*, Taipei, Taiwan, 2007, pp. 1817–1822.
12. S. Thuta, K.K. Mohapatra and N. Mohan: Matrix converter over-modulation using carrier-based control: maximizing the voltage transfer ratio, *IEEE-PESC*, Rhodes, Greece, 2008, pp. 1727–1733.
13. A. Hirota and M. Nakaoka: A low noise three phase matrix converter introducing delta-sigma modulation scheme, *37th IEEE – PESC*, Jeju, South Korea, 2006, pp. 1–6.
14. A. Hirota, S. Nagai, B. Saha and M. Nakaoka: Fundamental study of a simple control AC-AC converter introducing delta-sigma modulation approach, *IEEE-ICIT*, Chengdu, China, 2008, pp. 1–5.

15. A. Hirota, S. Nagai and M. Nakaoka: Suppressing noise peak single phase to three phase AC-AC direct converter introducing delta-sigma modulation technique, *IEEE-PESC*, Rhodes, Greece, 2008, pp. 3320–3323.

16. N.P.R. Iyer: Modelling, simulation and real time implementation of a three phase AC to AC matrix converter, Ph.D. thesis, Ch. 11 and Ch. 14, Department of ECE, Curtin University, Perth, WA, Australia, February 2012.

17. R. Schreier and G.C. Temes: *Understanding Delta-Sigma Data Converters*, IEEE Press, NJ, USA, 2005, pp. 21–38.

18. The Mathworks Inc.: www.mathworks.com, MATLAB/SIMULINK user manual, R2017b, 2017.

8

Single-Phase AC to Three-Phase AC Matrix Converter

8.1 Introduction

For single-phase AC to three-phase AC power conversion, three-phase indirect matrix converter (IMC) method is used where the single-phase AC is first rectified to DC and this DC-link voltage is inverted to AC using three-phase inverter. A control and designing method of the capacitance of the compensation capacitor for a direct single-phase AC to three-phase AC matrix converters (MCs) with application to a variable speed induction motor (IM) is reported in the literature [1]. The amplitude of the compensation capacitor voltage is controlled to absorb the single-phase power fluctuation along with the load power. A method is proposed to decide the input side parameters such as the capacitance of the compensation capacitor, considering the input voltage and the power of the IM [1]. A circuit configuration with a compensation capacitor in the input side is introduced which absorbs the fluctuating power due to the single-phase AC power. Here the single-phase AC source is connected to an LC filter and compensating capacitor [1]. The two nodes of the LC filter and the compensating capacitor end terminals are connected to nine bidirectional switches [1]. Indirect virtual rectifier-inverter analysis is used. Derivations for modulation ratio and output power flow are presented [1]. A method of designing and selecting the compensating capacitor is given in terms of the output phase voltage, output power flow and input frequency [1]. In this chapter, a detailed analysis of the single-phase AC to three-phase AC MC is presented. A model of this single-phase AC to three-phase AC MC is developed in SIMULINK [2]. Simulation results of the single-phase AC to three-phase AC MC are presented for both R–L load and three-phase IM load.

8.2 Analysis of Single-Phase AC to Three-Phase AC Matrix Converter

The single-phase AC to three-phase AC MC is shown in Figure 8.1a. Here the junction of the input filter inductor and capacitor is connected to input phase A, the junction of the input filter capacitor and compensation capacitor is connected to input phase B and the end of the compensation capacitor is connected to input phase C of the 3 × 3 MC consisting of nine bidirectional switches [1]. The equivalent circuit of Figure 8.1a is shown in Figure 8.1b which is a virtual indirect three-phase rectifier-inverter circuit [1]. The relationship between switches of Figure 8.1a,b is given by equation 5.1 in Chapter 5. The following analysis for the design of compensation capacitor C_C is based on Ref. [1].

8.2.1 Control of Virtual Rectifier

The rectifier stage in Figure 8.1b is a current source converter (CSC). Referring to Figure 8.1b, the input voltage is expressed as follows [1]:

$$v_S = \sqrt{2}V_S * \sin(\omega_i t)$$

(8.1)

$$V_S = \text{RMS value of input voltage}$$

Assuming unity input power factor, the reference input current i_S^* is given below:

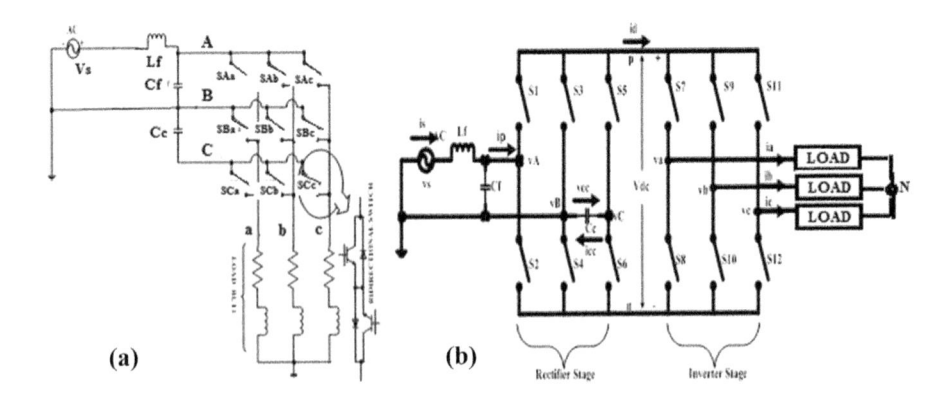

(a)   **(b)**

FIGURE 8.1
(a) Single-phase AC to three-phase AC matrix converter and (b) virtual indirect rectifier-inverter circuit.

$$i_S^* = \sqrt{2}I_S * \sin(\omega_i t)$$

$$I_S = \text{RMS value of input current} \tag{8.2}$$

Modulation signal m_S for switches S1 and S2 in Figure 8.1b is given below:

$$m_S = M_S * \sin(\omega_i t)$$

$$M_S = \text{Modulation ratio of modulation signal } m_S \tag{8.3}$$

The fundamental component of the current i_{pf} of current i_p in Figure 8.1b is given below:

$$i_{pf} = M_S * i_d * \sin(\omega_i t)$$

$$i_d = \text{DC link current} \tag{8.4}$$

When the current $i_{pf} = i_S^*$, we have the following equation by equating equations 8.2 and 8.4:

$$I_S = \frac{M_S * i_d}{\sqrt{2}} \tag{8.5}$$

Input power p_i is obtained by multiplying equations 8.1 and 8.2 which is given below [1]:

$$p_i = 2V_S * I_S * \sin^2(\omega_i t)$$

$$= V_S * I_S - V_S * I_S * \cos(2\omega_i t) \tag{8.6}$$

Compensating capacitor current reference i_{CC}^* is given below [1]:

$$i_{CC}^* = \sqrt{2}I_{CC} * \sin\left(\omega_i t + \frac{\pi}{4}\right)$$

$$I_{CC} = \text{RMS value of compensating capacitor current} \tag{8.7}$$

A modulation signal m_C for switches S5 and S6 in Figure 8.1b to control i_{CC} is given below [1]:

$$m_C = M_C * \sin\left(\omega_i t + \frac{\pi}{4}\right)$$

$$M_C = \text{Modulation ratio of modulation signal } m_C \tag{8.8}$$

Fundamental frequency component of compensation capacitor current i_{CCf} is given below [1]:

$$i_{CCf} = M_C * i_d * \sin\left(\omega_i t + \frac{\pi}{4}\right)$$

(8.9)

$$i_d = \text{DC link current}$$

Compensation capacitor voltage v_{CC} is given below:

$$v_{CC} = \frac{1}{C_C} * \left(\int i_{CCf} * dt\right) = -\frac{M_C * i_d}{\omega_i * C_C} * \cos\left(\omega_i t + \frac{\pi}{4}\right)$$

(8.10)

$$i_d = \text{DC link current}$$

Power input to compensating capacitor is obtained by multiplying equations 8.9 with 8.10 which is given below [1]:

$$p_C = -\frac{M_C^2 * i_d^2}{\omega_i * C_C} * \sin\left(\omega_i t + \frac{\pi}{4}\right) * \cos\left(\omega_i t + \frac{\pi}{4}\right)$$

$$= -\frac{M_C^2 * i_d^2}{2\omega_i * C_C} * \cos(2\omega_i t)$$

(8.11)

When the fluctuating power is compensated, then the second term on the RHS of equation 8.6 is equal to 8.11. Thus, equating the second term of equation 8.6 with equation 8.11 and using equation 8.5 for the value of I_S, we have the following equation for M_C [1]:

$$V_S * I_S * \cos(2\omega_i t) = \frac{M_C^2 * i_d^2}{2\omega_i * C_C} * \cos(2\omega_i t)$$

$$V_S * I_S = \frac{M_C^2 * i_d^2}{2\omega_i * C_C}$$

$$\frac{V_S * M_S * i_d}{\sqrt{2}} = \frac{M_C^2 * i_d^2}{2\omega_i * C_C}$$

(8.12)

$$M_C = \sqrt{\frac{\sqrt{2}\omega_i * C_C * M_S * V_S}{i_d}}$$

8.2.2 Control of Virtual Inverter

Referring to Figure 8.1b, when any one of switches S1, S3 and S5 of the virtual rectifier is ON, the p rail of DC link is $+V_{T\,max}$ and when any one of switches S2, S4 and S6 is ON, the n rail of DC link is $-V_{T\,max}$ where V_T is the RMS value of the output phase voltage of the virtual rectifier. Assume that the neutral N of the load connected to the virtual inverter is grounded. Then assuming that the modulation ratio of the switches of the virtual inverter is M_T, with reference the neutral point N of the load connected to the virtual inverter, output phase voltage can be expressed as follows [1]:

$$\left[+V_{T\,max} - \left(-V_{T\,max}\right)\right] = V_{dc} * M_T$$

$$V_{T\,rms} = V_T = \frac{V_{dc} * M_T}{2\sqrt{2}}$$

(8.13)

8.2.3 Calculation of Modulation Ratio

Balancing the input power and output power is required for the proper operation of the single-phase AC to three-phase AC MC. The virtual DC-link voltage is obtained from equation 8.13, as given below [1]:

$$V_{dc} = \frac{2\sqrt{2} * V_T}{M_T}$$

(8.14)

The sum of the constant and fluctuating component of the virtual DC-link voltage V_{dc} is given by multiplying equation 8.1 with equation 8.3, as given below [1]:

$$V_{dcf} = \sqrt{2}V_S * M_S * \sin^2\left(\omega_i t\right)$$

$$= \left(\frac{V_S * M_S}{\sqrt{2}}\right) - \left(\frac{V_S * M_S * \cos\left(2\omega_i t\right)}{\sqrt{2}}\right)$$

(8.15)

When the fluctuating component of DC-link voltage V_{dcf} is compensated fully, the second term on the RHS of equation 8.15 is zero and under this circumstance, equating the first term of equations 8.15 with 8.14, the modulation ratio M_S of switches S1 and S2 can be expressed as follows [1]:

$$\frac{V_S * M_S}{\sqrt{2}} = \frac{2\sqrt{2} * V_T}{M_T}$$

$$M_S = \frac{4V_T}{V_S * M_T}$$

(8.16)

If the single-phase fluctuating input power is compensated, then the output power P_O is a constant and is equal to the virtual DC-link voltage V_{dc} multiplied by the DC-link current i_d of the rectifier. This is given below [1]:

$$P_O = V_{dc} * i_d = \frac{2\sqrt{2} * V_T}{M_T} * i_d$$

$$i_d = \frac{P_O * M_T}{2\sqrt{2} * V_T}$$

(8.17)

Using equations 8.16 and 8.17 in equation 8.12, the modulation ratio M_C can be expressed as follows [1]:

$$M_C = \sqrt{\frac{\sqrt{2}\omega_i * C_C * 4V_T * V_S * 2\sqrt{2} * V_T}{P_O * M_T * V_S * M_T}}$$

$$= \sqrt{\frac{16\omega_i * C_C * V_T^2}{P_O * M_T^2}}$$

(8.18)

The modulation ratio M_C of switches S5 and S6 is a function of output power P_O and the virtual inverter output phase voltage V_T. The DC-link voltage V_{dc} is proportional to the ratio of V_T to M_T. The RMS compensation capacitor voltage V_{CC} is dependent on output power P_O. By designating RMS output phase voltage of the virtual inverter as V_T^*, the three-phase reference output voltage of the inverter for an output frequency of ω_o can be expressed as follows [1]:

$$v_a^* = \sqrt{2}V_T^* * \sin(\omega_o t)$$

$$v_b^* = \sqrt{2}V_T^* * \sin\left(\omega_o t - \frac{2\pi}{3}\right)$$

$$v_c^* = \sqrt{2}V_T^* * \sin\left(\omega_o t - \frac{4\pi}{3}\right)$$

(8.19)

The output power P_O is calculated as follows:

$$P_O = v_a^* i_a + v_b^* i_b + v_c^* i_c$$

(8.20)

The modulation function M_S is calculated by substituting V_T^* to V_T in equation 8.16 and M_C is calculated by using equation 8.18 and substituting P_O from equation 8.20. All modulation ratios M_S, M_C and M_T lie in the range of 0–1 [1].

8.3 Design of Compensation Capacitor

The capacitance of the compensating capacitor C_C is an important factor for the proper operation of single-phase AC to three-phase AC MC. The compensating power in the compensation capacitor is dependent on output power and the output voltage of MC is restricted by the compensation capacitor voltage. The compensation capacitor C_C decides the characteristics of the output power versus output phase voltage in a single-phase AC to three-phase AC MC. The compensation capacitor C_C is designed based on the relation between output power P_O and the output phase voltage V_T of the single-phase AC to three-phase AC MC with IM load to drive inertial load at constant acceleration. The maximum value of input phase voltage for the IM $V_{T\max}$ is decided by the maximum values of modulation ratios M_C and M_T given by equation 8.18. Output power P_O is determined by motor torque T_{em} and rotor angular speed ω_{re}. Using IM parameters and the IM RMS input phase voltage reference V_T^*, the IM torque T_{em} can be calculated. Using this T_{em} and rotor inertia J and damping constant B, the theoretical rotor speed ω_{ret} can be calculated.

Using RMS input voltage reference to the IM, the compensation capacitor C_C can be calculated using equation 8.18, where M_C and M_T are the maximum value unity. Thus, C_C and V_T^* and $V_{T\max}$ are related as follows [1]:

$$V_T^* \leq V_{T\max} = \frac{1}{4}\sqrt{\frac{P_O}{\omega_i * C_C}} \tag{8.21}$$

$$C_C \leq \frac{P_O}{16 * V_T^{*2} * \omega_i} \tag{8.22}$$

Using equations 8.10, 8.17 and 8.21 and for the maximum values of M_C and M_T, the RMS value of the compensation capacitor voltage V_{CC} can be expressed as follows [1]:

$$V_{CC} = \sqrt{\frac{P_O}{\omega_i * C_C}} \tag{8.23}$$

From equations 8.23 and 8.21, it is clear that the RMS value of the compensation capacitor voltage V_{CC} is four times greater than the maximum output phase voltage $V_{T\max}$.

8.4 Model Development

The model of the single-phase AC to three-phase AC MC-fed IM drive is developed in SIMULINK [2,3]. The parameter of the three-phase squirrel cage IM rated 15 kVA, 400 V, 50 Hz, 4 poles and 1,460 rpm is shown in Table 8.1 [2].

To design the compensating capacitor value, the three-phase IM is first driven on no load at rated input voltage. The model of the three-phase IM and its no-load characteristics are shown in Figures 8.2 and 8.3a,b, respectively [3]. From Figure 8.3a, it is seen that the time taken by the three-phase IM to reach a no-load speed of 152.9 mech. rad/s corresponding to 1,460 rpm is 45 ms. Corresponding to this time of 45 ms, the electromagnetic torque of the rotor from Figure 8.3b is 164.8 N m and the calculated power output P_O is 2,571.2 W. For this time of 45 ms, it is seen from Figure 8.4 that the RMS value of stator voltage per phase rises to 326.6 V. Thus, a V_T^* of 326.6 V is assumed. The stator applied voltage is calculated from its d-q component using an Fcn block with the formula $V_s = \sqrt{V_d^2 + V_q^2}$. Using equation 8.22, the value of the

TABLE 8.1

Three-Phase Cage IM Model Parameters

Sl. No.	Parameter	Value	Units
1	Stator resistance R_s	0.2147	Ω
2	Rotor resistance $R_{r'}$	0.2205	Ω
3	Stator leakage inductance L_{ls}	0.000991	H
4	Rotor leakage inductance $L_{lr'}$	0.000991	H
5	Mutual inductance L_{lm}	0.06419	H
6	Rotor inertia	0.102	kg m^2
7	Damping constant	0.009541	N m s

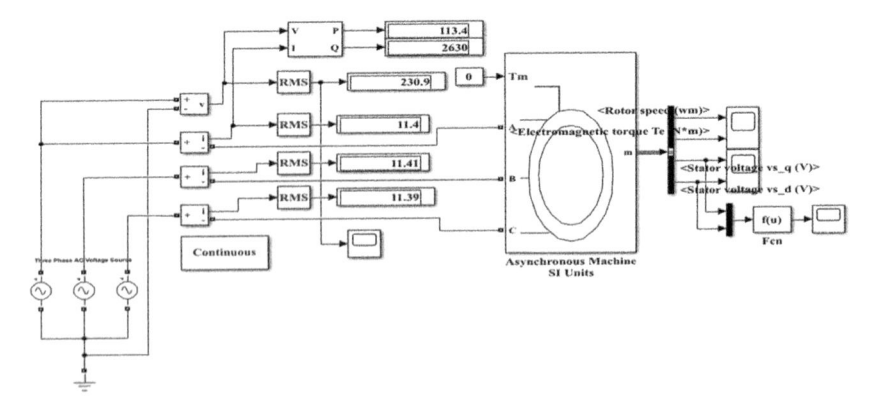

FIGURE 8.2
No-load test set-up for a three-phase cage induction motor.

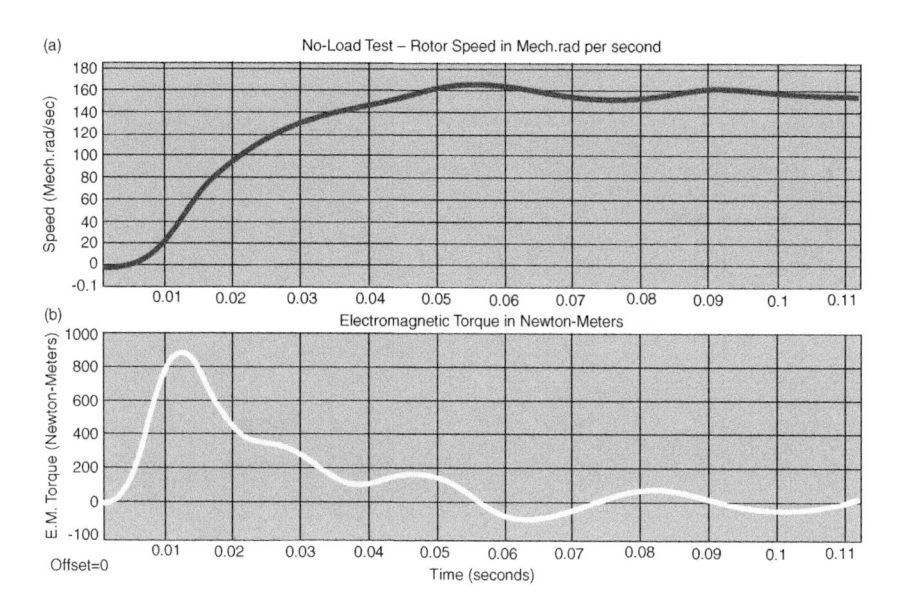

FIGURE 8.3
No-load characteristics of three-phase cage IM: (a) rotor speed in mechanical radians per second and (b) electromagnetic torque in Newton-metres.

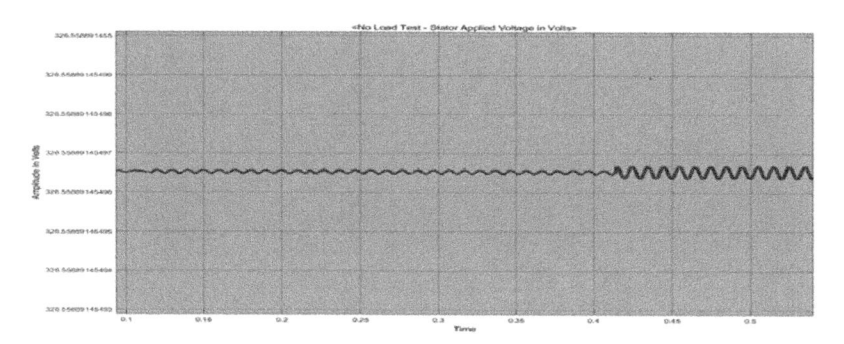

FIGURE 8.4
No-load test on three-phase cage IM – sector applied voltage in volts.

Compensating capacitor C_C is found to be 4.7955E–6 F. The input filter inductor L_f and filter capacitor C_f are taken to be 1E–3 H and 8E–6 F, respectively.

8.4.1 Model of Single-Phase AC to Three-Phase AC Matrix-Converter-Fed Induction Motor Drive

The model of the single-phase AC to three-phase AC MC driving the IM is shown in Figure 8.5 [3]. Here the gate pulses for the nine bidirectional switches are generated using Venturini modulation algorithm assuming unity input phase displacement factor given by equation 2.17 in Section 2.3 of Chapter 2.

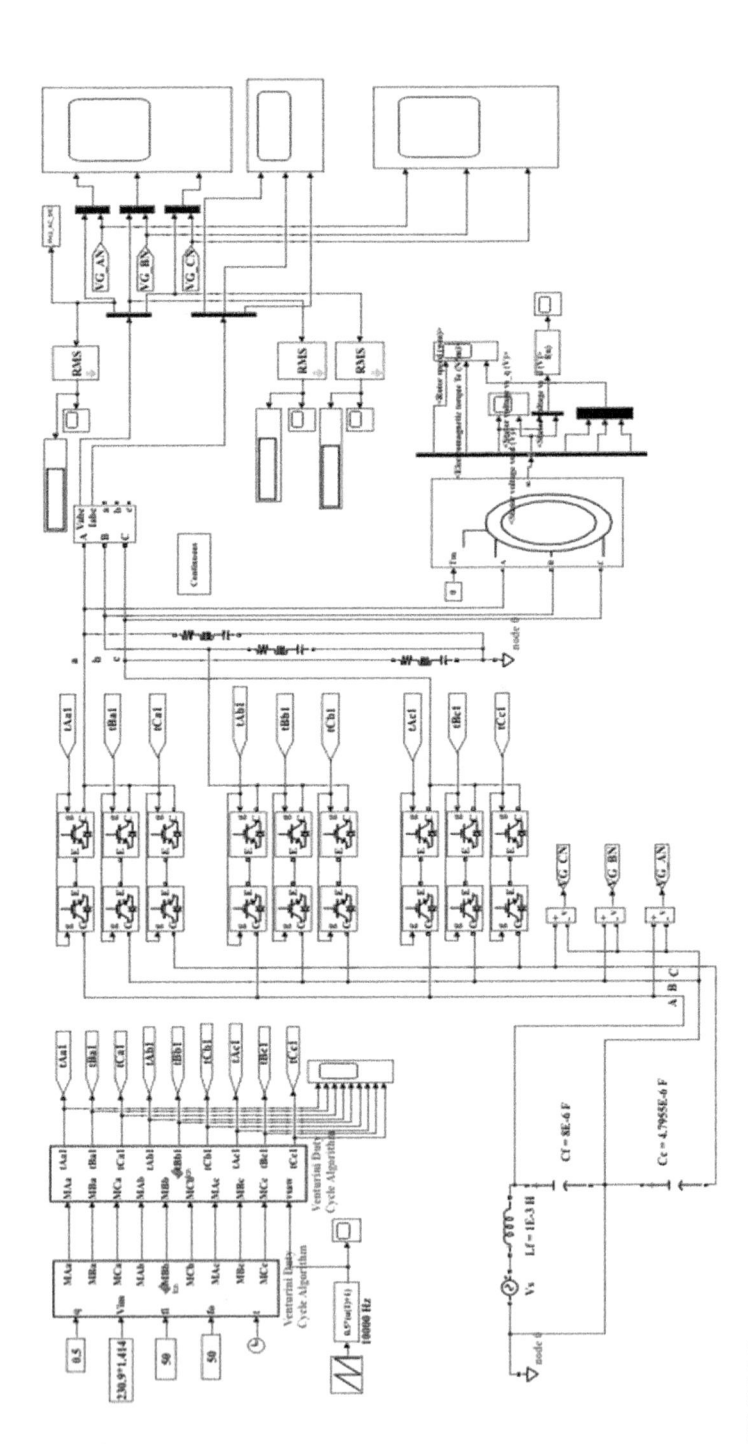

FIGURE 8.5
Model of single-phase AC to three-phase AC matrix-converter-fed IM drive.

In this model, a saw-tooth carrier frequency of 10 kHz is used. The single-phase AC input voltage source has an RMS value of 230.9 V and frequency 50 Hz. Two Embedded MATLAB Function blocks are used to generate the gate pulses. The method of generating these gate pulse and the source code used are already explained in Section 2.4.1 of Chapter 2. The gate pulses drive the nine bidirectional switches of the MC. The voltage source is a single-phase AC source with peak value $\sqrt{2} * 326.6$ V and frequency 50 Hz R_f and L_f forms the input filter and C_C forms the compensation capacitor.

8.4.2 Simulation Results

The simulation of the single-phase AC to three-phase AC MC driving the IM is carried out in SIMULINK [2]. The ode23tb (stiff/TR-BDF2) solver is used. The simulation parameters are given in Table 8.2. An output R–L–C filter to resonate at 10 kHz carrier switching frequency is used. The simulation results for the three-phase 50 Hz line-to-line output voltage and line-to-ground input voltage are shown in Figure 8.6. The stator current, rotor speed and electromagnetic

TABLE 8.2

Model Parameters

Sl. No.	Parameter	Value	Units
1	Single phase RMS line to ground input voltage V_s	231	V
2	Input frequency f_i	50	Hz
3	Output frequency f_o	50	Hz
4	Saw-tooth carrier frequency f_{saw}	10	kHz
5	Modulation index	0.5	-
6	Compensation capacitor	5.017e–6	F
7	Input filter inductor	1e–3	H
8	Input filter capacitor	8e–6	F

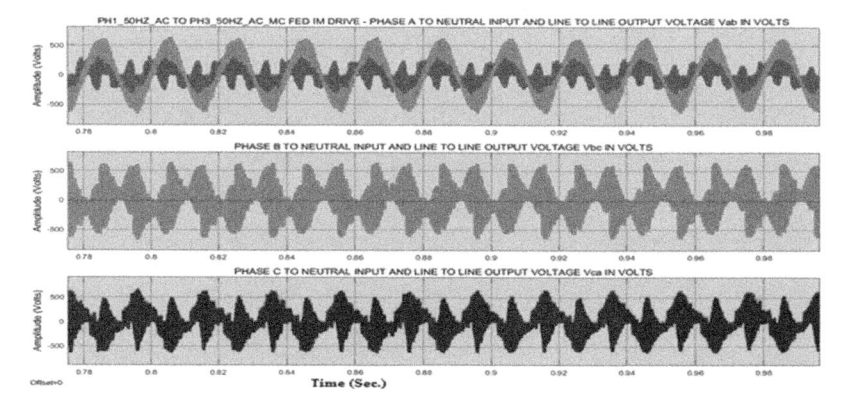

FIGURE 8.6

Three-phase AC line-to-line output voltage and line-to-neutral input voltage.

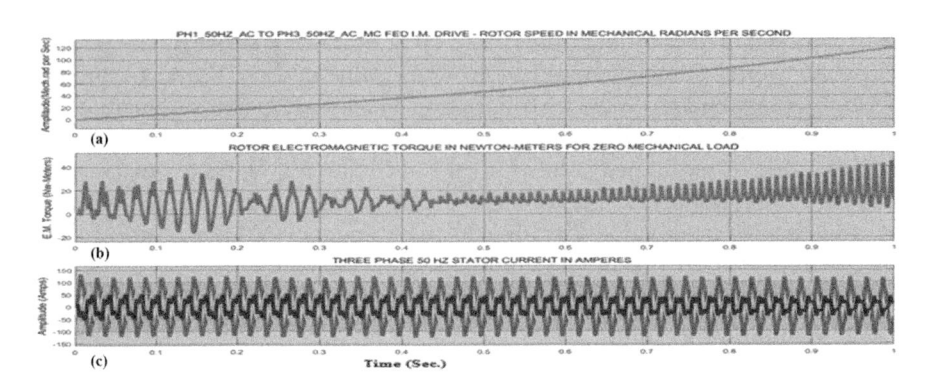

FIGURE 8.7
Single-phase 50 Hz AC to three-phase 50 Hz AC MC-fed cage IM drive – simulation results.
(a) Rotor speed in mechanical radians per second, (b) electromagnetic torque in Newton-
metres and (c) three-phase stator current in amperes.

torque of the three-phase cage IM are shown in Figure 8.7 for zero externally
applied mechanical load.

8.5 Conclusions

The model of a novel technique for single-phase AC to three-phase AC MC
is presented. Simulation results for three-phase cage IM show conversion
of single-phase AC voltage to three-phase AC voltage. However, the three-
phase output voltages and the load current or stator currents are found to
have magnitude unbalance by model simulation. The speed pick up and the
time to reach no-load rated speed for the three-phase cage IM are very slow
as compared with its performance by no-load test.

References

1. K. Iino, K. Kondo and Y. Sato: An experimental study on induction motor drive
   with a single phase-three phase matrix converter, *13th European Conference on
   Power Electronics and Applications, EPE,* Barcelona, Spain, 2009, pp. 1–9.
2. The Mathworks Inc.: www.mathworks.com, MATLAB/SIMULINK user manual, R2017b, 2017.
3. N.P.R. Iyer: Modelling, simulation and real time implementation of a three
   phase AC to AC matrix converter, Ph.D. thesis, Ch. 12 and Ch. 14, Department
   of ECE, Curtin University, Perth, WA, Australia, February 2012.

9

A Novel Single-Phase and Three-Phase AC to Single-Phase and Three-Phase AC Converter Using a DC Link

9.1 Introduction

A novel method of generating a pulse-width-modulated (PWM) single-phase and three-phase variable-frequency, variable-voltage AC from the normal 220/400 V, single-phase/three-phase 50 Hz AC supply mains using a step-down transformer and an intermediate DC link is presented here [1]. In this method, the DC-link voltage is converted to two square-wave AC voltages having 180° phase difference with frequency equal to that of the mains supply frequency which is then converted to variable-frequency variable-magnitude PWM square-wave AC voltage using semiconductor bidirectional switches.

9.2 Single-Phase AC to Single-Phase AC Converter Using a DC Link

The block diagram of the single-phase AC to single-phase AC converter is shown in Figure 9.1 [1]. The block diagram is explained below [1]:

- The supply mains is a 220 V, 50 Hz single-phase AC.
- The step-down transformer 230/6 V is connected to the supply mains.
- The secondary 6 V is connected to the operational amplifier zero crossing comparator (ZCC).
- The output of the ZCC forms the gate drive input to all the four MOSFET switches forming the bipolar dual output inverter.

- The bipolar dual output inverter consists of two legs. Each leg is a pair of N- and P-type MOSFETs connected in series. The two parallel legs of the inverter are connected to the DC source which forms the DC link. The two outputs of this inverter are $V_{dc}/\_0$ and $V_{dc}/\_{-180}$ with a frequency of 50 Hz, where $2V_{dc}$ is the DC-link voltage. These two voltage outputs are fed to the IGBT bidirectional switches.

- The IGBT bidirectional switches form a two-row, one-column arrangement. The four gates of the two rows of IGBTs are cross connected. This is different from the mode of connection used in conventional matrix converter. The output of this bidirectional switches is connected to the load.

- The clock generator output is a square pulse having a frequency of f_o Hz with 50% duty cycle. This frequency f_o Hz is variable.

- The carrier generator output V_c is a square pulse having a fixed frequency of f_c Hz with a variable duty cycle. The carrier frequency f_c Hz must be at least two times greater than the desired output frequency f_o Hz.

- The load voltage V_o is a PWM square-wave AC with a peak to peak value of $2V_{dc}$ in volts and frequency f_o Hz. By varying the duty cycle of the square pulse carrier V_c, the magnitude of the output voltage V_o can be controlled.

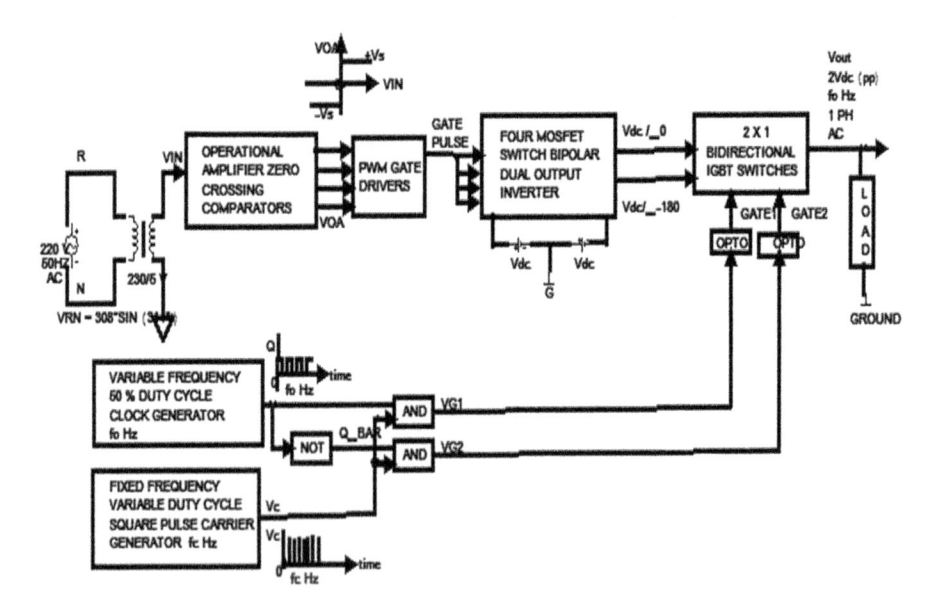

FIGURE 9.1
Block diagram of a single-phase AC to AC converter using a DC link.

9.3 Model of a PWM Single-Phase AC to Single-Phase AC Converter

The model of the novel PWM single-phase AC to single-phase AC converter is shown in Figure 9.2. The model subsystems are shown in Figure 9.3a–h. The subsystems are explained below:

The step-down transformer is shown in Figure 9.3a. This steps down the 220 V (RMS), 50 Hz single-phase AC voltage to 5 V (RMS). This transformer output is given to four operational amplifier Zero Crossing comparators (ZCC) as shown in Figure 9.3b. The output of ZCC is given to four PWM gate drives shown in Figure 9.3c. The square pulse output of these PWM gate drives forms the gate switching pulse for the dual MOSFET inverter shown in Figure 9.3d. This dual MOSFET inverter is an NMOSFET–PMOSFET pair combination connected as shown in Figure 9.3d. The two square-wave 50 Hz output of the dual MOSFET inverter with respect to ground are V_A and V_B, respectively, which differ in phase by 180°. The DC-link voltage connected to the dual inverter is V_{dc} with midpoint grounded as shown in Figure 9.2. Operational amplifier ZCC and logic gates are shown in Figure 9.3e which forms the gate drive for the IGBT bidirectional switch pair shown in Figure 9.3g. The sine wave (Figure 9.2) with amplitude 1 V and frequency f_o Hz is given to the ZCC in Figure 9.3e whose square pulse output has a frequency f_o Hz and amplitude 5 V. This ZCC and its inverted output are given to two separate two input AND gates. A variable duty-cycle square pulse generator (Figure 9.2) output with amplitude 5 V and frequency f_c Hz is given to the two AND gates for pulse width modulation, as shown in Figure 9.3e. Figure 9.3f provides isolation of PWM gate drive for IGBT bidirectional switches in Figure 9.3g. The output of this two AND gates is given to the gates of two IGBT bidirectional switches shown in Figure 9.3g via PWM gate driver in Figure 9.3f. The gates of the individual IGBTs in Figure 9.3g are cross connected. V_A and V_B from dual inverter are given to the collector and emitter of the IGBT bidirectional switch pair shown in Figure 9.3g. The load resistor and output voltage measurement unit are shown in Figure 9.3h.

9.3.1 Principle of Operation

Referring to Figure 9.3d, the working of the dual MOSFET inverter using N- and P-type MOSFETs in series is explained below [1]:

The 220 V single-phase 50 Hz sine-wave AC input voltage and the 5 V transformer secondary output voltage are shown in Figure 9.4. During the positive half cycle of the transformer secondary input voltage, the output OA1 to OA4 of the four operational amplifier ZCCs are driven to positive saturation and it is +10 V and during the negative half cycle of this input voltage, these ZCC outputs are driven to negative saturation and it is −10V.

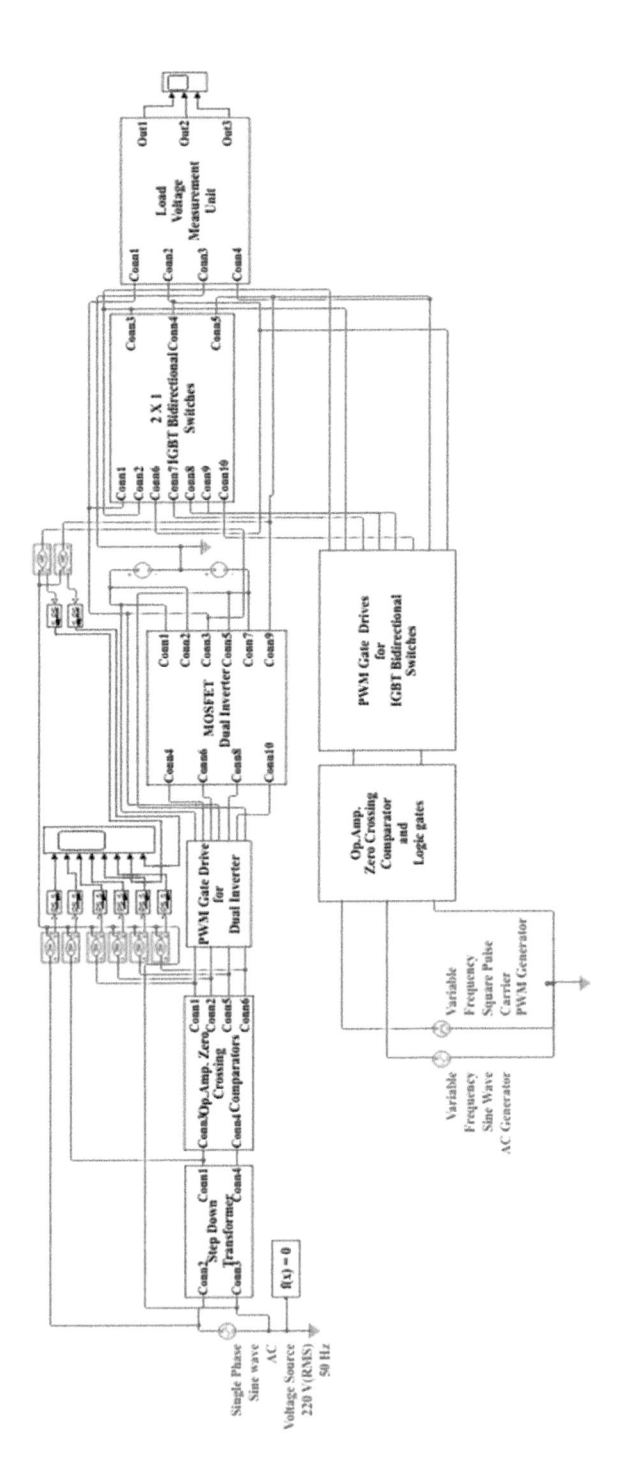

FIGURE 9.2
Model of a single-phase PWM AC to AC converter using a DC link.

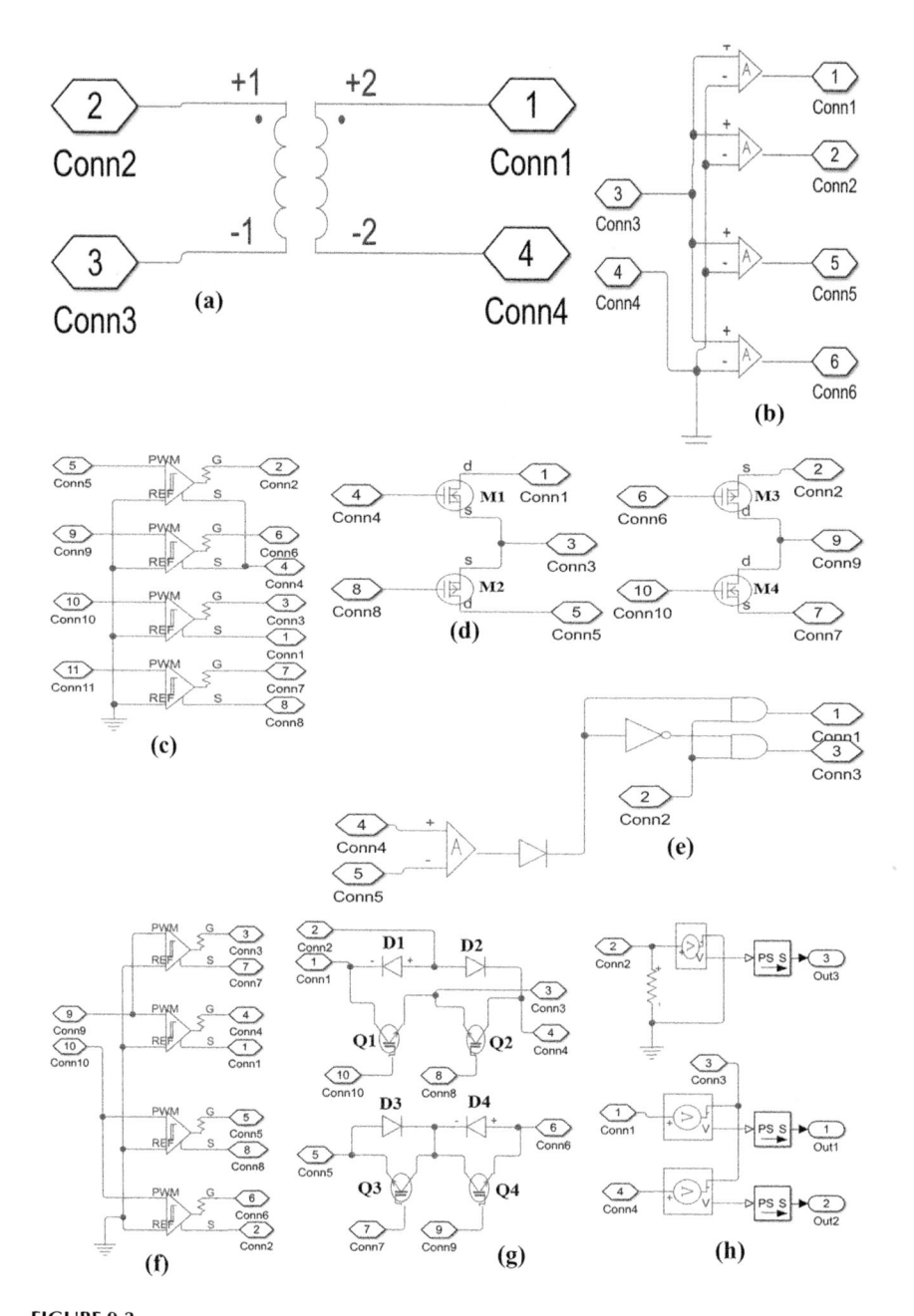

FIGURE 9.3
Single-phase AC to AC converter using a DC-link model subsystem: (a) step-down transformer, (b) operational amplifier zero crossing comparators, (c) PWM gate drives for dual inverter, (d) dual inverter, (e) operational amplifier zero crossing comparator and logic gates, (f) PWM gate drives for IGBT bidirectional switches, (g) IGBT bidirectional switches and (h) load voltage measurement unit.

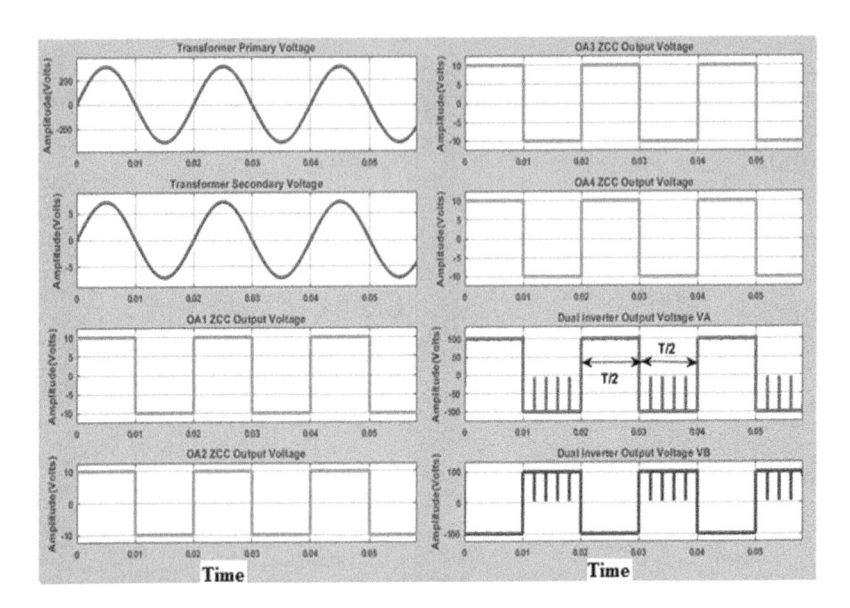

FIGURE 9.4
Transformer input and output voltage, four operational amplifier ZCC output voltage and dual MOSFET inverter output voltage V_A and V_B in volts.

This is shown in Figure 9.4. During the positive half cycle of the ZCC output OA1 to OA4, N-type MOSFETs M1 and M4 conducts, the output V_A and V_B with respect to DC-link ground reference are, respectively, $+V_{dc}/2$ (+100 V) and $-V_{dc}/2$ (–100 V), respectively, where V_{dc} is the DC-link voltage. Similarly during the negative half cycle of the ZCC output OA1 to OA4, P-type MOSFETs M2 and M3 conducts, the output VA and VB with respect to DC-link ground are, respectively, $-V_{dc}/2$ (–100 V) and $+V_{dc}/2$ (+100 V), respectively. This is shown in Figure 9.4.

Referring to Figures 9.2 and 9.3g, the working principle of the two IGBT bidirectional switch converters is given below [1].

Now consider that the period of the input voltage V_A and V_B is T and that of the clock switching the IGBT bidirectional switches is T_o with 50% duty cycle. Assume that there is no square pulse carrier PWM involved. This situation is created by replacing the square carrier pulse generator with a 5 V DC voltage source in Figure 9.2. Referring to Figures 9.2 and 9.3g, the following analysis holds good.

Case A: Let $T \gg T_o$. Consider interval $0 < t \le T_o/2$. Gate pulses for Q1 and Q4 are HIGH and those for Q2 and Q3 are LOW. Then the following argument is valid:

- Input voltage V_A is $+V_{dc}/2$ (+100 V) volts and V_B is $-V_{dc}/2$ (–100 V) volts.

- IGBT Q1 is ON and Q2, Q3 and Q4 are OFF.
- IGBT Q1 and the diode D2 across Q2 conducts.
- Output voltage V_{OUT} is $+V_{dc}/2$ (+100 V) volts.

Case B: Let $T \gg T_o$. Consider interval $T_o/2 < t \leq T_o$. Gate pulses for Q2 and Q3 are HIGH and those for Q1 and Q4 are LOW. Then the following argument is valid:

- Input voltage V_A is $+V_{dc}/2$ (+100 V) volts and V_B is $-V_{dc}/2$ (–100 V) volts.
- IGBT Q3 is ON and Q1, Q2 and Q4 are OFF.
- IGBT Q3 and the diode D4 across Q4 conducts.
- Output voltage V_{OUT} is $-V_{dc}/2$ (–100 V) volts.

Case C: Let $T \gg T_o$. Consider interval $0 < t \leq T_o/2$. Gate pulses for Q1 and Q4 are HIGH, and those for Q2 and Q3 are LOW. Then the following argument is valid:

- Input voltage V_A is $-V_{dc}/2$ (–100 V) volts and V_B is $+V_{dc}/2$ (+100 V) volts.
- IGBT Q4 is ON and Q1, Q2 and Q3 are OFF.
- IGBT Q4 and the diode D3 across Q3 conducts.
- Output voltage V_{OUT} is $+V_{dc}/2$ (+100 V) volts.

Case D: Let $T \gg T_o$. Consider interval $T_o/2 < t \leq T_o$. Gate pulses for Q2 and Q3 are HIGH and those for Q1 and Q4 are LOW. Then the following argument is valid:

- Input voltage V_A is $-V_{dc}/2$ (–100 V) volts and V_B is $+V_{dc}/2$ (+100 V) volts.
- IGBT Q2 is ON and Q1, Q3 and Q4 are OFF.
- IGBT Q2 and the diode D1 across Q1 conducts.
- Output voltage V_{OUT} is $-V_{dc}/2$ (–100 V) volts.

Period of output voltage V_{OUT} is T_o. Cases A to D are shown in Figure 9.5.

Case E: Let $T \ll T_o$. Consider interval $0 < t \leq T_o/2$. Gates pulse for Q1 and Q4 are HIGH and those for Q2 and Q3 are LOW. Then the following argument is valid:

- When V_A is $+V_{dc}/2$ (+100 V) volts and V_B is $-V_{dc}/2$ (–100 V) volts, IGBT Q1 and the diode D2 across Q2 conducts. Output V_{OUT} is $+V_{dc}/2$ (+100 V) volts.
- When V_A is $-V_{dc}/2$ (–100 V) volts and V_B is $+V_{dc}/2$ (+100 V) volts, IGBT Q4 and the diode D3 across Q3 conducts. Output V_{OUT} is $+V_{dc}/2$ (+100 V) volts.

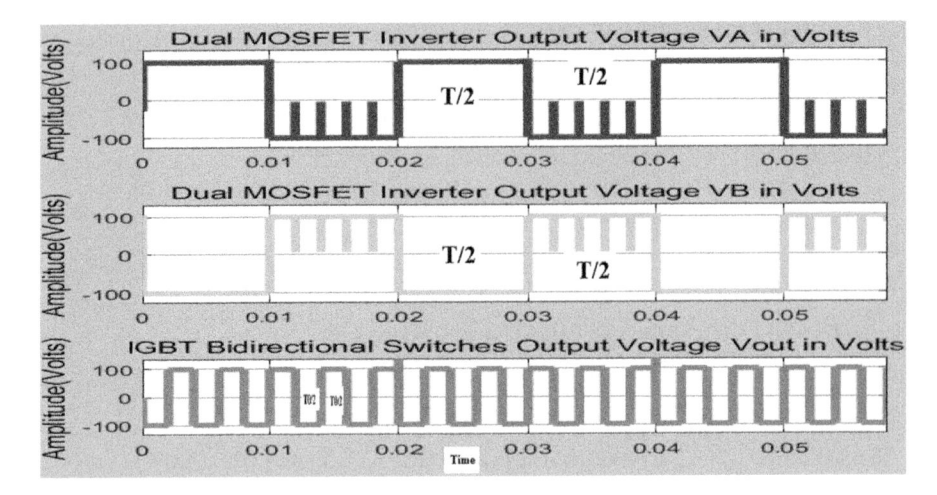

FIGURE 9.5
IGBT bidirectional switches input and output voltage waveforms.

Case F: Let $T \ll T_o$. Consider interval $T_o/2 < t \le T_o$. Gate pulses for Q1 and Q4 are LOW, and those for Q2 and Q3 are HIGH. Then the following argument is valid:

- When V_A is $+V_{dc}/2$ (+100 V) volts and V_B is $-V_{dc}/2$ (−100 V) volts, IGBT Q3 and the diode D4 across Q4 conducts. Output V_{OUT} is $-V_{dc}/2$ (−100 V) volts.
- When V_A is $-V_{dc}/2$ (−100 V) volts and V_B is $+V_{dc}/2$ (+100 V) volts, IGBT Q2 and the diode D1 across Q1 conducts. Output V_{OUT} is $-V_{dc}/2$ (−100 V) volts.

Case G: Let $T \ll T_o$. Consider interval $0 < t \le T_o/2$. Gate pulses for Q1 and Q4 are LOW and those for Q2 and Q3 are HIGH. Then the following argument is valid:

- When V_A is $+V_{dc}/2$ (+100 V) volts and V_B is $-V_{dc}/2$ (−100 V) volts, IGBT Q3 and the diode D4 across Q4 conducts. Output V_{OUT} is $-V_{dc}/2$ (−100 V) volts.
- When V_A is $-V_{dc}/2$ (−100 V) volts and V_B is $+V_{dc}/2$ (+100 V) volts, IGBT Q2 and the diode D1 across Q1 conducts. Output V_{OUT} is $-V_{dc}/2$ (−100 V) volts.

Case H: Let $T \ll T_o$. Consider interval $T_o/2 < t \le T_o$. Gate pulses for Q1 and Q4 are HIGH and those for Q2 and Q3 are LOW. Then the following argument is valid:

- When V_A is $+V_{dc}/2$ (+100 V) volts and V_B is $-V_{dc}/2$ (−100 V) volts, IGBT Q1 and the diode D2 across Q2 conducts. Output V_{OUT} is $+V_{dc}/2$ (+100 V) volts.

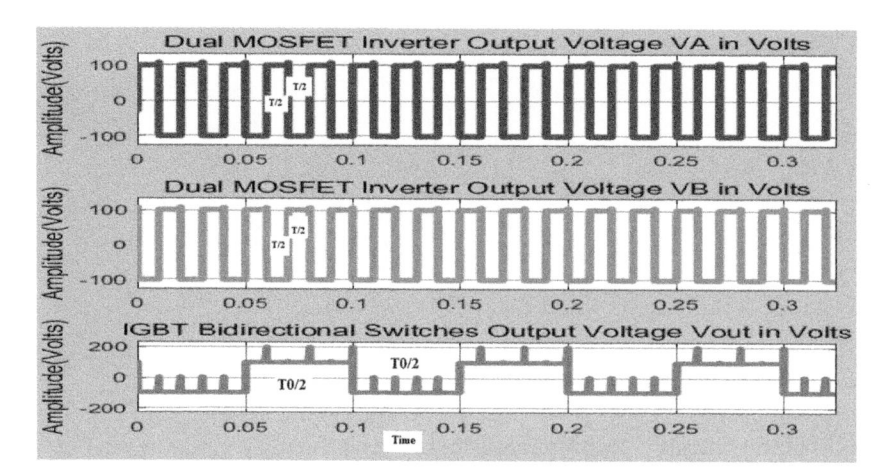

FIGURE 9.6
IGBT bidirectional switches input and output voltage waveforms.

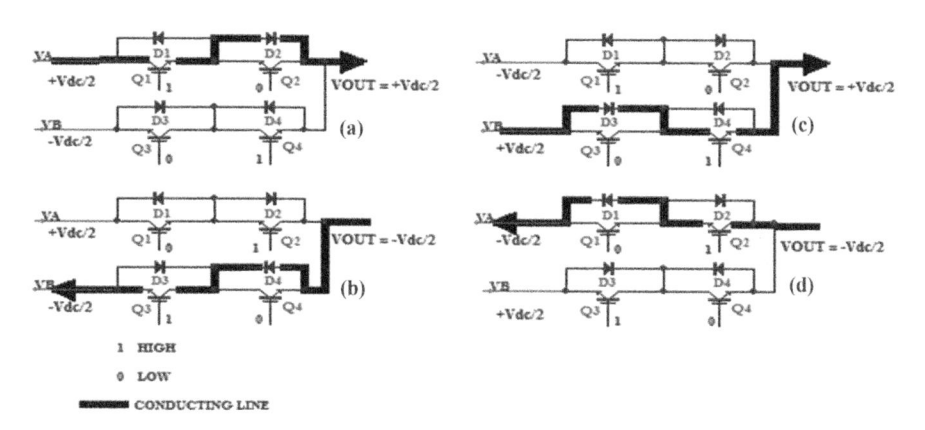

1 HIGH
0 LOW
■■■■■ CONDUCTING LINE

FIGURE 9.7
Bidirectional IGBT switches conducting modes.

- When V_A is $-V_{dc}/2$ ($-100\,$V) volts and V_B is $+V_{dc}/2$ ($+100\,$V) volts, IGBT Q4 and the diode D3 across Q3 conducts. Output V_{OUT} is $+V_{dc}/2$ ($+100\,$V) volts.

The period of the output voltage V_{OUT} is T_0. Cases E to H are shown in Figure 9.6. Figure 9.7a–d correspond to cases A to H discussed above.

9.3.2 RMS Output Voltage

A typical output voltage waveform of the above converter is shown in Figure 9.8. This is uniform pulse-width modulation. Let D_C be the duty cycle of the square pulse carrier and $T_C = 1/f_c$ be its period. Let $T_o = 1/f_o$ be the

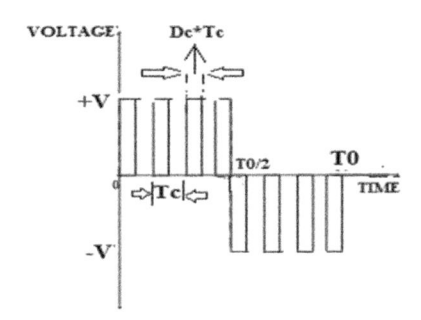

FIGURE 9.8
Square PWM output voltage waveform.

period of the output voltage waveform. Let V be the peak value of the output PWM voltage waveform. The following derivation for RMS output voltage waveform holds good [1,2].

Number of PWM square pulses for one half cycle of the PWM output voltage is given below:

$$n_p = \frac{T_o}{2T_C} \tag{9.1}$$

Let V_{rms} be the RMS value of the output voltage and T_1 be the time of starting of the first pulse measured from the origin.

$$V_{rms}^2 = \frac{2n_p}{T_o} * \left[\int_{T_1}^{T_1 + D_C T_C} V^2 dt \right] \tag{9.2}$$

$$V_{rms}^2 = \frac{2V^2 n_p D_C T_C}{T_o} \tag{9.3}$$

$$V_{rms} = V\sqrt{\frac{2n_p D_C T_C}{T_o}} \tag{9.4}$$

$$V_{rms} = V\sqrt{2n_p D_C T_C f_o} \tag{9.5}$$

Equations 9.4 and 9.5 give the RMS value of the PWM output voltage waveform.

9.3.3 Simulation Results

To study the behaviour of the proposed single-phase DC–AC–AC converter, a model of the same is developed using SIMULINK [3]. This is presented in Figure 9.2. The ode23t (mod.stiff/Trapezoidal) solver is used. Here,

TABLE 9.1

DC–AC–AC Converter Model Parameters

Sl. No.	Parameter	Value	Units
1	RMS line-to-neutral input voltage	220	V
2	Input frequency f_i	50	Hz
3	DC-link voltage	200	V
4	Output frequency f_o	10, 250	Hz
5	PWM carrier switching frequency	10	kHz
6	PWM carrier duty cycle	0.2, 0.8	-
7	Load resistance	100	kΩ

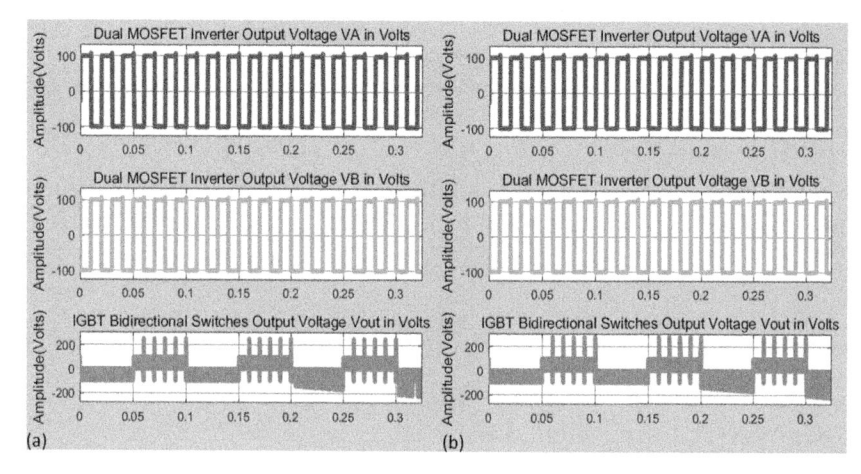

(a)   (b)

FIGURE 9.9
Simulation results for 10 Hz AC output voltage for a carrier duty cycle of 0.2 (a) and simulation results for 10 Hz AC output voltage with a carrier duty cycle of 0.8 (b).

simulation results are presented for the parameters given in Table 9.1. Figure 9.9 shows the simulation results for a 10 Hz AC output voltage with a PWM carrier duty cycle of 0.2 and 0.8, respectively. Figure 9.10 shows the simulation for a 250 Hz AC output voltage with a PWM carrier duty cycle of 0.2 and 0.8, respectively.

9.4 Discussion of Results

From the simulation results, it is seen that a variable-frequency PWM variable-voltage output is obtained at the load terminal from a constant 50 Hz frequency AC input voltage. The variable frequency at the load terminal is obtained by selecting the appropriate clock frequency to drive the

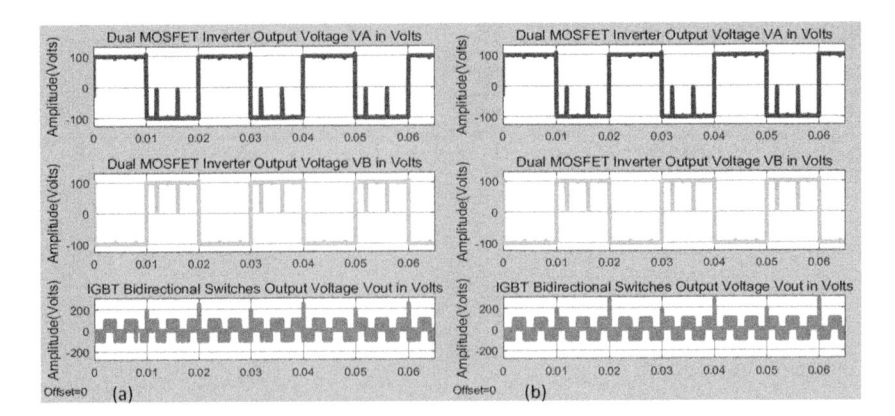

FIGURE 9.10
Simulation results for 250 Hz AC output voltage with duty cycle 0.2 (a) and simulation results for 250 Hz AC output voltage with duty cycle 0.8 (b).

IGBT bidirectional switches. The variable-voltage magnitude at the load terminal is obtained by varying the duty cycle of the square pulse carrier.

9.5 Three-Phase AC to Three-Phase AC Converter Using a DC Link

The block diagram of the three-phase fixed frequency AC to three-phase variable-frequency AC converter using a DC link is shown in Figure 9.11 [1]. The block diagram is explained below [1]:

- The supply mains is a 220 V (line to neutral), 50 Hz three-phase AC.
- The step-down transformer 230/6 V is connected to each phase of the supply mains.
- The secondary 6 V of each phase is connected to operational amplifier ZCC.
- The output of the ZCC forms the gate drive input to all the four MOSFET switches forming the bipolar dual output inverter in each phase.
- The bipolar dual output inverter in each phase consists of two legs. Each leg is a pair of N- and P-type MOSFETs connected in series. The two parallel legs of the inverter are connected to the DC source which forms the DC link. The two outputs of this inverter in each phase are $V_A = V/\_0$, $V_A = V/\_{-}180$, $V_B = V/\_{-}120$, $V_B = V/\_{-}300$ and $V_C = V/\_{-}240$, $V_C = V/\_{-}420$ with a frequency of 50 Hz, where 2 V is the DC-link

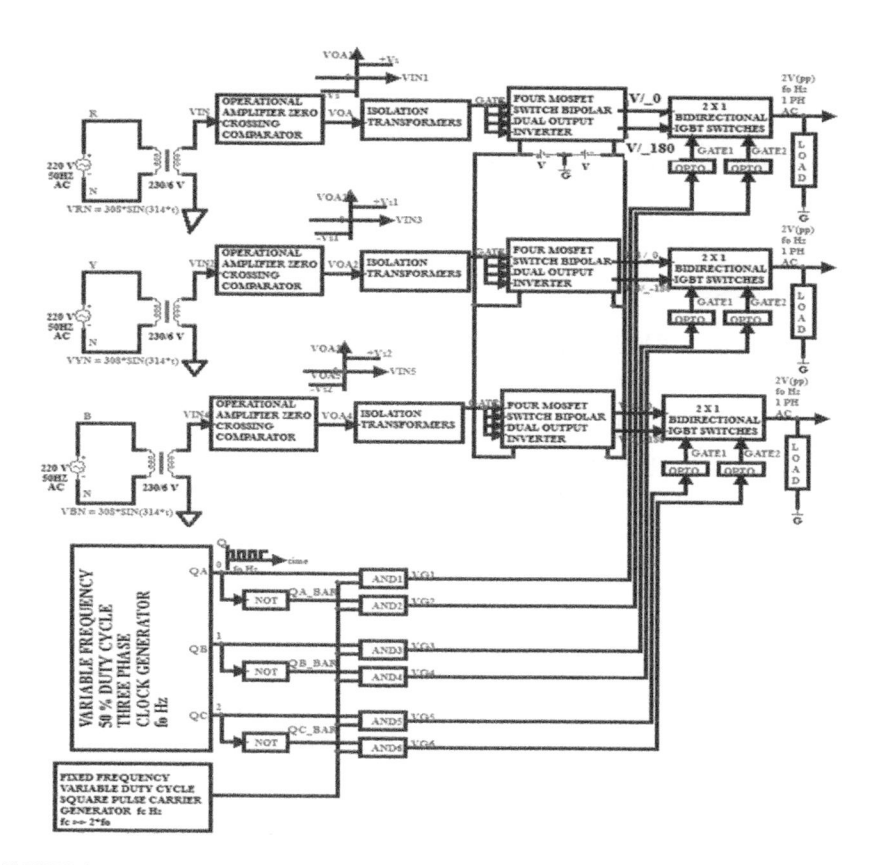

FIGURE 9.11
Block diagram of a PWM three-phase AC to three-phase AC converter.

voltage. These two outputs of inverter in each phase are fed to the IGBT bidirectional switches corresponding to each output phase.

- The IGBT bidirectional switches in each output phase form a two-row, one-column arrangement. The four gates of the two rows of IGBTs are cross connected. This is different from the mode of connection used in conventional matrix converter. The output of this bidirectional switches is connected to the three-phase load.

- The three-phase clock generator output is a square pulse for each phase having a frequency of f_o Hz *with 50% duty cycle*. The clocks QA, QB and QC corresponding to each output phase are separated from each other by a phase difference of $T_o/3$ s, where $T_o = 1/f_o$. This frequency f_o Hz is variable.

- The carrier generator output is a square pulse having a pre-calculated fixed frequency of f_c Hz with a variable duty cycle. The carrier frequency f_c Hz must be at least two times greater than the desired output frequency of f_o Hz.

- The load voltage V_o is a PWM square-wave AC with a peak to peak value of 2 V and frequency f_o Hz.
- By varying the duty cycle of the square pulse carrier generator, the magnitude of the output voltage V_o can be controlled.

9.6 Model of a PWM Three-Phase AC to Three-Phase AC Converter

The SIMULINK model schematic of the novel PWM DC to three-phase AC to three-phase AC converter is shown in Figure 9.12. The model subsystems are the same as shown in Figure 9.3a–h except that three single-phase step-down transformers for each input phase along with three sets of four operational amplifier ZCC and PWM gate drivers are used for each input phase. A total of six NMOSFET–PMOSFET pairs are used to generate fixed input frequency three-phase AC voltage. To generate variable-frequency three-phase AC output voltage, three low-voltage sine-wave operational amplifier ZCCs are used along with logic gates and a square pulse carrier. Three sets of two IGBT bidirectional switch pairs are used to generate variable-frequency three-phase AC output voltage. Three balanced load is connected to each output phase. The subsystems for the three-phase dual NMOSFET–PMOSFET inverter, low-voltage operational amplifier ZCC for three-phase gate drive for IGBT bidirectional switches and the IGBT bidirectional switches for each output phase are shown in Figure 9.13a–c. Three-phase MOSFET dual inverter shown in Figure 9.13a generates three-phase dual square-wave voltages at the input frequency with phase angle 0°, –180°, –120°, –300°, –240°, –420° (–60°), respectively, with peak value $+V_{dc}/2$ and $-V_{dc}/2$ volts where V_{dc} is the DC-link voltage of dual inverter. Operational amplifier ZCCs in each of the three-phase shown in Figure 9.13b are driven by a three-phase sine-wave voltage source with peak value 1 V and frequency f_o Hz which is the desired output frequency and forms the gate drive for IGBT bidirectional switches in each output phase. Three-phase IGBT bidirectional switch pairs generate three-phase AC output voltage with frequency f_o Hz. The pair of IGBTs in each output phase has their gates cross connected. The principle of operation three-phase AC to AC converter using a DC link is the same as explained in Section 9.3.1.

9.6.1 Simulation Results

To study the behaviour of the proposed three-phase DC–AC–AC converter, a model of the same is developed using SIMULINK [3]. This is presented in Figure 9.12. Parameters are given in Table 9.1. The ode15s (stiff/NDF) solver is

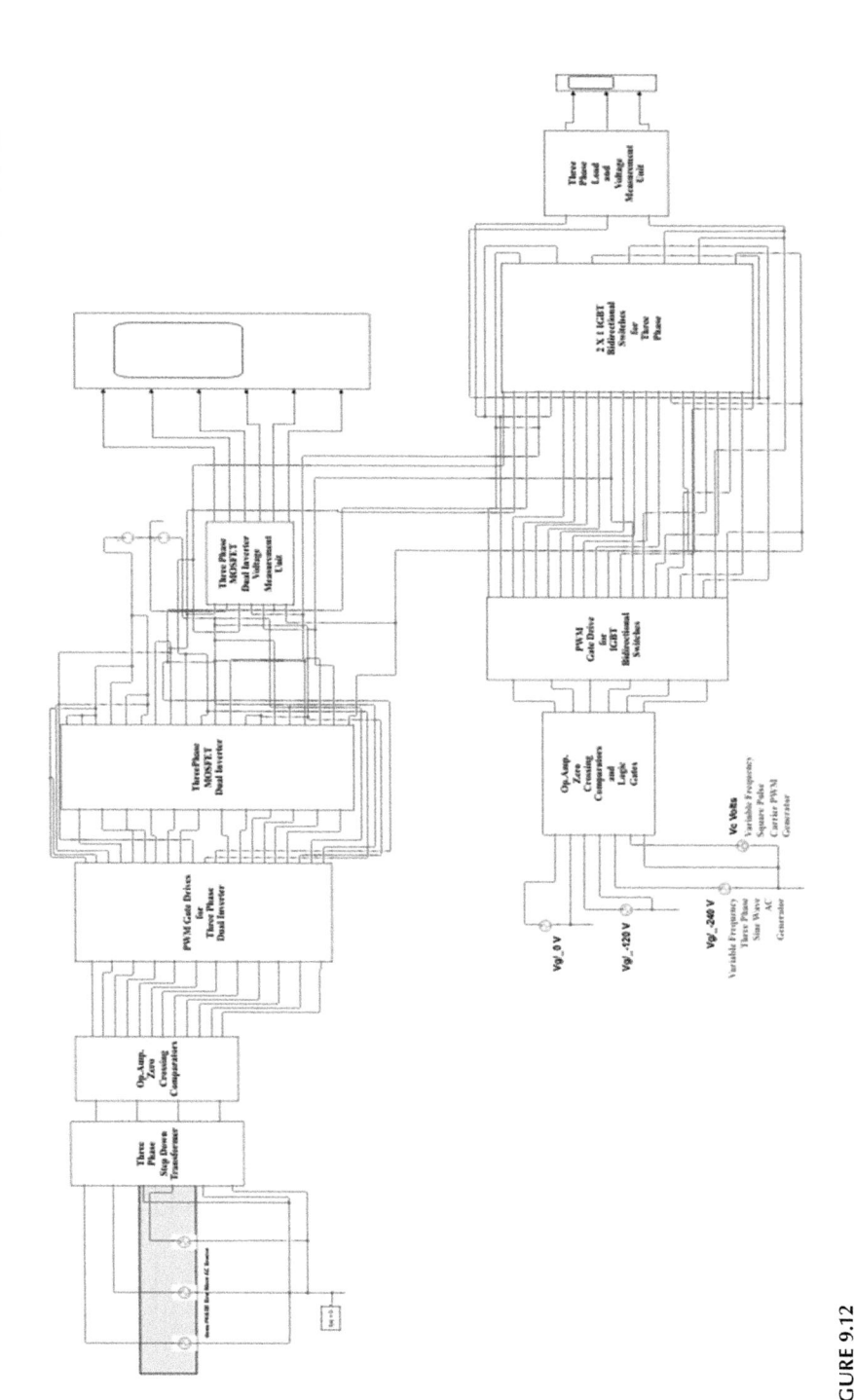

FIGURE 9.12

Model of a PWM three-phase AC to three-phase AC converted using a DC link.

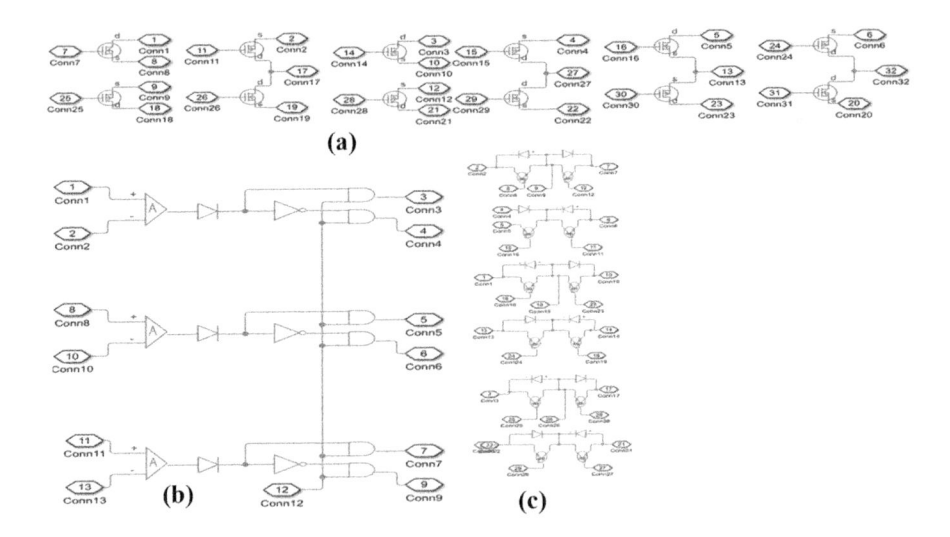

FIGURE 9.13
Three-phase PWM AC to AC converter using a DC-link model subsystem: (a) three-phase MOSFET dual inverter, (b) operational amplifier ZCC and logic gates drive for three-phase IGBT bidirectional switches and (c) three-phase IGBT bidirectional switch pair.

used. The three-phase 50 Hz dual MOSFET inverter output voltage is shown in Figure 9.14. The three-phase PWM 10 Hz output voltage for a carrier duty cycle of 0.2 and 0.8 is shown in Figure 9.15a,b, respectively, and that for a 250 Hz AC output voltage for a carrier duty cycle of 0.2 and 0.8 is shown in Figure 9.16a,b.

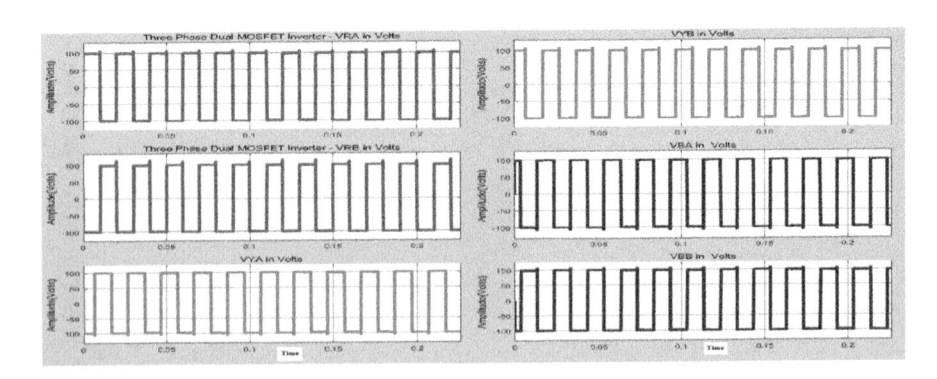

FIGURE 9.14
Three-phase dual MOSFET inverter output voltage.

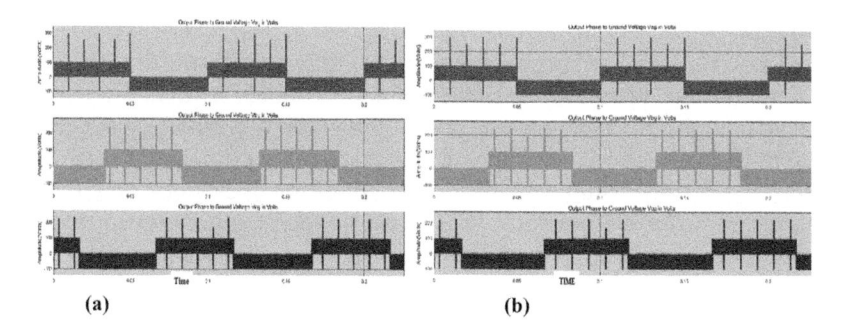

FIGURE 9.15
Three-phase PWM AC to AC converter using a DC link: (a) three-phase 10 Hz output voltage for a carrier duty cycle of 0.2 and 0.8 (b).

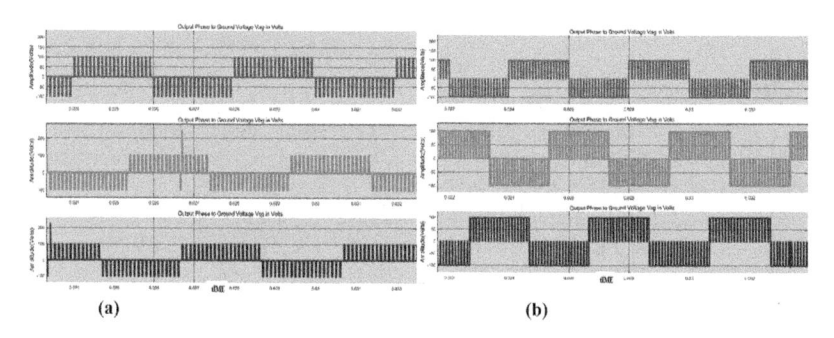

FIGURE 9.16
Three-phase PWM AC to AC converter using a DC link: (a) three-phase 250 Hz output voltage for a carrier duty cycle of 0.2 and 0.8 (b).

9.7 Discussion of Results

From the simulation results, it is seen that a variable-frequency PWM three-phase variable-voltage output is obtained at the load terminal from a constant 50 Hz frequency three-phase AC input voltage. The variable frequency at the load terminal is obtained by selecting the appropriate clock frequency to drive the IGBT bidirectional switches. The variable-voltage magnitude at the load terminal is obtained by varying the duty cycle of the carrier switching pulse.

9.8 Conclusions

A novel method of generating a variable-frequency variable-magnitude single-phase and three-phase AC output voltage from a single-phase and three-phase 50 Hz AC input voltage is presented. The PWM scheme used is well known. Both the single-phase AC to single-phase AC and the three-phase AC to three-phase AC variable-voltage variable-frequency conversion can be achieved. This is a new method in the sense the topology of connecting the IGBT bidirectional switches and its gate connections are much different from the conventional MC. The above topology presents a new method of obtaining variable-frequency AC output voltage from a constant supply frequency input voltage. The gate drive can be easily implemented using analogue and digital ICs. Sine-wave AC output voltage can also be obtained by connecting an appropriate sine-wave filter to the output of IGBT bidirectional switches.

References

1. N.P.R. Iyer: Modelling, simulation and real time implementation of a three phase AC to AC matrix converter, Ph.D. thesis, Ch. 13, Department of ECE, Curtin University, Perth, WA, Australia, February 2012.
2. M.H. Rashid: *Power Electronics*, Prentice Hall Inc., Upper Saddle River, NJ, 2017.
3. The Mathworks Inc.: www.mathworks.com, MATLAB/SIMULINK user manual, R2017b, 2017.

10

Real-Time Hardware-in-the-Loop Simulation of a Three-Phase AC to Single-Phase AC Matrix Converter

10.1 Introduction

In this chapter, real-time hardware-in-the-loop (HIL) simulation of a three-phase AC to single-phase AC matrix converter (MC) is presented. HIL is a technique used in the development and testing of complex real-time embedded Systems. HIL simulation provides an effective platform by adding the complexity of the plant under control to the test platform. The complexity of the plant under control is included in test and development by adding a mathematical representation or model of all related system under consideration [1]. These mathematical representations are referred to as the 'plant simulation'. The embedded system to be tested interacts with this plant simulation [1]. The ability to design and automatically test power electronics systems with HIL simulations will reduce development cycle, increase efficiency, and improve reliability and safety of these systems for a large number of applications [1]. There are at least three strong reasons for using HIL simulation for power electronics, namely, (a) reduction of development cycle, (b) demand to extensively test control hardware and software in order to meet safety and quality requirements, and (c) need to prevent costly and dangerous failures [1].

10.2 Model of Three-Phase AC to Single-Phase AC Matrix Converter

The model of the three-phase AC to single-phase AC MC is developed using the Venturini modulation algorithm assuming unity input phase displacement factor [2–6]. A schematic of the three-phase AC to single-phase AC MC is shown in Figure 10.1. There are three bidirectional switches connected to each input phase A, B and C, and all these switches are connected to one

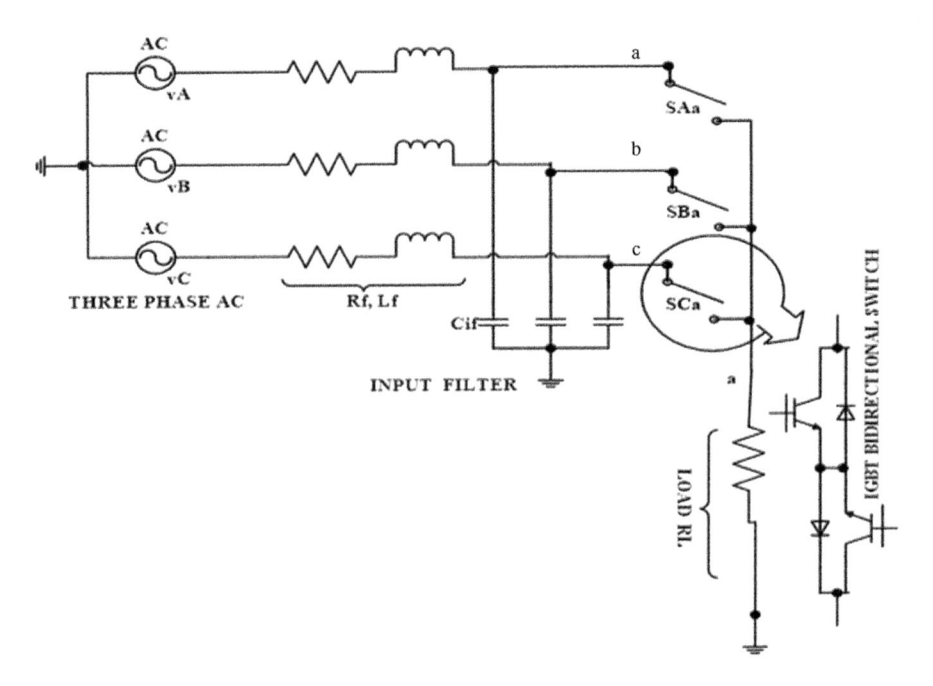

FIGURE 10.1
Three-phase AC to single-phase AC matrix converter.

output phase a in series with a resistive load. The 3×1 MC shown in Figure 10.1 connects the three-phase AC source to the single-phase load. The switching function for a 3×1 MC can be defined as follows [2,3]:

$$S_{Kj} = \begin{cases} 1 \text{ when } S_{Kj} \text{ is closed} \\ 0 \text{ when } S_{Kj} \text{ is open} \end{cases} \tag{10.1}$$

$$K \in \{A, B, C\} \text{ and } j \in \{a\}$$

The above constraint can be expressed in the following form:

$$S_{Aj} + S_{Bj} + S_{Cj} = 1 \tag{10.2}$$

$$j \in \{a\}$$

Here the input-output voltage relation simplifies to:

$$v_a = \begin{bmatrix} S_{Aa} & S_{Ba} & S_{Ca} \end{bmatrix} * \begin{bmatrix} v_A \\ v_B \\ v_C \end{bmatrix} \tag{10.3}$$

Similarly, the input-output current relation can be expressed as follows:

$$\begin{bmatrix} i_A \\ i_B \\ i_C \end{bmatrix} = \begin{bmatrix} S_{Aa} \\ S_{Ba} \\ S_{Ca} \end{bmatrix} * i_a \tag{10.4}$$

where v_a, i_a and i_A, i_B, i_C are the output voltage, output current and input currents, respectively.

To determine the behaviour of the MC at output frequencies well below the switching frequency, a modulation duty cycle can be defined for each switch. The modulation duty cycle M_{Kj} for the switch S_{Kj} in Figure 10.1 is defined as follows:

$$M_{Kj} = \frac{t_{Kj}}{T_s} \tag{10.5}$$

$$K \in \{A, B, C\} \text{ and } j \in \{a\}$$

where t_{Kj} is the on time for the switch S_{Kj} between input phase $K \in \{A, B, C\}$ and $j \in \{a\}$ and T_s is the period of the pulse-width-modulated (PWM) switching signal or sampling period. In terms of the modulation duty cycle, equations 10.2–10.4 can be rewritten as

$$v_a = \begin{bmatrix} M_{Aa} & M_{Ba} & M_{Ca} \end{bmatrix} * \begin{bmatrix} v_A \\ v_B \\ v_C \end{bmatrix} \tag{10.6}$$

$$\begin{bmatrix} i_A \\ i_B \\ i_C \end{bmatrix} = \begin{bmatrix} M_{Aa} \\ M_{Ba} \\ M_{Ca} \end{bmatrix} * i_a \tag{10.7}$$

$$M_{Aj} + M_{Bj} + M_{Cj} = 1$$
$$j \in \{a\} \tag{10.8}$$

The Venturini modulation algorithm for a three-phase AC to three-phase AC MC is presented in Section 2.2 of Chapter 2. For unity input phase displacement factor, the modulation function can be expressed as follows:

$$M_{Kj} = \frac{t_{Kj}}{T_s} = \begin{bmatrix} \frac{1}{3} + \frac{2v_K.v_j}{3V_{im}^2} \end{bmatrix} \tag{10.9}$$

$$\text{for } K = A, B, C \text{ and } j = a$$

10.3 Model Development

To study the behaviour of the three-phase AC to single-phase AC MC using the Venturini modulation algorithm assuming unity input phase displacement factor, a model of this MC is developed in SIMULINK [7,8]. The data shown in Table 10.1 are used to develop the model. The MC switching is developed based on equation 10.9. A three-phase square-wave AC voltage is used.

10.3.1 Model of Three-Phase AC to Single-Phase AC MC Using the Venturini Algorithm

The model of the three-phase AC to single-phase AC MC is developed using the Venturini second method discussed in Section 2.3.3 of Chapter 2. Unity input phase displacement factor is assumed. The model is developed using SIMULINK [7,8]. The data shown in Table 10.1 are used to develop the model. The model simulation is carried out for three-phase square-wave input voltages for the parameters in Table 10.1. The model is shown in Figure 10.2 [8]. The gate pulses for the three bidirectional switches are developed using Embedded MATLAB Function, saw-tooth carrier and logic gates. Embedded MATLAB Function is used to realize equations 2.9 and 2.11 of Chapter 2 and the three modulation functions for output phase a, defined in equation 10.9. Using these three modulation functions and a 5 kHz saw-tooth carrier as inputs, gate timing pulses are generated using comparators and logic gates for the three bidirectional switches, as shown in Figure 10.2. The three-phase sine-wave input has an RMS value of 123 V between line and neutral and frequency 50 Hz. In addition, model is shown to generate three-phase 123 V (RMS), 50 Hz square-wave input, using three threshold switches. The threshold switches have zero as the threshold value and output first input u(1) when their second input u(2) is greater than or equal to zero, else their output is the third input u(3). All these switches have u(1) of +123 V and u(3) of −123 V. The second input $u(2)$ to each of the threshold switches is three-phase 50 Hz sine-wave voltages. When $u(2)$ crosses zero and goes positive and negative, the output of the threshold switch goes +123 and −123 V, respectively, giving the three-phase line to ground output voltages. The modulation index and output frequency are set to 0.5 and 50 Hz, respectively.

TABLE 10.1

Real-Time HIL Simulation – Model Parameters

Sl. No.	Modulation Index q	RMS Line to Neutral Input Voltage (V)	Input Frequency (Hz)	Output Frequency (Hz)	Carrier Frequency (kHz)	Load Resistance (Ω)
1	0.5	123	50	50	5	4,700

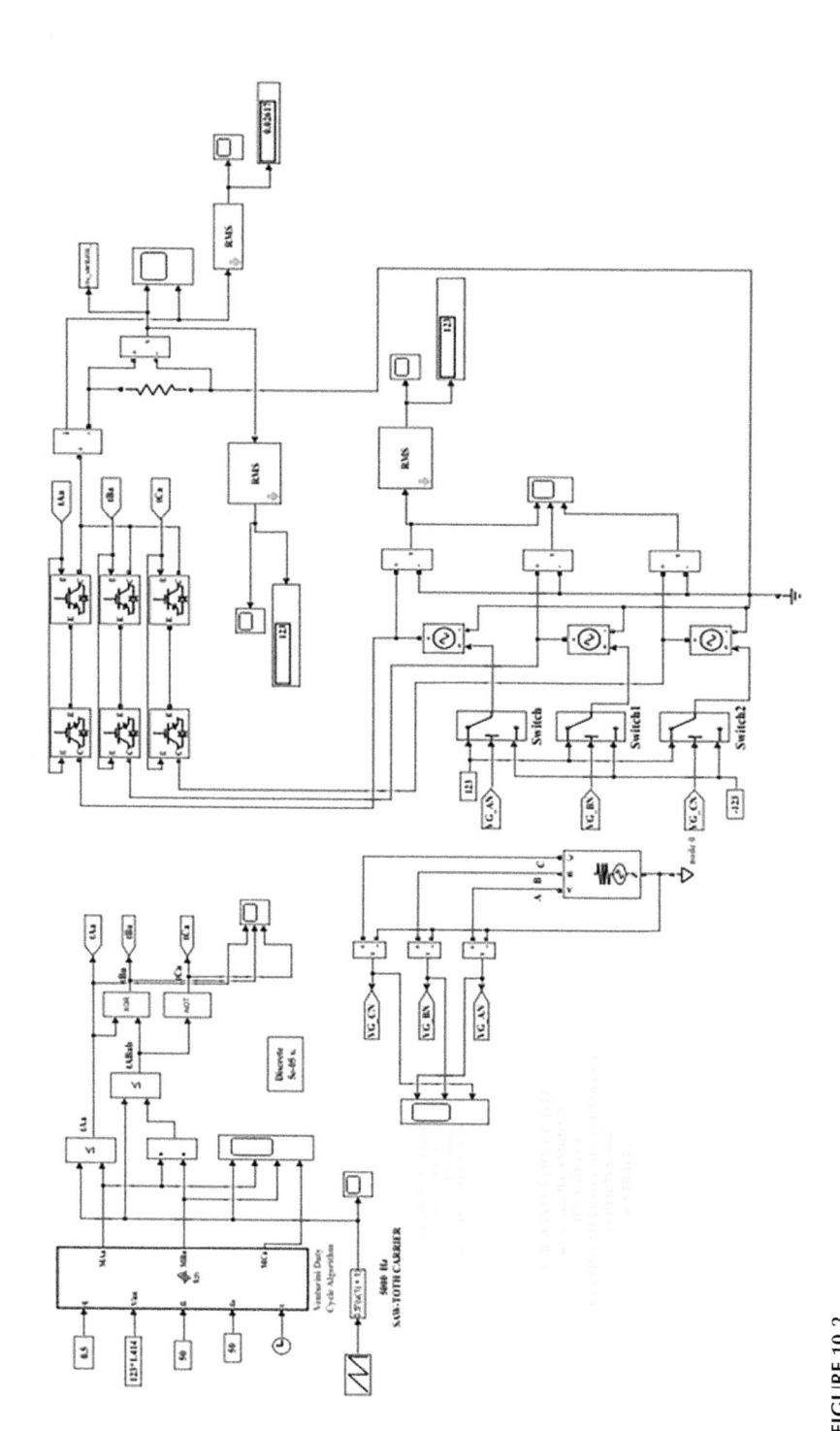

FIGURE 10.2
SIMULINK model for three-phase AC to single-phase AC matrix converter for three-phase square-wave input.

10.3.2 Simulation Results

The simulation of the above model was carried out in SIMULINK [7]. The ode15s (Stiff/NDF) solver is used in both cases. Simulation results for three-phase 50 Hz square-wave input voltage are discussed below.

The simulation result of the output voltage, load current and the RMS value of output voltage for the three-phase 123 V (RMS), 50 Hz square-wave input voltage is shown in Figure 10.3a,b. The output voltage across the load resistor is found to be 123 V (RMS) which is shown in Figure 10.3b. The three-phase 50 Hz square-wave input voltage and its RMS value are shown in Figure 10.4a,b respectively. Simulation results are tabulated in Table 10.2.

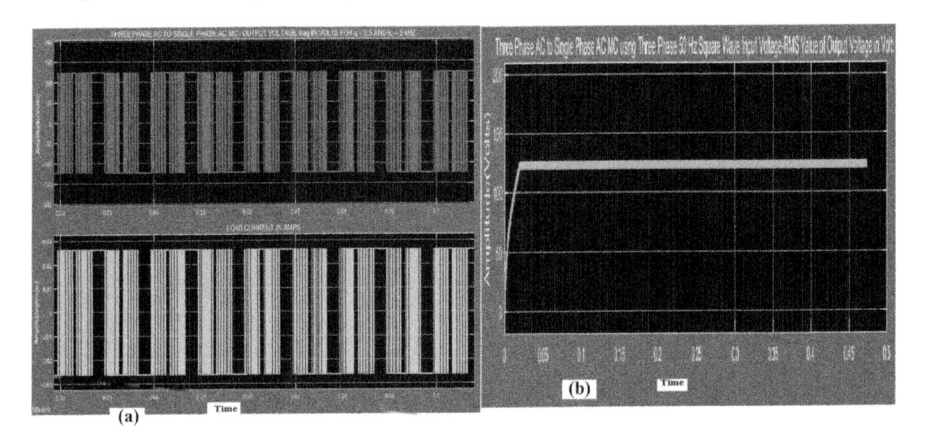

FIGURE 10.3
Three-phase AC to single-phase AC MC: (a) output voltage (top) and load current (bottom) and (b) RMS value of output voltage.

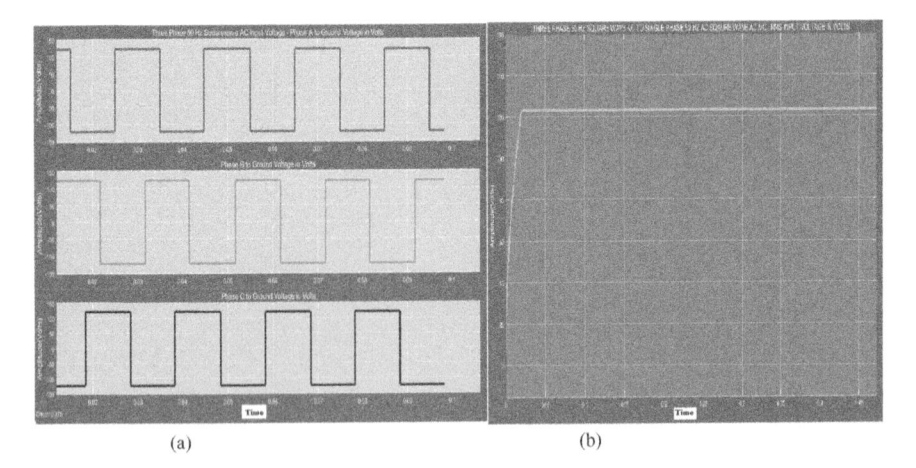

FIGURE 10.4
Three-phase AC to single-phase AC MC: (a) three-phase 50 Hz square-wave AC input voltage and (b) RMS value of line to ground input voltage.

TABLE 10.2

Three-Phase AC to Single-Phase AC MC – Simulation Results

Sl. No.	Input Voltage			Output Voltage			Modulation Index
	Nature	Frequency	RMS Value	Nature	Frequency	RMS Value	
1	Three-phase square-wave AC	50 Hz	123 V (L to N)	Single-phase square-wave AC	50 Hz	123 V (L to G)	0.5

10.4 Experimental Verification Using dSPACE Hardware Controller Board

The first experiment is developed using dSPACE DS1104 hardware controller board to study the gate drive waveforms for three-phase 50 Hz AC to single-phase 50 Hz AC MC [9]. The experimental set-up is shown in Figure 10.5. The method of interfacing the dSPACE hardware controller board to the SIMULINK model of three-phase AC to single-phase AC MC is shown in Figure 10.6. The DS1104 has a main processor MPC8240 and a slave DSP TMS320F240 [9]. In Figure 10.6, the Embedded MATLAB Function generates the three modulation functions M_{Aa}, M_{Ba} and M_{Ca}. The comparator block, A ≤ B provides output HIGH when saw-tooth carrier V_{saw} input given to node A is less than or equal to the input given to node B. The comparator connected to V_{saw} and M_{Aa} produces output t_{Aa} and that connected to V_{saw} and $(M_{Aa}+M_{Ba})$ produces output t_{ABa}. The outputs t_{Aa} and t_{ABa} are connected to EXCLUSIVE-OR gate to obtain t_{Ba} and the output t_{ABa} is inverted using a NOT gate to obtain t_{Ca}. t_{Aa}, t_{Ba} and t_{Ca} form the gate pulses for the bidirectional switches connected between phase A-a, B-a and C-a, respectively, in Figure 10.1. The waveforms of the saw-tooth

FIGURE 10.5
dSPACE experiment set-up for the gate drive of 3×1 matrix converter.

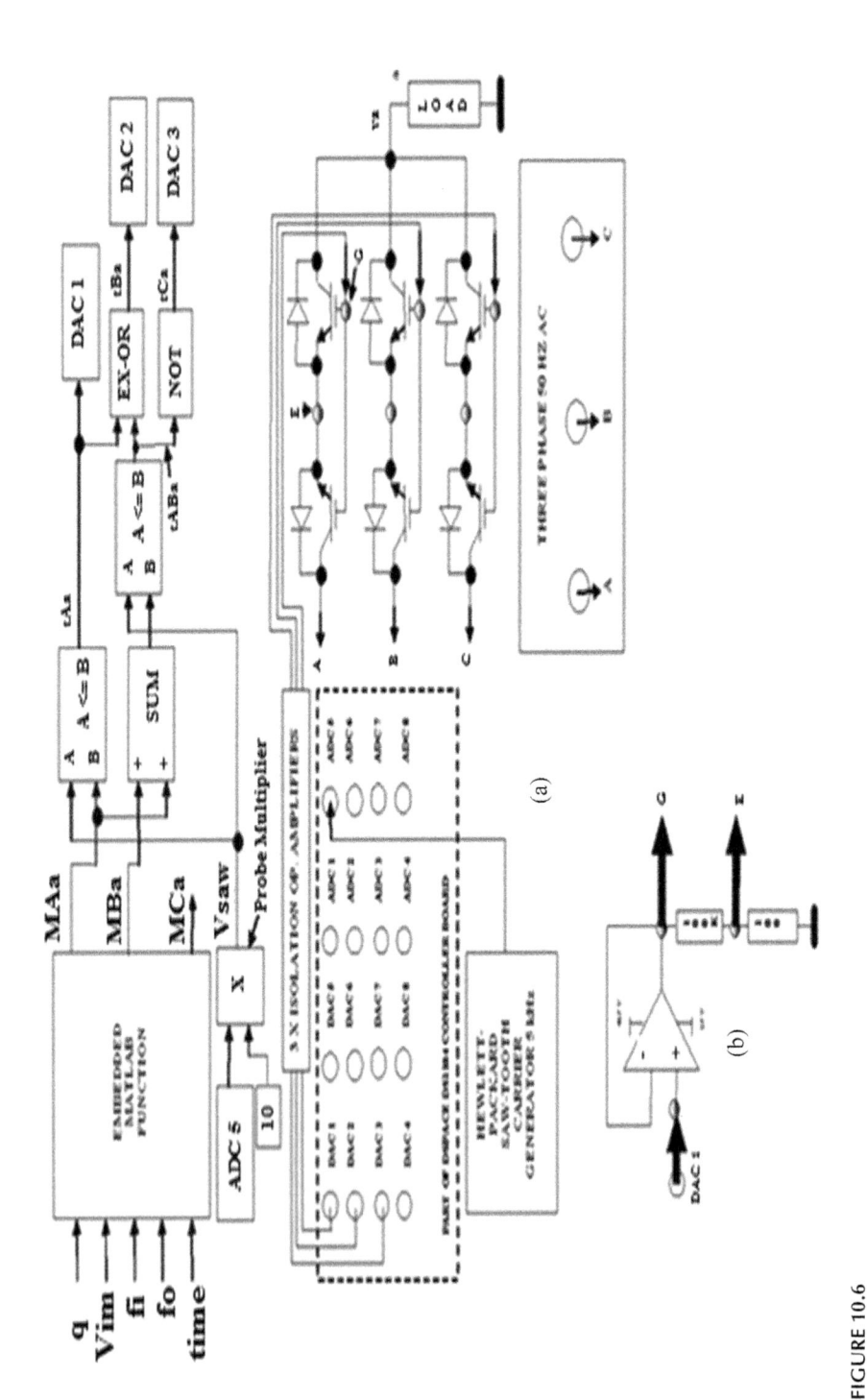

FIGURE 10.6
(a) and (b) dSPACE DS1104 hardware controller board interface to three-phase AC to single-phase AC matrix converter SIMULINK model.

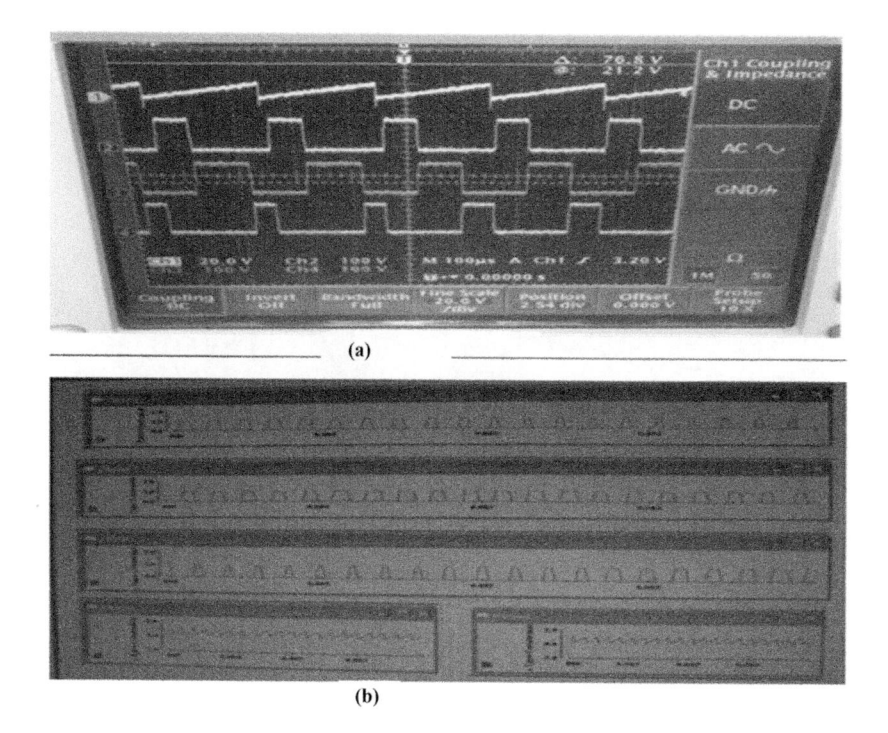

FIGURE 10.7
Saw-tooth carrier and gate drive waveform for a 3×1 MC: (a) using oscilloscope and (b) using control desk.

carrier gate drive obtained for the above experiment using oscilloscope and control desk are shown in Figure 10.7a,b respectively [8].

The second experiment is developed to observe the performance of an actual three-phase 50 Hz AC to single-phase AC MC when switching of bidirectional switches is done using the above gate pulse drive. The dSPACE hardware set-up for this experiment is shown in Figure 10.8a and the 3×1 MC board is shown in Figure 10.8b. Figure 10.6b, which is a high input impedance operational amplifier, shows the method of connecting the output of DAC1 of DS1104 hardware controller board to the bidirectional switch. Similar connections apply to DAC2 and DAC3 of DS1104 hardware controller board. All IGBTs used in the experiment are STGP14NC60KD with a voltage and current rating of 600 V, 25 A. The Texas Instruments quad TL084 operational amplifier is used for isolation as this has a high input impedance of the order of 10^{12} Ω and slew rate of the order of 13 V/μs. Although the Venturini algorithm is developed for three-phase sine-wave AC input voltage only, in this experiment, a three-phase 123 V (RMS) line-to-neutral, 50 Hz square-wave AC input voltage is used.

The waveform of this input voltage for the reference phase A is shown in Figure 10.9a in both the channels. The gate drive pattern obtained is shown in Figure 10.7a,b. The output voltage observed across the load resistor is found to be 86 V (RMS) and this waveform is shown in Figure 10.9b [8]. In this experiment, a modulation index of 0.5 is used. The dSPACE hardware controller board experimental result is tabulated in Table 10.3.

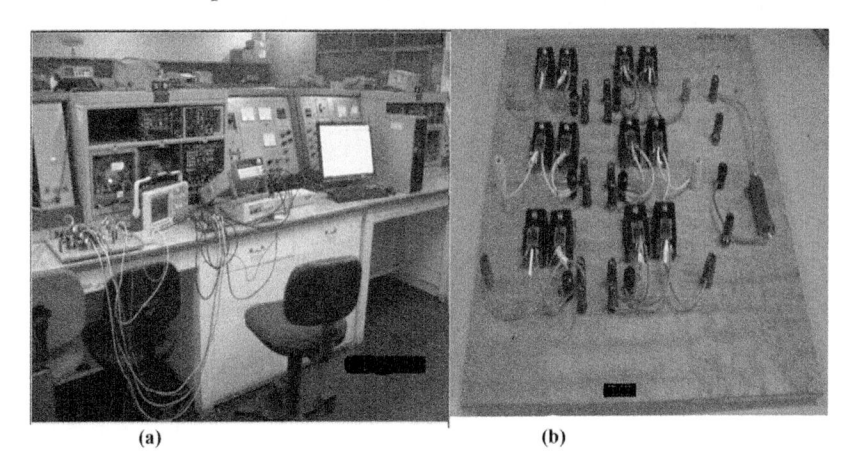

(a)   (b)

FIGURE 10.8
(a) dSPACE experimental set-up for a 3×1 MC and (b) a 3×1 MC board.

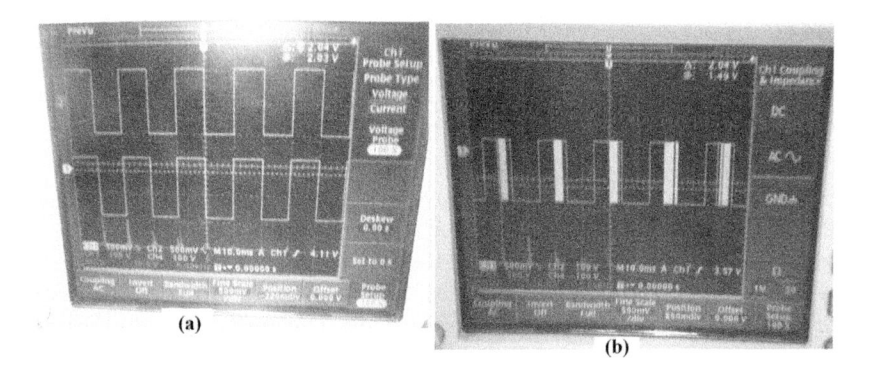

(a)   (b)

FIGURE 10.9
(a) Reference phase A input voltage of three-phase square-wave AC voltage (both channels) and (b) output voltage of 3×1 MC across the load.

TABLE 10.3

The 3×1 MC dSPACE Hardware Experimental Results

| SI. No. | Input Voltage | | | Output Voltage | | | Modulation Index |
	Nature	Frequency	RMS Value	Nature	Frequency	RMS Value	
1	Three-phase square-wave AC	50 Hz	123 V (L to N)	Single-phase square-wave AC	50 Hz	86 V (L to G)	0.5

10.5 Discussion of Results

The real-time HIL simulation of the above three-phase AC to single-phase AC MC is carried out using dSPACE [9]. The real-time platform is developed using dSPACE DS1104 hardware controller board. In this dSPACE experiment, a three-phase 50 Hz square-wave AC voltage from a three-phase inverter is used as the input. A resistive load is used for this experiment. The experimental results obtained are shown. The experimental results are verified by simulation. The Venturini modulation algorithm assuming unity input phase displacement factor is used for the real-time experiment on the above MC and also for the verification by simulation.

10.6 Conclusions

The waveforms observed using oscilloscope almost well agree with the one obtained by simulation of the model. However, the RMS value of the output voltage by experiment has a variation of 30% compared to the one obtained by simulation of the model. The bidirectional IGBT switch models use RC snubber for convergence of simulation, whereas no such snubber is present in the IGBTs used for the experiment.

References

1. Wikipedia: https://en.wikipedia.org/wiki/Hardware-in-the-loop_simulation.
2. A. Alesina and M.G.B. Venturini: Solid state power conversion: a Fourier analysis approach to generalized transformer synthesis, *IEEE Transactions on Circuits and Systems*, Vol. CAS-28, No. 4, 1981, pp. 319–330.
3. A. Alesina and M. Venturini: Intrinsic amplitude limits and optimum design of 9-switches direct PWM AC-AC converters, *IEEE-PESC*, Kyoto, Japan, Vol. 2, 1988, pp. 1284–1291.
4. A. Alesina and M. Venturini: Analysis and design of optimum amplitude nine-switch direct AC-AC converters, *IEEE Transactions on Power Electronics*, Vol. 4, 1989, pp. 101–112.
5. P. Wheeler, J. Rodriguez, J.C. Clare, L. Empringham and A. Weinstein: Matrix converters – a technology review, *IEEE Transactions on Industrial Electronics*, Vol. 49, No. 2, 2002, pp. 276–288.
6. P. Wheeler, J. Clare, L. Empringham, M. Apap and M. Bland: Matrix converters, *Power Engineering Journal*, 2002, Vol. 16, No. 6, pp. 273–282.

7. The Mathworks Inc.: www.mathworks.com, MATLAB/SIMULINK user manual, MATLAB R2017b, 2017.
8. N.P.R. Iyer: Modelling, simulation and real time implementation of a three phase AC to AC matrix converter, Ph.D. thesis, Ch. 6, Department of ECE, Curtin University, Perth, WA, Australia, February 2012.
9. dSPACE GmbH: www.dspace.com, dSPACE users' manual, 2010.

11

Three-Phase Z-Source Matrix Converter

11.1 Introduction

The traditional three-phase AC to three-phase AC matrix converter (MC) has an inherent limitation in their output voltage gain which is limited to 0.866 [1–5]. While switching MCs, care must be taken not to close two or more switches connected to the same output phase simultaneously as this causes dangerous short-circuit current [1–5]. Also any one switch in each output phase must remain closed to provide a current path with inductive loads. With Z-source MC all the bidirectional switches connected to the same output phase can be closed simultaneously providing a shoot-through (ST) state [6]. During this ST state, the three-phase AC mains supply is switched off. Depending on the ST time interval, the output voltage can be boosted to a value higher than the input voltage. Longer the ST time interval, higher the output voltage [6]. During non-shoot through (NST) state, the three-phase AC mains supply is switched on and the bidirectional switches are turned on as in a conventional MC.

In this chapter, models are developed for three-phase voltage-fed Z-source direct matrix converter (ZSDMC) and that for three-phase quasi Z-source indirect matrix converter (QZSIMC). The voltage boost phenomenon is studied by model simulation and the results are presented.

11.2 Three-Phase Voltage-Fed Z-Source Direct Matrix Converter

The three-phase voltage-fed ZSDMC configuration is shown in Figure 11.1 [6]. The input is a three-phase AC voltage source. The three-phase input filter is a capacitor connected between each of the three input lines. The S0 forms the switch used to open during ST state and close during NST state. The L–C network comprising L_1–C_1, L_2–C_2 and L_3–C_3 forms the Z-source network for each input phase. The nine bidirectional switches form the 3×3 MC, and the

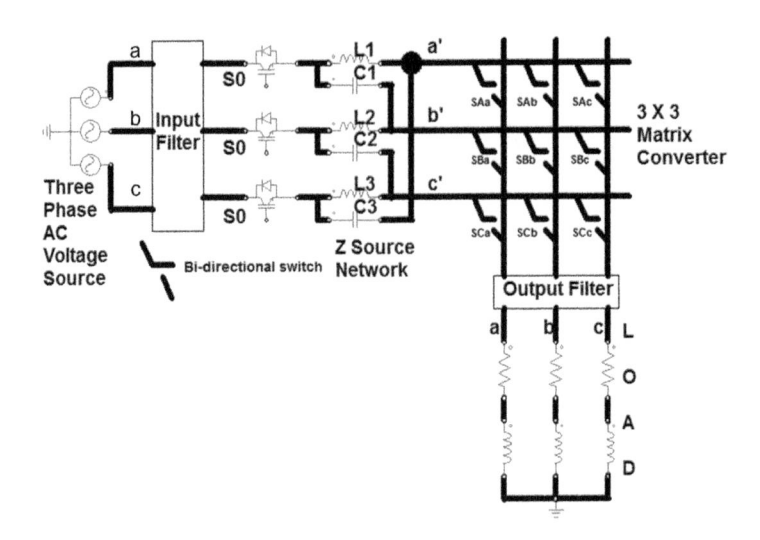

FIGURE 11.1
Three-phase voltage-fed Z-source matrix converter.

three-phase load is a series combination of resistor and inductor. A low-pass R–L–C circuit forms the output filter.

11.2.1 Principle of Operation and Analysis – Simple Boost Control

The equivalent circuit of the three-phase voltage-fed ZSDMC is shown in Figure 11.2a for the ST state and in Figure 11.2b for NST state. Let T_s be the carrier switching period. Let T_0 and T_1 be the ST and NST period. Then $T_s = (T_0 + T_1)$. Let $D = (T_0/T_s)$ be the ST duty cycle.

Referring to Figure 11.2a, the following equation can be written [6]:

$$\begin{bmatrix} v_{C1} \\ v_{C2} \\ v_{C3} \end{bmatrix} = \begin{bmatrix} v_{L1} \\ v_{L2} \\ v_{L3} \end{bmatrix} \tag{11.1}$$

Referring to Figure 11.2b, the following equations can be written:

$$\begin{bmatrix} v_{AB} \\ v_{BC} \\ v_{CA} \end{bmatrix} = \begin{bmatrix} v_{C1} \\ v_{C2} \\ v_{C3} \end{bmatrix} - \begin{bmatrix} v_{L2} \\ v_{L3} \\ v_{L1} \end{bmatrix} \tag{11.2}$$

$$\begin{bmatrix} v_{a'b'} \\ v_{b'c'} \\ v_{c'a'} \end{bmatrix} = \begin{bmatrix} v_{C1} \\ v_{C2} \\ v_{C3} \end{bmatrix} - \begin{bmatrix} v_{L1} \\ v_{L2} \\ v_{L3} \end{bmatrix} \tag{11.3}$$

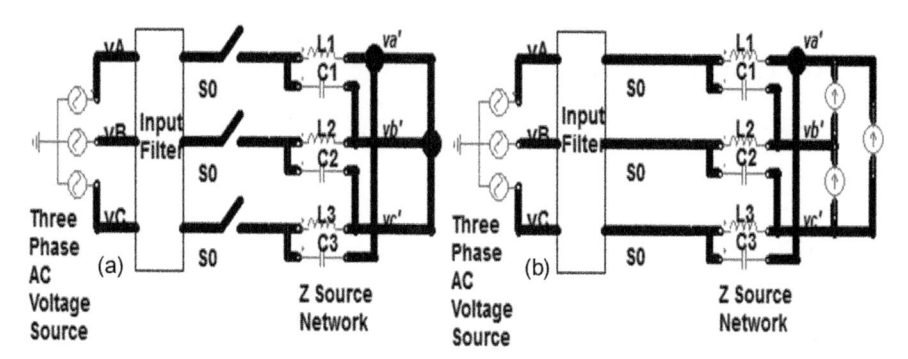

FIGURE 11.2
Voltage fed Z-source MC: (a) shoot through and (b) non-shoot through states.

For high carrier switching frequency compared to frequency of operation of the ZSDMC, the average voltage across the inductor over one switching cycle is zero. Thus, from equations 11.1 and 11.2, this can be expressed as follows:

$$D * v_{C1} + (1-D) * (v_{C3} - v_{CA}) = 0$$
$$D * v_{C2} + (1-D) * (v_{C1} - v_{AB}) = 0 \tag{11.4}$$
$$D * v_{C3} + (1-D) * (v_{C2} - v_{BC}) = 0$$

Similarly, from equations 11.1 and 11.3, the following equation can be written:

$$v_{C1} - (1-D) * v_{a'b'} = 0$$
$$v_{C2} - (1-D) * v_{b'c'} = 0 \tag{11.5}$$
$$v_{C3} - (1-D) * v_{c'a'} = 0$$

Let the three-phase line-to-line input voltage be defined in the following form:

$$v_{jk} = V_i * \sin(\omega t + \varphi_{jk}) \tag{11.6}$$

where $\varphi_{jk} = 0, -120$ and $+120$ for $jk = AB, BC$ and CA, respectively, and V_i and ω are the amplitude and angular frequency of the input voltage source.

Let the voltage across the capacitors C_1, C_2 and C_3 lead the respective input voltage by φ_C so that their respective voltage can be expressed as follows:

$$v_{Cn} = V_C * \sin(\omega t + \varphi_C + \varphi_n) \tag{11.7}$$

where $\varphi_n = 0, -120$ and $+120$ for $n = 1, 2$ and 3, respectively, and V_C is the amplitude of the capacitor voltage.

Assume that the voltages across the ZSDMC bridge lead the respective input voltage by φ_m so that their voltages can be expressed as follows:

$$v_{pq} = V_m * \sin\left(\omega t + \varphi_m + \varphi_{pq}\right) \tag{11.8}$$

where $\varphi_{pq} = 0, -120$ and $+120$ for $pq = a'b', b'c'$ and $c'a'$, respectively, and V_m is the amplitude of the Z-source network voltage.

Expanding, using and rearranging equations 11.4, 11.6 and 11.7, we have

$$V_C * \begin{bmatrix} D*\sin\left(\omega t + \varphi_C\right) + \left(1-D\right)*\sin\left(\omega t + \varphi_C + 120\right) \\ D*\sin\left(\omega t + \varphi_C - 120\right) + \left(1-D\right)*\sin\left(\omega t + \varphi_C\right) \\ D*\sin\left(\omega t + \varphi_C + 120\right) + \left(1-D\right)*\sin\left(\omega t + \varphi_C - 120\right) \end{bmatrix}$$
$$\tag{11.9}$$
$$= \left(1-D\right)*V_i * \begin{bmatrix} \sin\left(\omega t + 120\right) \\ \sin\left(\omega t\right) \\ \sin\left(\omega t - 120\right) \end{bmatrix}$$

Solving equation 11.9, the following equation can be obtained:

$$\frac{V_C}{V_i} = \frac{\left(1-D\right)}{\sqrt{\left(3D^2 - 3D + 1\right)}} \tag{11.10}$$

Equation 11.10 is proved in the Appendix A.

Similarly, from equations 11.5, 11.7 and 11.8, we have

$$V_C = \left(1-D\right)*V_m \tag{11.11}$$

From equations 11.10 and 11.11, the line-to-line output voltage of the Z-source network which is also the MC bridge voltage at the input side can be expressed as follows:

$$\frac{V_m}{V_i} = B = \frac{1}{\sqrt{\left(3D^2 - 3D + 1\right)}} \tag{11.12}$$

where B represents the boost factor. By proper choice of D desired boost factor can be obtained. Boost factor B depends on the type of control used. In a simple boost control, ST duty cycle D will be constant and its maximum value is $D \leq \left[\dfrac{2 - 4q}{3}\right]$, where q is the modulation index [6].

11.2.2 Simple Boost Control Strategy

The simple boost control is shown in Figure 11.3. Two ST reference voltages are used as defined below, one for the top and the other for the bottom input voltage envelope [6]:

$$v_{\text{str\_pos}} = \left[\left(e_{i\_\max} - e_{i\_\min} \right) * n + e_{i\_\max} + e_{i\_\min} \right] \Big/ 2 \qquad (11.13)$$

$$v_{\text{str\_neg}} = \left[\left(e_{i\_\max} - e_{i\_\min} \right) * (-n) + e_{i\_\max} + e_{i\_\min} \right] \Big/ 2 \qquad (11.14)$$

where n determines the ST duty cycle lying in the range $1 \geq n \geq (1 + 4q)/3$, q is the modulation index and $e_{i\_\max}$ and $e_{i\_\min}$ are the maximum and minimum of three-phase input voltage reference.

Value of q is less than 0.5 for simple boost control. The ST interval from top reference for one full carrier period is given below [6]:

$$\frac{T_0}{T_s} = D = (1 - n) \qquad (11.15)$$

In Figure 11.3, the three-phase input voltages v_A, v_B, v_C, three-phase output voltages v_a, v_b and v_c, ST references $v_{\text{str\_pos}}$ and $v_{\text{str\_neg}}$ and the Data Type

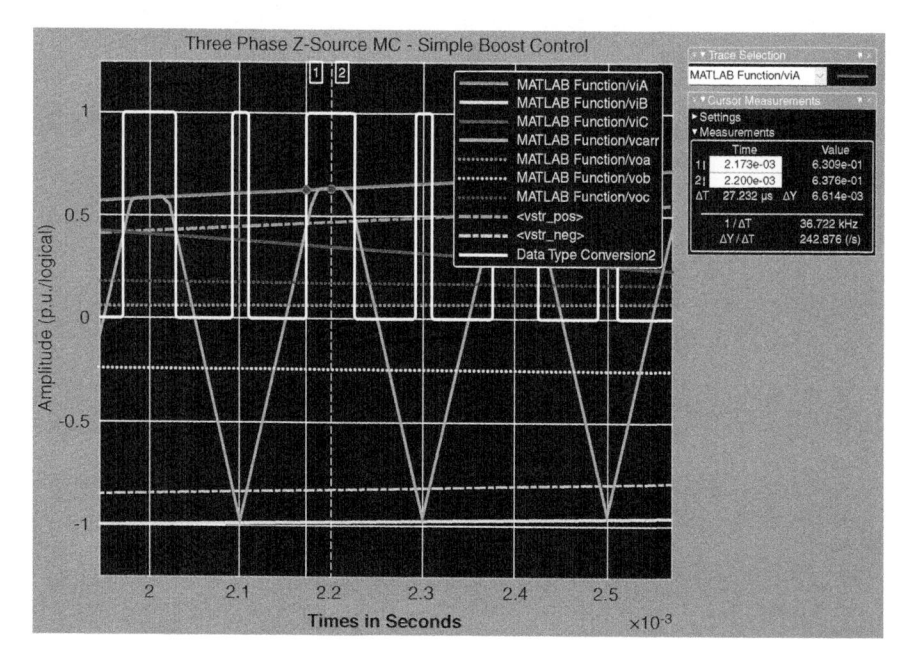

FIGURE 11.3
Three-phase Z-source MC – simple boost control.

Conversion 2 output which is the shoot-through pulse with period T_0 are shown. The ST state has no effect on the output voltage waveform. The pulse-width-modulated (PWM) pulses S_a, S_b and S_c are created by comparing the output voltage references v_a, v_b and v_c with the triangle carrier v_{carr}. If v_a is greater than or equal to v_{carr}, S_a is LOW or zero and HIGH or 1 otherwise. Similarly PWM pulses S_b and S_c are developed by comparing v_b and v_c with v_{carr} in the same way as for output voltage reference v_a. The PWM waveforms S_a, S_b and S_c are shown in Figure 11.4.

The PWM pulses S_a, S_b and S_c have to be distributed to nine bidirectional switches of the MC.

For this purpose, six additional logic signals S_{x1}, S_{x2}, S_{y1}, S_{y2}, S_{z1} and S_{z2} are created. This is shown in Figure 11.5. Signals S_{x1}, S_{y1} and S_{z1} are the top envelope indicators for the input voltage references v_A, v_B and v_C. During positive half cycle if v_A is the largest compared to v_B and v_C, then S_{x1} remains HIGH

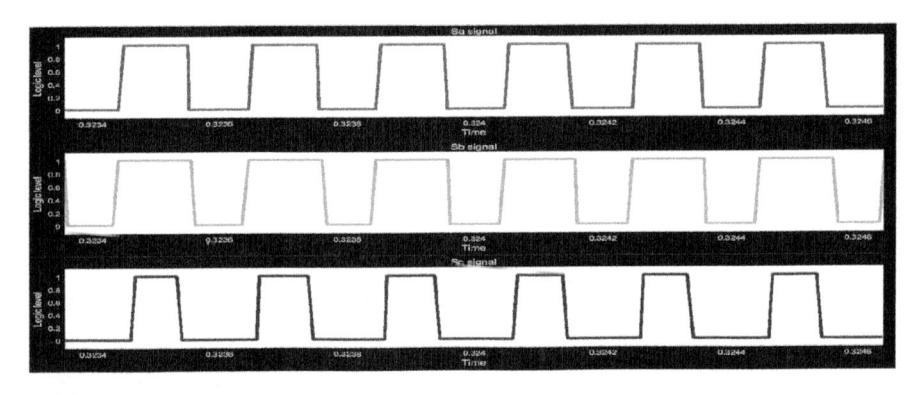

FIGURE 11.4
Three-phase Z-source MC – PWM signals S_a, S_b and S_c.

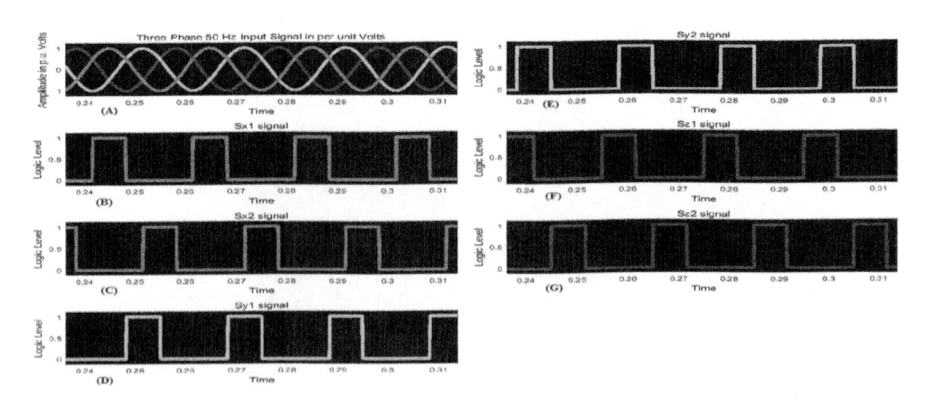

FIGURE 11.5
Three-phase Z-source MC – six input voltage envelop indicators.

or logic 1, else its value is LOW or logic 0. Similarly, during negative half cycle if v_A is the lowest compared to v_B and v_C then S_{x2} is HIGH else its value is LOW. The nine switching signals for the bidirectional switches of the MC are generated using the following logic [6]:

$$S_{Aa} = \left(S_{x1} \cap S_a\right) \cup \left(S_{x2} \cap \bar{S}_a\right)$$

$$S_{Ba} = \left(S_{y1} \cap S_a\right) \cup \left(S_{y2} \cap \bar{S}_a\right)$$

$$S_{Ca} = \left(S_{z1} \cap S_a\right) \cup \left(S_{z2} \cap \bar{S}_a\right)$$

$$S_{Ab} = \left(S_{x1} \cap S_b\right) \cup \left(S_{x2} \cap \bar{S}_b\right)$$

$$S_{Bb} = \left(S_{y1} \cap S_b\right) \cup \left(S_{y2} \cap \bar{S}_b\right) \tag{11.16}$$

$$S_{Cb} = \left(S_{z1} \cap S_b\right) \cup \left(S_{z2} \cap \bar{S}_b\right)$$

$$S_{Ac} = \left(S_{x1} \cap S_c\right) \cup \left(S_{x2} \cap \bar{S}_c\right)$$

$$S_{Bc} = \left(S_{y1} \cap S_c\right) \cup \left(S_{y2} \cap \bar{S}_c\right)$$

$$S_{Cc} = \left(S_{z1} \cap S_c\right) \cup \left(S_{z2} \cap \bar{S}_c\right)$$

where $\cap$ represents the AND logic and $\cup$ represents the OR logic operation.

11.2.3 Model Development

The model of the three-phase Z-source MC with simple boost control is developed using SIMULINK [7]. The model is shown in Figure 11.6. The model parameters are shown in Table 11.1. The three-phase input reference voltage with 1 p.u. amplitude and frequency 50 Hz, three-phase output reference voltage with amplitude 0.25 p.u. and frequency 20 Hz are generated using Embedded MATLAB Function. An external triangle carrier V_{tri} with peak value +1/−1 V and frequency 5 kHz is given as input to the Embedded MATLAB Function and this value of V_{tri} is saved in V_{carr}. The output voltage reference for each phase is compared with V_{carr} and the PWM signals S_a, S_b and S_c are generated. The input voltage reference for each phase is compared with V_{carr} to generate six input voltage envelope indicators S_{x1}, S_{x2}, S_{y1}, S_{y2}, S_{z1} and S_{z2}. The input voltage reference for each phase is connected to Min and Max modules to determine `eimax` and `eimin`. This `eimax`, `eimin` and n value are used as inputs to the Embedded MATLAB Function to generate `vstr_pos` and `vstr_neg` and the `st_sig` signals. The source code for generating these signals is shown in Program segment 11.1.

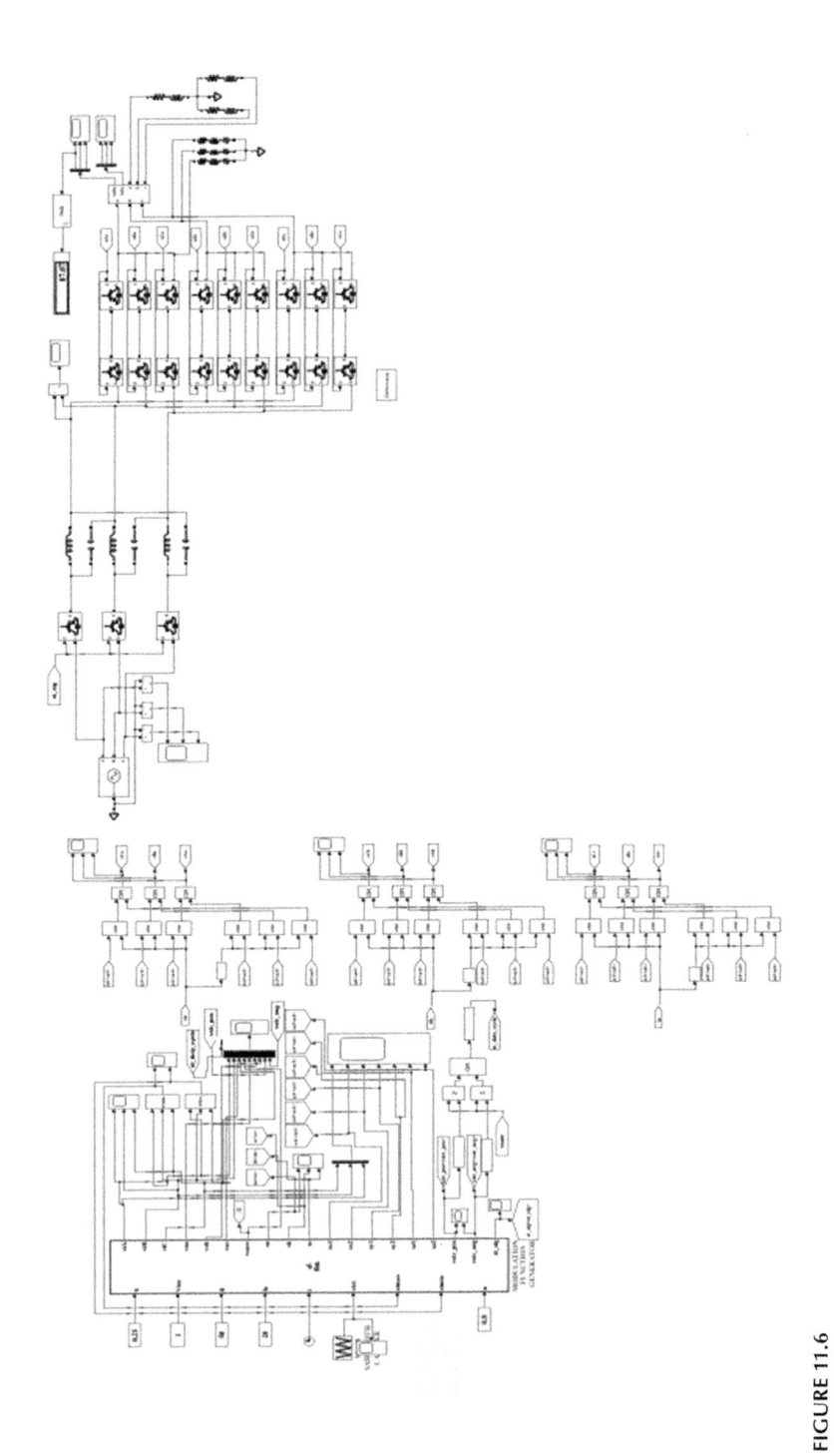

FIGURE 11.6
Three-phase voltage-fed Z-source matrix converter model with simple boost control.

TABLE 11.1

Three-Phase Z-Source MC Model Parameters

Sl. No.	Parameter	Value	Units
1	ZS inductor	600e–6	H
2	ZS capacitor	168.87e–6	F
3	Output filter resistor	10	Ω
4	Output filter inductor	800e–6	H
5	Output filter capacitor	1.2665e–6	F
6	Load resistor	30	Ω
7	Load inductor	0.1	H
8	Carrier switching frequency	5	kHz
9	RMS line-to-neutral input voltage	100	V
10	Input frequency	50	Hz
11	Output frequency	20	Hz
12	Modulation index q	0.25	-
13	n value	0.8	-

Program Segment 11.1

```
function
[viA,viB,viC,voa,vob,voc,vcarr,sa,sb,sc,sx1,sx2,sy1,sy2,sz1,
sz2,vstr_pos,vstr_neg,st_sig] = fcn(q,Vim,fi,fo,t,vtri,eimax,
eimin,n)
%%Three Phase Voltage Fed Z-Source Direct Matrix Converter -
Simple Boost Control.
%% Dr. Narayanaswamy P R Iyer.
viA = Vim*sin(2*pi*fi*t);
viB = Vim*sin(2*pi*fi*t - 2*pi/(3));
viC = Vim*sin(2*pi*fi*t + 2*pi/(3));
%%
voa = q*Vim*sin(2*pi*fo*t);
vob = q*Vim*sin(2*pi*fo*t - 2*pi/(3));
voc = q*Vim*sin(2*pi*fo*t + 2*pi/(3));
%%
vcarr = vtri;
if vcarr >= eimax
vcarr = eimax;
 end
if vcarr <= eimin
vcarr = eimin;
end
if vcarr >= voa
    sa = 0;
else
    sa = 1;
end
if vcarr >= vob
```

```
    sb = 0;
else
    sb = 1;
end
if vcarr >= voc
    sc = 0;
else
    sc = 1;
end
if viA >= 0.5
    sx1 = 1;
else
    sx1 = 0;
end
if viB >= 0.5
    sy1 = 1;
else
    sy1 = 0;
end
if viC >= 0.5
    sz1 = 1;
else
    sz1 = 0;
end
if viA <= -0.5
    sx2 = 1;
else
    sx2 = 0;
end
if viB <= -0.5
    sy2 = 1;
else
    sy2 = 0;
end
if viC <= -0.5
    sz2 = 1;
else
    sz2 = 0;
end
vstr_pos = (( eimax - eimin )*n + eimax + eimin )/2;
vstr_neg = (( eimax - eimin )*(-n) + eimax + eimin )/2;
if vcarr <= vstr_pos || vcarr >= vstr_neg
    st_sig = 1;
else
    st_sig = 0;
end
```

The above six input reference voltage envelope indicators along with the three output voltage PWM signals are given to AND, OR and NOT logic gates as per equation 11.16 to generate gate pulse for the nine bidirectional switches of the MC. This is shown in Figure 11.6. The three-phase AC input voltage generator, Z-source network, switch S0, 3×3 MC, output filter and load are shown in Figure 11.6.

11.2.4 Simulation Results

The simulation of the three-phase ZSDMC with simple boost control for the parameters shown in Table 11.1 is carried out using ode23tb (stiff/ TR-BDF2) solver. The simulation results for the output line-to-neutral voltage, line-to-line voltage, output voltage across Z-source network and load current are shown in Figure 11.7a–d, respectively. Also, the ST period T_0 is measured for one carrier period from the simulated waveform shown in Figure 11.3 at three different locations each commencing at 2e–3, 2.091e–3 and 2.173e–3 s. The respective values for T_{01}, T_{02} and T_{03} are found to be 0.029e–3, 0.018e–3 and 0.027e–3 s, respectively. The total T_0 is found to be 0.074 ms. The simulation results and theoretical values are tabulated in Table 11.2. The value of ST duty cycle D is found to be 0.37 by model simulation. This value of D by calculation is 0.2 using equation 11.15.

11.2.5 Discussion of Results

The ST duty cycle D by calculation and by model simulation differs by 80%. However, the variation of boost factor B by calculation and by model simulation varies only by a narrow margin. The value of Z-source network peak output voltage V_m by calculation for a D of 0.2 closely agrees with that by model simulation. The peak line-to-line output voltage by model simulation closely agrees with the value of Z-source network output voltage by calculation for a D of 0.2.

11.2.6 Maximum Boost Control Strategy

The maximum boost control is shown in Figure 11.8. Two ST reference voltages are used as defined below, one for the top and another for the bottom input voltage envelope [6]:

$$v_{str\_pos} = \left[\left(e_{i\_max} - e_{i\_min}\right) * q + e_{i\_max} + e_{i\_min}\right]/2 \tag{11.17}$$

$$v_{str\_neg} = \left[\left(e_{i\_max} - e_{i\_min}\right) * \left(-q\right) + e_{i\_max} + e_{i\_min}\right]/2 \tag{11.18}$$

where q is the modulation index and e_{i_max} and e_{i_min} are the maximum and minimum of three-phase input voltage reference. The ST interval from top reference for one full carrier period is given below [6]:

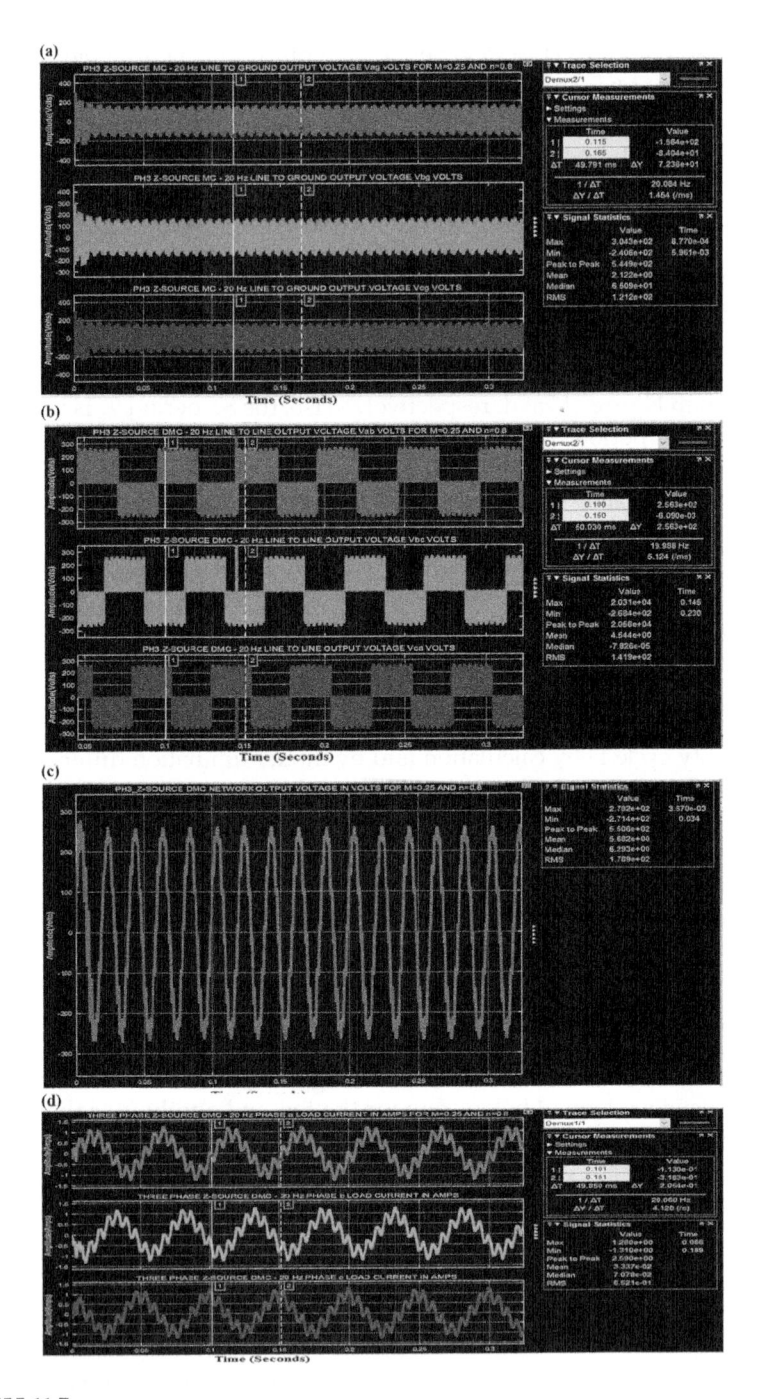

FIGURE 11.7
Simulation results for three-phase Z-source MC with simple boost control: (a) line-to-ground voltage, (b) line-to-line voltage, (c) Z-source source network voltage and (d) load current.

TABLE 11.2

Three-Phase ZSMC by Simple Boost Control – Simulation and Calculated Results

Sl. No.	ST Duty Cycle D	Boost Factor B	Peak Line-to-Line Input Voltage V_i (V)	Z-Source Network Output Voltage V_m (V)	Remarks
1	0.37	1.1489	244.9048	281.3657 (265)	Calculated (from model)
2	0.2	1.1094	244.9048	271.6975	Calculated

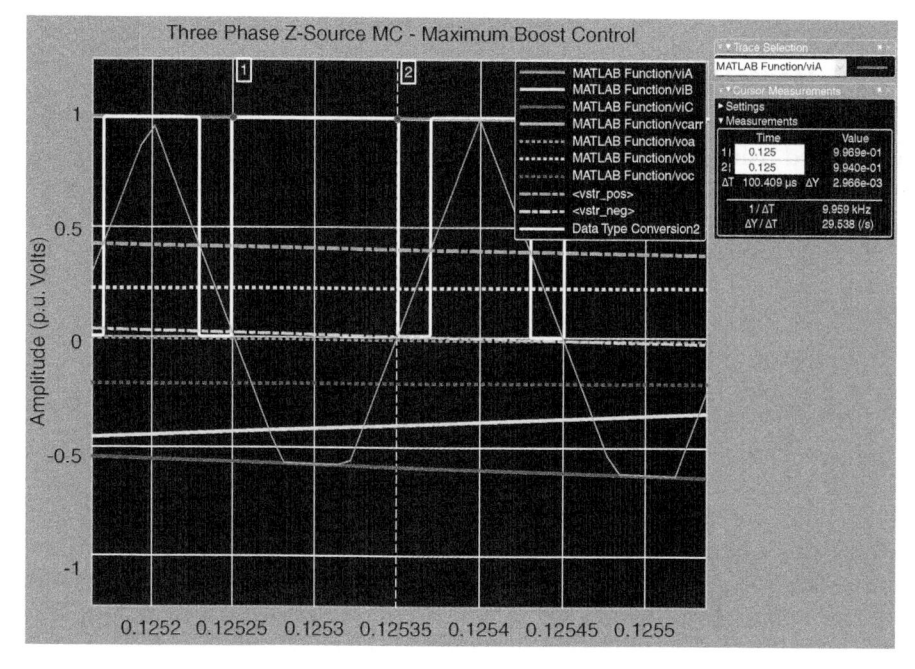

FIGURE 11.8
Three-phase Z-source MC – maximum boost control.

$$\frac{T_0}{T_s} = D = (1 - q) \tag{11.19}$$

By comparison with the derivation given in Section 11.2.1, we have

$$\frac{V_C}{V_i} = \frac{(1 - q)}{\sqrt{(3q^2 - 3q + 1)}} \tag{11.20}$$

$$\frac{V_m}{V_i} = B = \frac{1}{\sqrt{(3q^2 - 3q + 1)}} \tag{11.21}$$

The maximum boost control will provide a voltage gain G as given below:

$$G = B * q = \frac{q}{\sqrt{(3q^2 - 3q + 1)}} \qquad (11.22)$$

The method of generating the PWM waveforms S_a, S_b and S_c, three-phase input voltage envelope indicators and the gate pulses for the nine bidirectional switches are the same as presented in Section 11.2.2.

11.2.7 Model Development

The model of the three-phase Z-source MC with maximum boost control is developed using SIMULINK [7]. The model is the same as shown in Figure 11.6. The model parameters are shown in Table 11.1. The only difference is that n value is not used. Modulation index q is used to generate ST reference voltages vstr_pos and vstr_neg. The statements for vstr_pos and vstr_neg in Program segment 11.1 are replaced with equations 11.17 and 11.18.

11.2.8 Simulation Results

The simulation of the three-phase ZSDMC with maximum boost control for the parameters shown in Table 11.1 is carried out using ode23tb (stiff/TR-BDF2) solver. The simulation results for the output line-to-neutral voltage, line-to-line voltage, output voltage across Z-source network and load current are shown in Figure 11.9a–d, respectively. Also the ST period T_0 is measured for one carrier period from the simulated waveform shown in Figure 11.8 at three different locations each commencing at 125.2e–3, 125.25e–3 and 125.37e–3 s. The respective values for T_{01}, T_{02} and T_{03} are found to be 28.738e–6, 100.409e–6 and 30.064e–6 s, respectively.

The total T_0 is found to be 159.211e–6 s. The simulation results and theoretical values are tabulated in Table 11.3. The value of ST duty cycle D is found to be 0.796 by model simulation.

This value of D by calculation is 0.75 using equation 11.19.

11.2.9 Discussion of Results

The ST duty cycle D by calculation and by model simulation for the ZSDMC with maximum boost control differs by a narrow margin. Similarly, the variation of boost factor B by calculation and by model simulation is only a narrow percentage. The value of Z-source network peak output voltage V_m by calculation for a D of 0.75 and 0.796 closely agree.

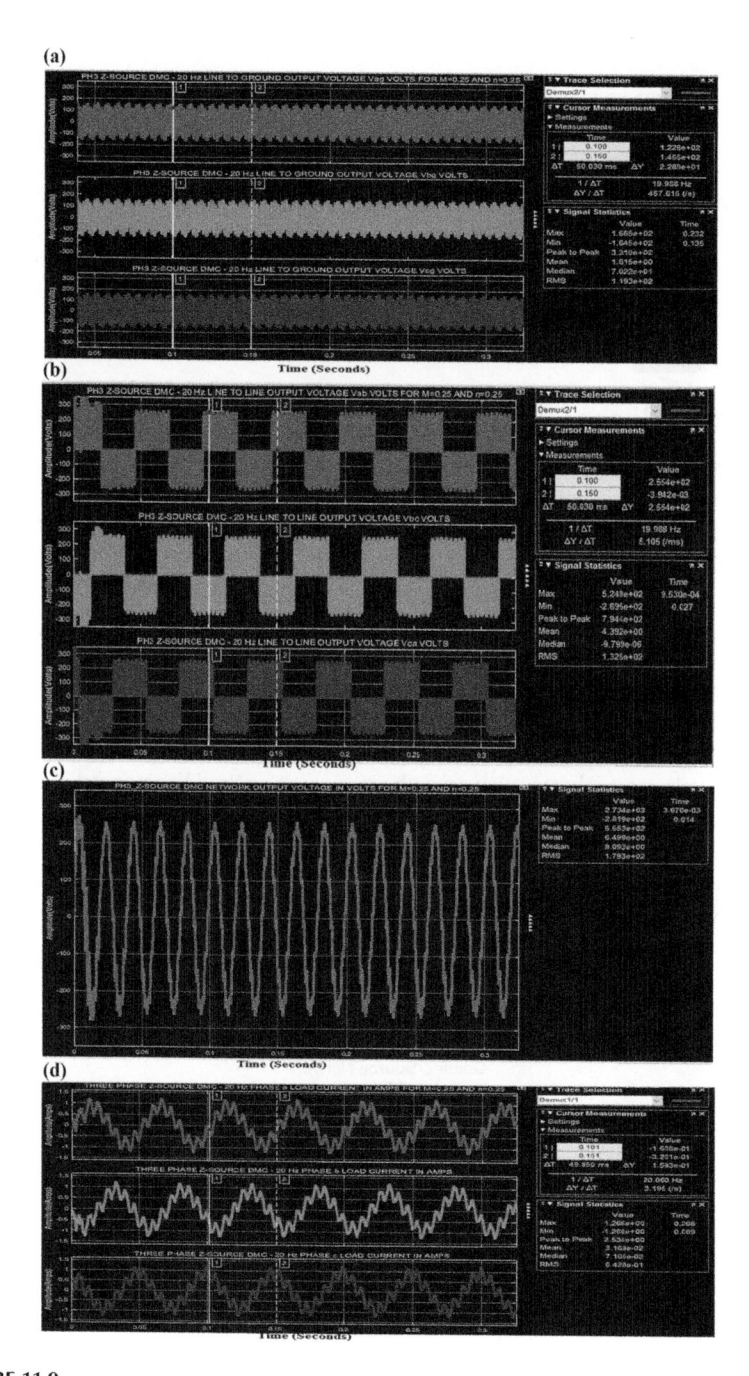

FIGURE 11.9
Three-phase ZSMC with maximum boost control: (a) line-to-neutral output voltage, (b) line-to-line output voltage, (c) Z-source network output voltage and (d) three-phase load current.

TABLE 11.3

Three-Phase ZSMC by Maximum Boost Control – Simulation and Calculated Results

Sl. No.	ST Duty Cycle D	Boost Factor B	Peak Line-to-Line Input Voltage V_i (V)	Z-Source Network Output Voltage V_m (V)	Remarks
1	0.796	1.3964	268.6	342 (278.6)	Calculated (from model)
2	0.75	1.5119	268.6	370.26	Calculated

11.3 Three-Phase Quasi Z-Source Indirect Matrix Converter

The conventional indirect matrix converter (IMC) avoids commutation switching problem compared to direct matrix converter (DMC) and their voltage gain is limited to 0.866. To improve the voltage gain of IMC above unity, QZSIMC topology is used [8].

Figure 11.10 shows the topology of a three-phase QZSIMC [8]. The Quasi Z-source network consists of two inductors L_1, L_2, two capacitors C_1, C_2 and a switch S_x connected between three-phase rectifier source at the front end and three-phase inverter source at the back end as shown in Figure 11.10. To improve the voltage gain above unity, an ST state is introduced. During ST state, switch S_x is open and inverter side upper and lower leg switches are closed simultaneously providing a short-circuit path. This period T_0 is called ST period. During NST state, switch S_x is closed and the gate pulses to the inverter switches are applied as for a conventional three-phase inverter. This NST period is designated as T_1. The total period $(T_1 + T_0)$ is the carrier switching period T. The equivalent circuits of the three-phase QZSIMC for the NST and ST period are shown in Figure 11.11a,b, respectively [8].

Let v_{C1}, v_{C2} be the voltage across capacitors C_1 and C_2 and i_{L1}, i_{L2} be the current through inductors L_1 and L_2. Let i_d be the DC-link current, v_{in} be the

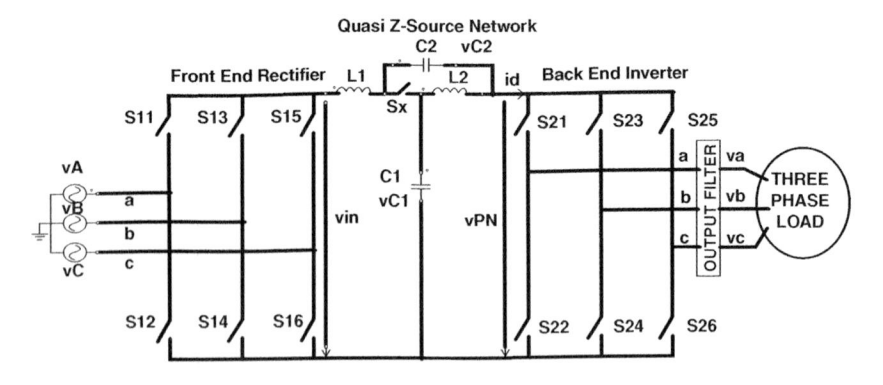

FIGURE 11.10

Three-phase quasi Z-source indirect matrix converter.

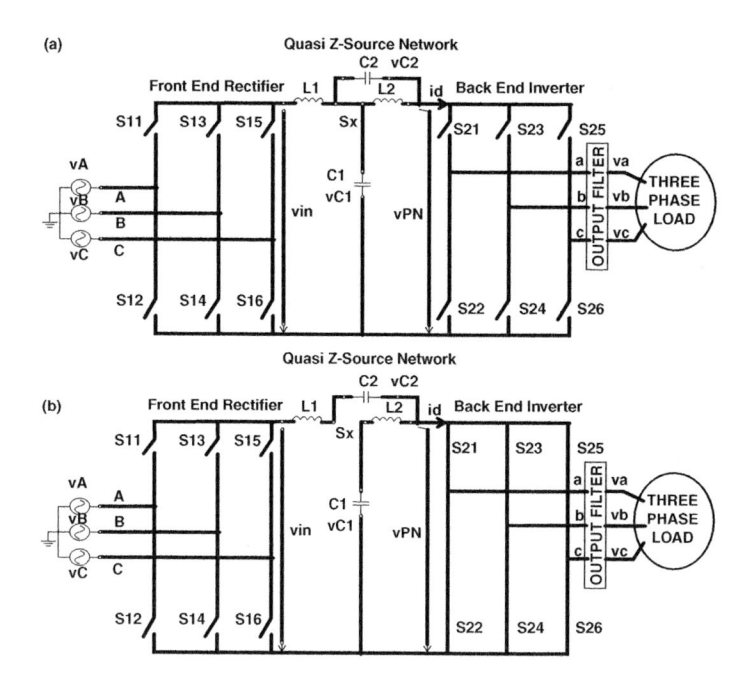

FIGURE 11.11
Equivalent circuit of three-phase QZSIMC: (a) non-shoot through state and (b) shoot through state.

output voltage of three-phase rectifier and v_{PN} be the DC-link voltage available at the input side of three-phase inverter. Referring to Figure 11.11a during NST state, the following state space equations apply:

$$
\begin{bmatrix}
\dfrac{dv_{C1}}{dt} \\[2mm]
\dfrac{dv_{C2}}{dt} \\[2mm]
\dfrac{di_{L1}}{dt} \\[2mm]
\dfrac{di_{L2}}{dt}
\end{bmatrix}
=
\begin{bmatrix}
0 & 0 & \dfrac{1}{C_1} & 0 \\[2mm]
0 & 0 & 0 & \dfrac{-1}{C_2} \\[2mm]
\dfrac{-1}{L_1} & 0 & 0 & 0 \\[2mm]
0 & \dfrac{1}{L_2} & 0 & 0
\end{bmatrix}
*
\begin{bmatrix}
v_{C1} \\[2mm]
v_{C2} \\[2mm]
i_{L1} \\[2mm]
i_{L2}
\end{bmatrix}
$$

$$
+ \; v_{in} *
\begin{bmatrix}
0 \\[2mm]
0 \\[2mm]
\dfrac{1}{L_1} \\[2mm]
0
\end{bmatrix}
+ i_d *
\begin{bmatrix}
\dfrac{-1}{C_1} \\[2mm]
\dfrac{1}{C_2} \\[2mm]
0 \\[2mm]
0
\end{bmatrix}
\tag{11.23}
$$

Similarly from Figure 11.11b during ST state, the following state space equations apply:

$$
\begin{bmatrix} \dfrac{dv_{C1}}{dt} \\[2mm] \dfrac{dv_{C2}}{dt} \\[2mm] \dfrac{di_{L1}}{dt} \\[2mm] \dfrac{di_{L2}}{dt} \end{bmatrix} = \begin{bmatrix} 0 & 0 & 0 & \dfrac{-1}{C_1} \\[2mm] 0 & 0 & \dfrac{1}{C_2} & 0 \\[2mm] 0 & \dfrac{-1}{L_1} & 0 & 0 \\[2mm] \dfrac{1}{L_2} & 0 & 0 & 0 \end{bmatrix} * \begin{bmatrix} v_{C1} \\[2mm] v_{C2} \\[2mm] i_{L1} \\[2mm] i_{L2} \end{bmatrix} + v_{in} * \begin{bmatrix} 0 \\[2mm] 0 \\[2mm] \dfrac{1}{L_1} \\[2mm] 0 \end{bmatrix} \qquad (11.24)
$$

The ST duty cycle D is defined as T_0/T. In steady state, the average voltage across the two inductors and the average current through two capacitors over one carrier switching period is zero. From equations 11.23 and 11.24, this can be expressed as follows:

$$
\frac{-v_{C1} * (1-D) * T}{L_1} + \frac{v_{in} * (1-D) * T}{L_1} + \frac{v_{C2} * D * T}{L_1} + \frac{v_{in} * D * T}{L_1} = 0 \quad (11.25)
$$

$$
\frac{-v_{C2} * (1-D) * T}{L_2} + \frac{v_{C1} * D * T}{L_2} = 0 \qquad (11.26)
$$

$$
\frac{i_{L1} * (1-D) * T}{C_1} - \frac{i_d * (1-D) * T}{C_1} - \frac{i_{L2} * D * T}{C_1} = 0 \qquad (11.27)
$$

$$
\frac{i_{L2} * (1-D) * T}{C_2} - \frac{i_d * (1-D) * T}{C_2} - \frac{i_{L1} * D * T}{C_2} = 0 \qquad (11.28)
$$

From equations 11.25 and 11.26,

$$
\frac{v_{C1}}{v_{C2}} = \frac{(1-D)}{D} \qquad (11.29)
$$

Using equations 11.25 and 11.29,

$$
\frac{v_{C1}}{v_{in}} = \frac{(1-D)}{(1-2D)} \qquad (11.30)
$$

From equations 11.29 and 11.30,

$$\frac{v_{C2}}{v_{in}} = \frac{D}{(1-2D)} \tag{11.31}$$

From equations 11.27 and 11.28,

$$i_{L1} = i_{L2} = \frac{P}{v_{in}} \tag{11.32}$$

where P is the power input from rectifier side. Also neglecting ST time interval being small compared to carrier period, the following equation applies:

$$v_{PN} = (v_{C1} + v_{C2}) = \frac{v_{in}}{(1-2D)} \tag{11.33}$$

$$B = \frac{1}{(1-2D)} \tag{11.34}$$

where B is the boost factor.

11.3.1 Model Development

The model of the three-phase QZSIMC was developed using SIMULINK [7]. This model is shown in Figure 11.12. The model parameters are given in Table 11.4.

The three-phase rectifier and inverter in Figure 11.12 are the ideal semiconductor switches S11 to S16 and S21 to S26, respectively, as shown in Figure 11.10. The six Pulse generator blocks from 'Pulse Generator' to 'Pulse Generator 5' in Figure 11.12 form the gate drive for the switches S11 to S16 of three-phase rectifier. The period of these gate drives corresponds to the period of the three-phase sine-wave input voltage, pulse width 50% of period and phase delays are, 0, π, $2\pi/3$, $5\pi/3$, $4\pi/3$ and $\pi/3$ rad, respectively. This DC output voltage of rectifier is filtered and forms the input to the Quasi Z-source network comprising inductors L_1, L_2 and capacitors C_1, C_2 and the diode which forms the switch S_x in Figure 11.10.

The triangle carrier has a frequency of 5 kHz, positive and negative peak of +1 and –1 V, respectively. Three-phase sine-wave modulating signal having a peak of 0.6 V and frequency 50 Hz is generated using three sine-wave AC function generator blocks. Triangle carrier and sine-wave modulating signals for each of the three phases are compared in a relational operator block which forms the comparator. The respective output of relational operator block for each phase forms the gate pulse for upper switches S21, S23 and S25 of the three-phase inverter and the gate pulse for lower switches S22, S24 and S26 for each phase is formed by inverting gate pulse for respective

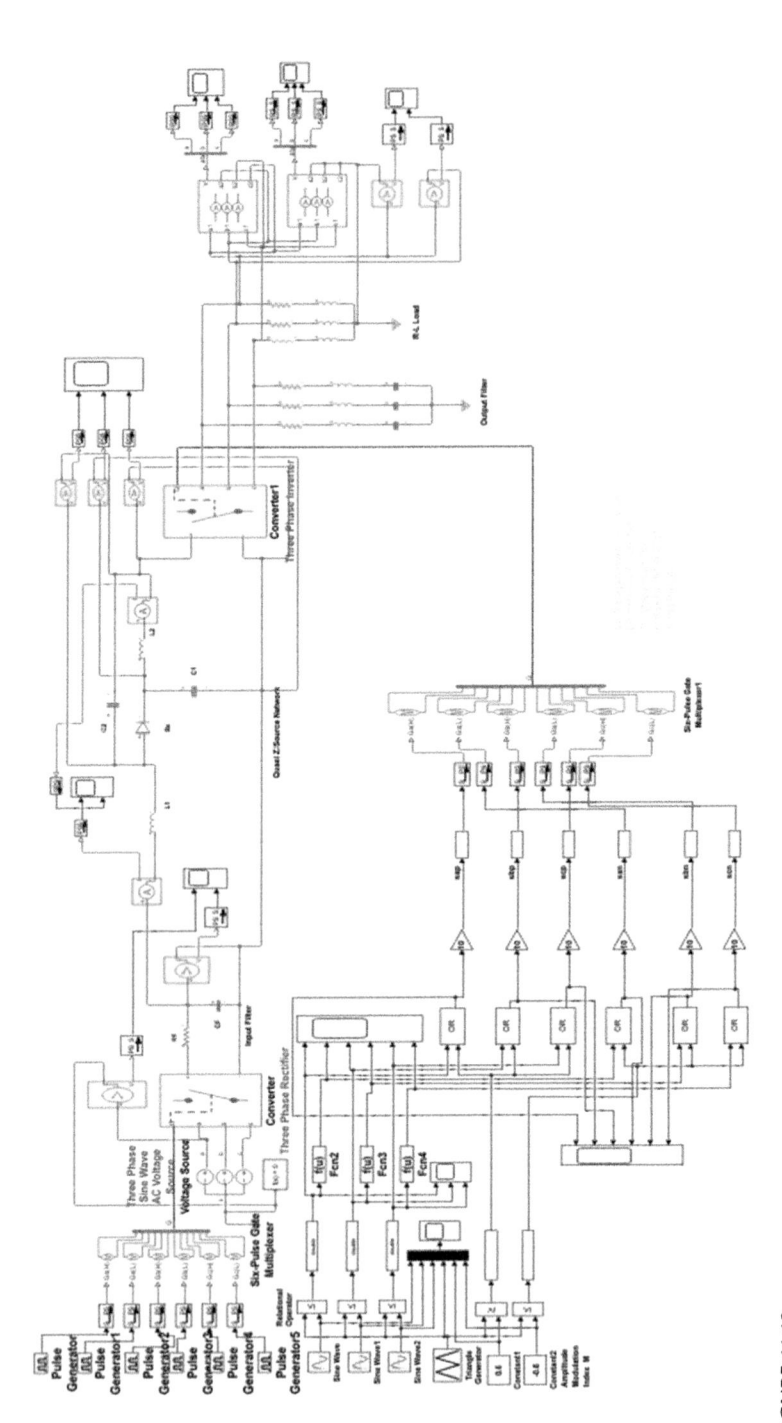

FIGURE 11.12
Model of three-phase AC to three-phase AC quasi Z-source indirect matrix converter.

TABLE 11.4

Three-Phase QZSIMC Model Parameters

	Three-Phase Sine-Wave AC Voltage Source		Quasi Z-Source Network					R–L–C	R-L
Sl. No.	Line-to-Neutral RMS Input Voltage (V)	Input Frequency (Hz)	L_1, L_2 (H)	C_1, C_2 (F)	Carrier Switching Frequency (Hz)	A.M. Index	Output Frequency (Hz)	Output Filter (Ω, H, F)	Load (Ω, H)
1	100	50	0.5e–3, 0.5e–3	150e–6, 150e–6	5e3	0.6	50	10, 800e–6, 1.2665e–6	30, 0.1

upper switches using three Fcn blocks marked Fcn2, Fcn3 and Fcn4 in Figure 11.12. This forms the gate pulse as for a three-phase conventional inverter. To generate ST zero switching state, two constant blocks each with value corresponding to positive and negative peak of sine-wave modulating signal are compared with triangle carrier in two relational operator blocks as shown in Figure 11.12. The respective outputs of gate pulse generated as for three-phase conventional inverter and that generated using two constant blocks and triangle carrier are then given to OR logic gates and then to gain blocks to generate gate pulses with required amplitude greater than the gate threshold value of the semiconductor switches used in the three-phase inverter shown in Figure 11.12. This gate pulse so generated provides ST zero switching state. An R–L–C filter and R–L load are connected to the output of the three-phase inverter.

11.3.2 Simulation Results

The simulation of the three-phase QZSIMC is carried out using fixed-step ode14x (extrapolation) solver. The data shown in Table 11.4 are used. Simulation results are shown in Figure 11.13a–f. The following calculations are made from the gate pulse waveform of three-phase inverter shown in Figure 11.13d.

From the gate pulse waveform for switches S21 and S22 of three-phase inverter, ST periods are found in two locations using scope cursor in the time interval 0.97–0.9702 s. These are, respectively, 33.048e–6 and 31.754e–6 s, giving a total time T_0 of 64.802e–6 s. The ST duty cycle D works out to 0.324. The output voltage of three-phase rectifier v_{in} is found from Figure 11.13a as 61.5 V. The ratio v_{c1}/v_{c2} from equation 11.29 works out to 2.0864. The values of v_{c1} and v_{c2} from Figure 11.13b are found to be 119.4 and 58.07 V, respectively, which gives a ratio of 2.0561. Using equations 11.30 and 11.31, theoretical values for v_{c1} and v_{c2} work out to 118.1080 and 56.6080 V, respectively. Similarly, RMS values of inductor currents i_{L1} and i_{L2} from Figure 11.13c are found to be 3.6790 A in both cases confirming

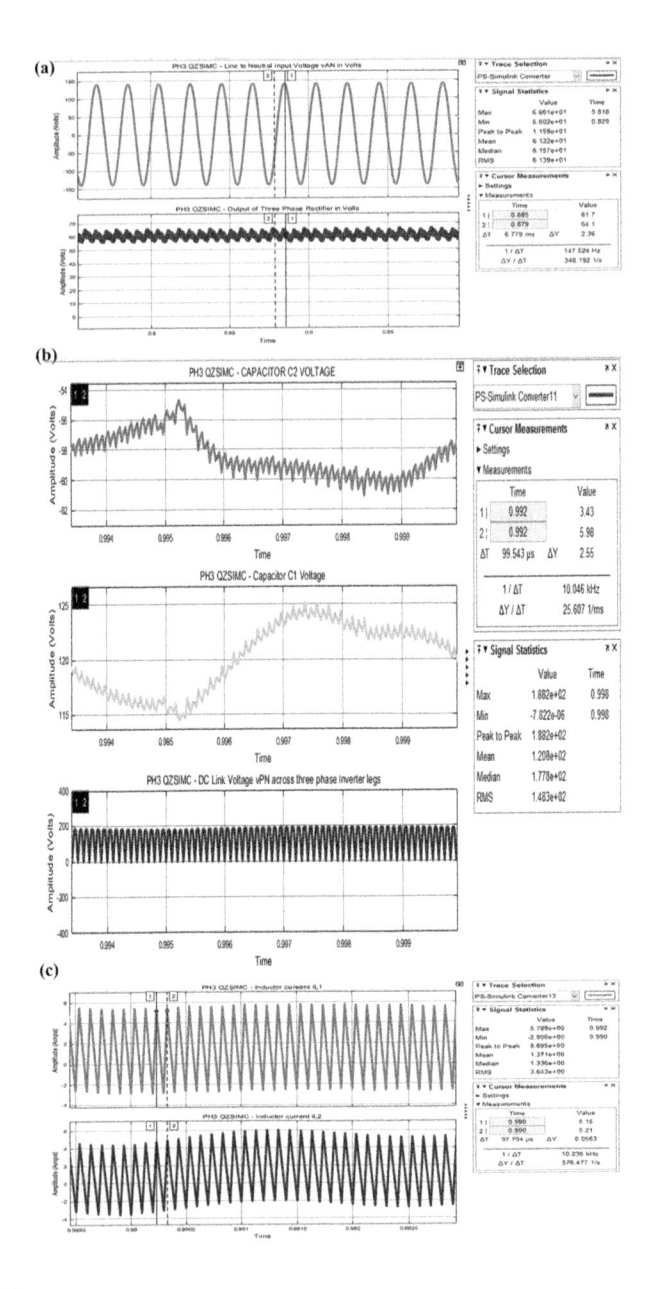

FIGURE 11.13

Three-phase QZSIMC simulation results: (a) line-to-neutral input voltage (top) and DC output voltage of rectifier (bottom); (b) capacitor C_2 voltage (top) and C_1 voltage (middle), voltage across inverter legs (bottom); (c) inductor current i_{L1} (top) and i_{L2} (bottom); (d) six gate pulse for the three-phase inverter; (e) three-phase line-to-line output voltage; and (f) three-phase line-to-ground output voltage.

(Continued)

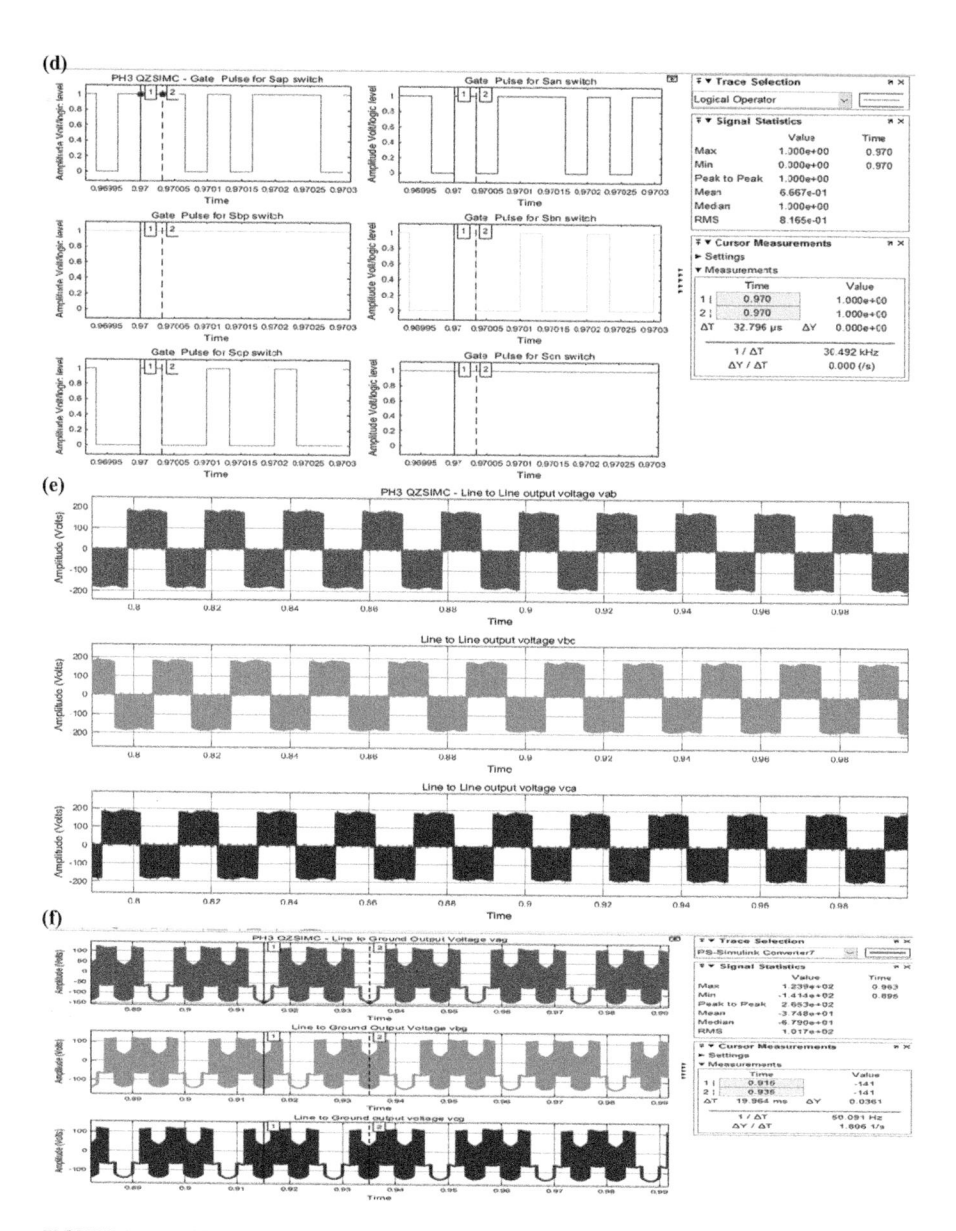

FIGURE 11.13 (CONTINUED)
Three-phase QZSIMC simulation results: (a) line-to-neutral input voltage (top) and DC output voltage of rectifier (bottom); (b) capacitor C_2 voltage (top) and C_1 voltage (middle), voltage across inverter legs (bottom); (c) inductor current i_{L1} (top) and i_{L2} (bottom); (d) six gate pulse for the three-phase inverter; (e) three-phase line-to-line output voltage; and (f) three-phase line-to-ground output voltage

TABLE 11.5

Three-Phase QZSIMC – Model Simulation Results

Sl. No.	ST Duty Cycle D	Boost Factor B	Three-Phase Rectifier Output Voltage V_{in} (V)	V_{c1} (V)	V_{c2} (V)	Ratio V_{c1}/V_{c2}	i_{L1}, i_{L2} (A) (RMS)	v_{PN} (V)	Output Voltage (V) (RMS) v_{LL}, v_{LG}	Remarks
1	0.324	2.8409	61.5	119.4 (118.1)	58.07 (56.6)	2.0561 (2.086)	3.6790, 3.6790	174.715 (174.716)	117.3, 102.2 (104.8)	Model results (theoretical values)

equation 11.32. Using equation 11.33, v_{PN} is found to be 174.7159 V, whereas by addition of v_{c1} and v_{c2}, v_{PN} works out to 174.716 V.

Three-phase 50 Hz line-to-line and line-to-ground output voltages are shown in Figure 11.13e,f, respectively. Their RMS values are found to be 117.3 and 102.2 V, respectively. Three-phase rectifier and inverter modulation indices are 1 and 0.6, respectively. The RMS line-to-ground output voltage is $v_{PN}*1*0.6$ which works out to 104.8295 V which closely agree with simulation result. The simulation results are tabulated in Table 11.5.

11.3.3 Discussion of Results

In the case of three-phase QZSIMC, ratio of v_{c1}/v_{c2}, values of i_{L1}, i_{L2}, voltage v_{PN} across back-end inverter legs, RMS value of line-to-ground output voltage by calculation and model simulation closely agree.

11.4 Conclusions

With a conventional three-phase MC, it is possible to achieve output voltage which is less than the input voltage by a factor corresponding to modulation index q and the maximum achievable voltage is limited to 86.6% of input voltage. By inserting a Z-source L–C network along with a series semiconductor switch, it is possible to boost the output voltage by a factor called the boost factor B which can go up to two with simple boost control. The recorded RMS and peak line-to-ground and line-to-line output voltage by model simulation indicate that their values have been boosted as compared to their respective input values. In the case of three-phase QZSIMC, also the output voltage is boosted by a factor B greater than unity. For three-phase QZSIMC, this factor is B multiplied by modulation index of front-end rectifier and back-end inverter.

References

1. A. Alesina and M.G.B. Venturini: Solid state power conversion: a Fourier analysis approach to generalized transformer synthesis, *IEEE Transactions on Circuits and Systems*, Vol. CAS-28, No. 4, 1981, pp. 319–330.
2. A. Alesina and M. Venturini: Intrinsic amplitude limits and optimum design of 9-switches direct PWM AC-AC converters, *IEEE-PESC*, Kyoto, Japan, 1988, pp. 1284–1291.
3. A. Alesina and M. Venturini: Analysis and design of optimum amplitude nine-switch direct AC-AC converters, *IEEE Transactions on Power Electronics*, Vol. 4, 1989, pp. 101–112.
4. P. Wheeler, J. Rodriguez, J.C. Clare, L. Empringham and A. Weinstein: Matrix converters – a technology review, *IEEE Transactions on Industrial Electronics*, Vol. 49, No. 2, 2002, pp. 276–288.
5. P. Wheeler, J. Clare, L. Empringham, M. Apap and M. Bland: Matrix converters, *Power Engineering Journal*, 2002, Vol. 16, No. 6, pp. 273–282.
6. B. Ge, Q. Lei, W. Qian and F.Z. Peng: A family of Z-source matrix converters, *IEEE Transactions on Industrial Electronics*, Vol. 59, No. 1, 2012, pp. 35–46.
7. The Mathworks Inc.: www.mathworks.com, MTALAB/SIMULINK user manual, MATLAB R2017b, 2017.
8. S. Liu, B. Ge, H.A. Rub, X. Jiang, and F.Z. Peng: A novel indirect Quasi-Z-source matrix converter applied to induction motor drives, *IEEE Energy Conversion Congress and Exposition (ECCE)*, Denver, CO, September 2013, pp. 2440–2444.

12

A Combined PWM Sine-Wave AC to AC and AC to DC Converter

12.1 Introduction

A combined pulse-width-modulated (PWM) single-phase/three-phase sine-wave AC to single-phase/three-phase variable-voltage variable-frequency sine-wave AC converter which can also be used as a sine-wave AC to DC converter with provision for reversing the polarity of the DC output voltage is reported in this chapter [1,2]. The model of this dual converter is developed using SIMULINK software. Simulation results indicate that both variable-frequency variable-voltage PWM AC output voltage in the AC mode and a PWM variable DC output voltage with polarity reversal capability in the DC mode are obtainable from the same converter, by suitably transferring the gate drive using a selector switch. The polarity reversal capability of DC output voltage in the DC mode renders this converter suitable for speed control and brake by plugging of separately excited DC (SEDC) motor and permanent magnet DC (PMDC) motor, thus opening new possibilities for applications in hybrid electric vehicles (HEVs) and electric traction (ET) with the former type and aircraft control applications with the later type motor.

12.2 Single-Phase PWM AC to AC and AC to DC Converter

The block diagram of a combined PWM sine-wave AC to AC and AC to DC converter is shown in Figure 12.1. Single-phase sine-wave AC supply having frequency f_s Hz is given to the primary. The two secondaries have equal voltages with a phase difference of 180°. The voltage across first secondary winding is connected to IGBT bidirectional switch with a common-emitter configuration. The voltage across the second secondary winding is connected to another IGBT bidirectional switch with a common-collector configuration. The gates of these two IGBT bidirectional switches are cross connected

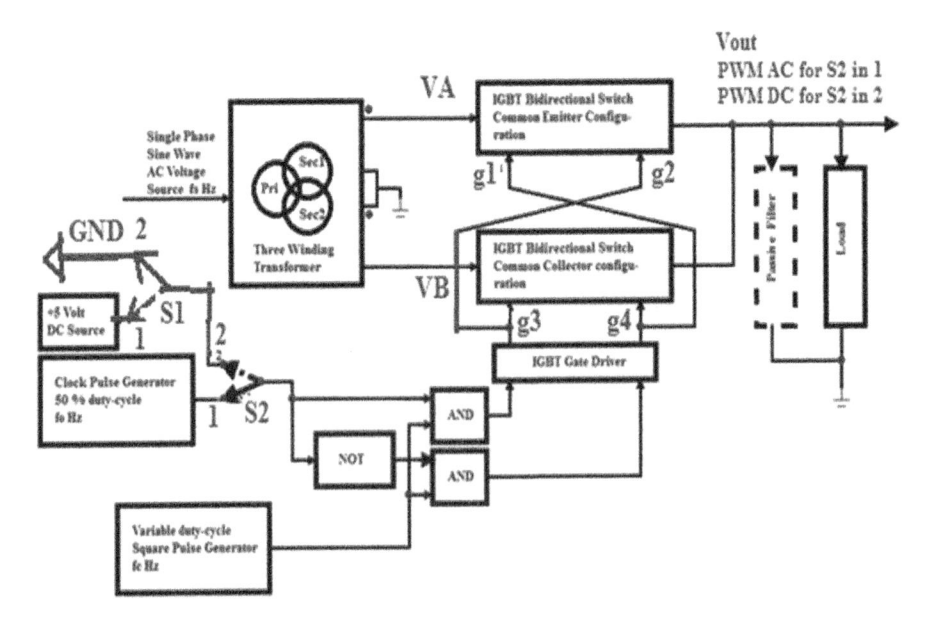

FIGURE 12.1
Block diagram of a combined PWM AC/AC and AC/DC converter.

[1,2]. The outputs of these two bidirectional switches are connected together which drives the load.

The gate drive consists of clock generator with frequency f_o Hz and duty cycle 50% for the AC output voltage mode and a +5 V DC source and zero volts or ground for the DC output voltage mode. The selector switch S2 in position 1 irrespective of S1 position selects the AC output voltage mode. With S2 in position 2, S1 in positions 1 and 2, respectively, selects the positive and negative DC output voltage mode. The variable duty-cycle square pulse generator with frequency f_c Hz forms the carrier signal. The gate drive output is inverted using a NOT gate. The gate drive output ANDed with the carrier signal forms the gate drive for the IGBTs whose gates are marked g2 and g3 [1,2]. The inverted gate drive ANDed with carrier signal forms the gate drive for the two IGBTS whose gates are marked g1 and g4 [1,2].

12.2.1 Model Development

To understand the operation of the above AC to AC cum AC to DC converter with polarity reversal capability, a SIMULINK model [3] is developed neglecting carrier PWM, as shown in Figure 12.2. In the AC mode, square pulse is selected as the gate drive. This square pulse having period T_0, frequency f_0 and duty cycle 50% forms the gate drive for bidirectional switches Q1 and Q4. This square pulse inverted using NOT gate forms the gate drive for Q2 and Q3. The three winding transformer primary is connected to

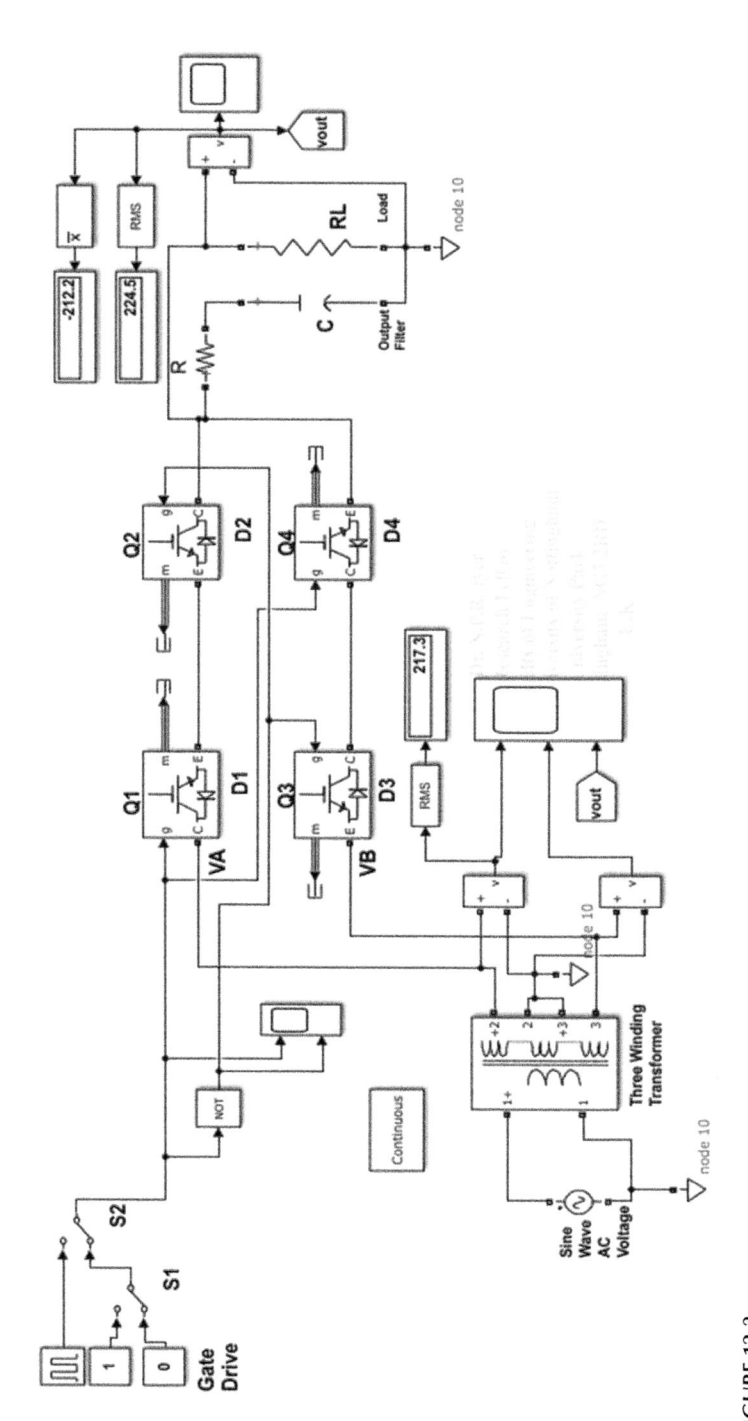

FIGURE 12.2

Basic model of a combined sine-wave AC to AC and AC to DC converter.

single-phase sine-wave AC voltage source having frequency f_s Hz and the two secondaries are connected to V_A and V_B inputs of IGBT bidirectional switches. A resistive load is used.

12.2.2 Principle of Operation

The principle of operation of the above converter for AC and DC modes is explained below. Referring to Figure 12.2, in the AC mode, switch S2 is connected to square pulse gate drive having period T_0 seconds and duty cycle 50% irrespective of the position of S1. DC mode has two gate drive inputs namely input HIGH or logic 1 (+5 V) or input LOW or logic 0 (GND). Switch S2 is disconnected from square pulse gate drive and is thrown to select S1 output for DC mode. The period of the input sine-wave AC voltage V_s is T $(1/f_s)$ s.

12.2.2.1 AC Mode

The three winding transformer has unity turns ratio with both secondaries. Assume that NO output filter and NO carrier modulation are used.

Given $T \ll T_0$. Cases E to H in Section 9.3.1 apply with the following modification.

The value $+V_{dc}/2$ for V_A, V_B and V_{OUT} is to be replaced with $+V_s$ where V_s is the time instantaneous value of the input sine-wave AC voltage. Similarly, the value $-V_{dc}/2$ is to be replaced with $-V_S$. The output voltage period is T_0. The simulation result for this case is shown in Figure 12.3.

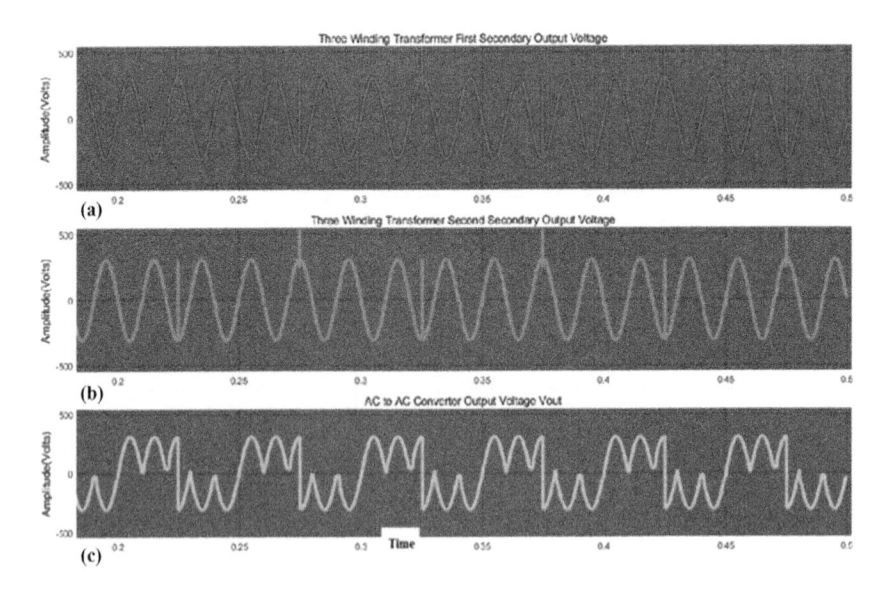

FIGURE 12.3
Sine-wave AC to AC and AC to DC converter – AC mode for $T \ll T_0$ – simulation results. First secondary input voltage (a), second secondary input voltage (b) and output voltage for $T \ll T_0$ (c).

Given $T \gg T_0$. Cases A to D in Section 9.3.1 apply with the modification as for Cases E to H mentioned above. The output voltage period is T_0. The simulation result for this case is shown in Figure 12.4. The bidirectional IGBT switch conducting pattern for AC mode is shown in Figure 12.5.

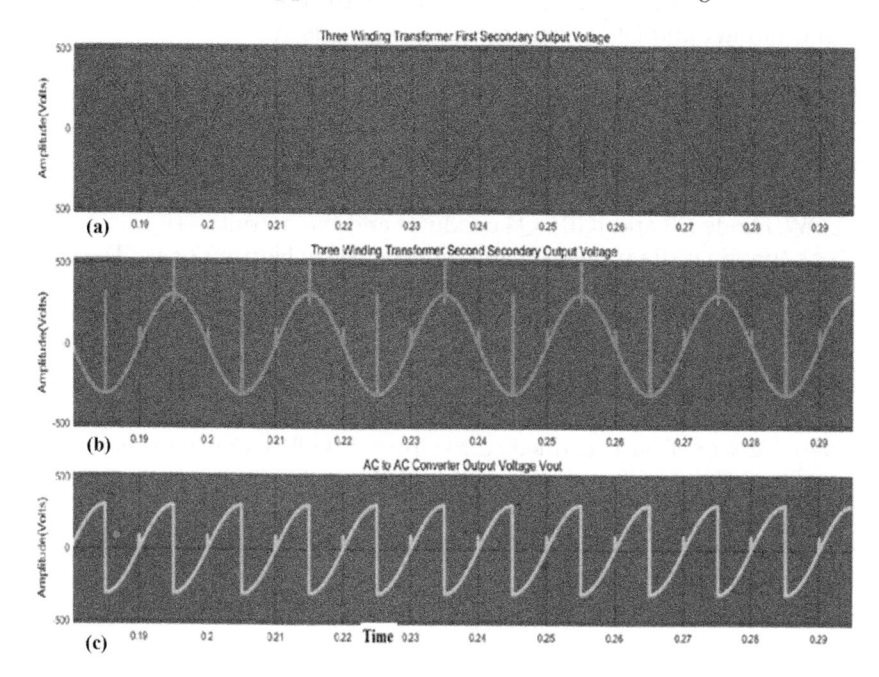

FIGURE 12.4
Sine-wave AC to AC and AC to DC converter – AC mode for $T \gg T_0$ – simulation results. First secondary input voltage (a), second secondary input voltage (b) and output voltage for $T \gg T_0$ (c).

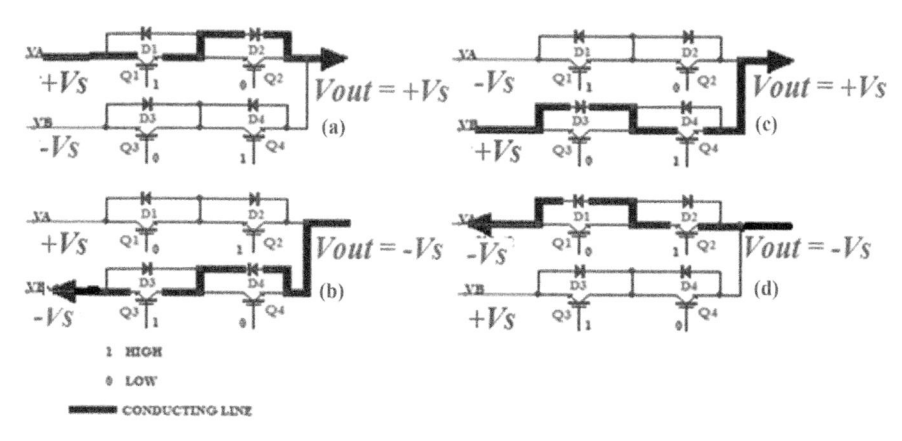

FIGURE 12.5
Bidirectional IGBT switches conducting pattern for AC mode.

12.2.2.2 DC Mode

In this mode, there are two types of gate drive inputs. Referring to Figure 12.2, switch S1 connected to +5 V (HIGH) and the other it is connected to ground or zero volts (LOW). Switch S2 is connected to the output of S1. Assume that NO output filter and NO carrier modulation are used.

Given $0 < t \leq T/2$ s, S1 is connected to +5 V (HIGH). During this time interval, V_A is $+V_s$ and V_B is $-V_s$. The gates of Q1 and Q4 are HIGH and those of Q2 and Q3 are LOW. IGBT Q1 and the diode D2 conducts and the output is $+V_s$.

Given $T/2 < t \leq T$, S1 is connected to +5 V (HIGH). During this interval, V_A is $-V_s$ and V_B is $+V_s$. The gates of Q1 and Q4 are HIGH and those of Q2 and Q3 are LOW. Diode D3 and IGBT Q4 conducts and the output is $+V_s$.

This situation is diagrammatically represented in Figure 12.5a,c. The simulation result are shown in Figure 12.6.

Given $0 < t \leq T/2$ s, S1 is connected to Ground (LOW). During this time interval, V_A is $+V_s$ and V_B is $-V_s$. The gates of Q1 and Q4 are LOW and those of Q2 and Q3 are HIGH. IGBT Q3 and the diode D4 conducts and the output is $-V_s$.

Given $T/2 < t \leq T$, S1 is connected to Ground (LOW). During this interval, V_A is $-V_s$ and V_B is $+V_s$. The gates of Q1 and Q4 are LOW and those of Q2 and Q3 are HIGH. Diode D1 and IGBT Q2 conducts and the output is $-V_s$.

This situation is diagrammatically represented in Figure 12.5b,d. The simulation results are shown in Figure 12.7.

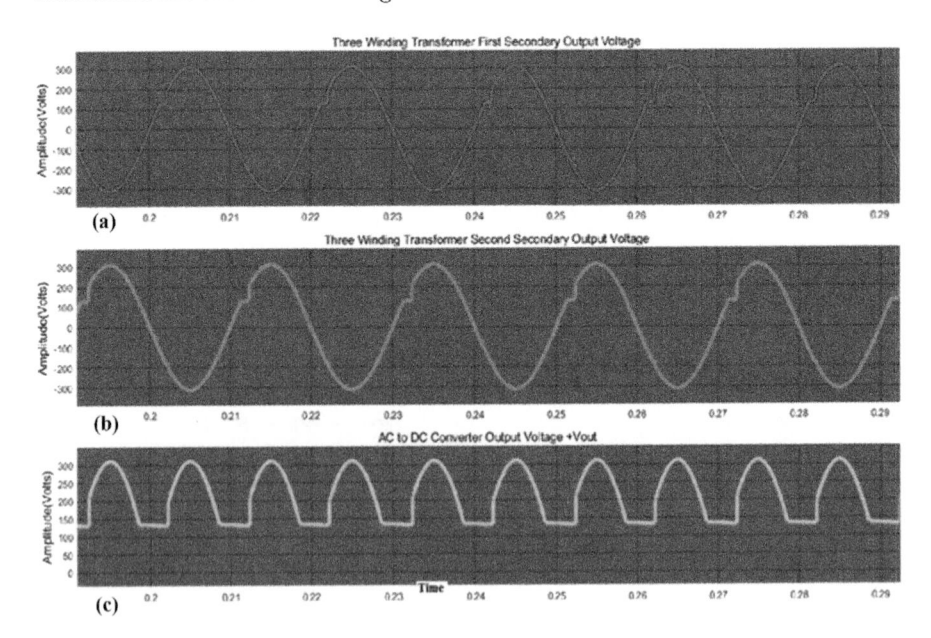

FIGURE 12.6
Sine-wave AC to AC and AC to DC converter – DC mode simulation results. First secondary input voltage (a), second secondary input voltage (b) and DC output voltage $+V_{OUT}$ (c).

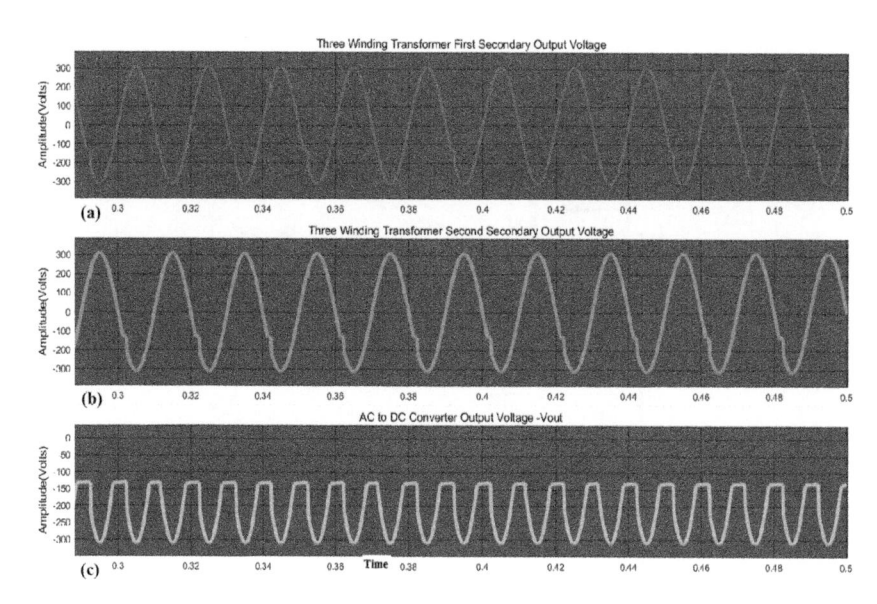

FIGURE 12.7
Sine-wave AC to AC and AC to DC converter – DC mode simulation results. First secondary input voltage (a), second secondary input voltage (b) and DC output voltage –V_{OUT} (c).

12.2.3 Single-Phase Sine-Wave PWM AC to AC and AC to DC Converter

The complete block diagram of the single-phase sine-wave PWM AC to AC and AC to DC converter is shown in Figure 12.1 [2]. Based on this block diagram, a model of the same is developed using SIMULINK [3]. This model is shown in Figure 12.8. The model is explained below.

The AC gate drive is a square pulse generator with appropriate amplitude and frequency f_0 Hz corresponding to desired output frequency and 50% duty cycle. The DC gate drives are +5 and 0 V constant DC voltage signals. Pulse-width modulation is achieved by a square pulse carrier generator having frequency f_c Hz and a variable duty cycle. By varying the duty cycle of square pulse carrier, the output voltage magnitude can be controlled. Single-phase sine-wave AC voltage is given to the primary of the three winding transformer. The two secondary windings have appropriate turns ratio as per output voltage requirements. The two secondary windings are connected in series and the common point of the secondary is grounded. The output AC voltage of the two secondary windings V_A and V_B differ in phase by 180°. The IGBTs Q1–D1 and Q2–D2 form one bidirectional switch pair and so also Q3–D3 and Q4–D4. The gates of Q1 and Q4 are interconnected and so also the gates of Q2 and Q3. The collector of Q2 and emitter of Q4 are tied together which forms the output terminal. The two secondary voltages V_A and V_B are connected to the collector of Q1 and emitter of Q3, respectively. The load is connected to the output terminal. Output filter can be used and this connection is not shown.

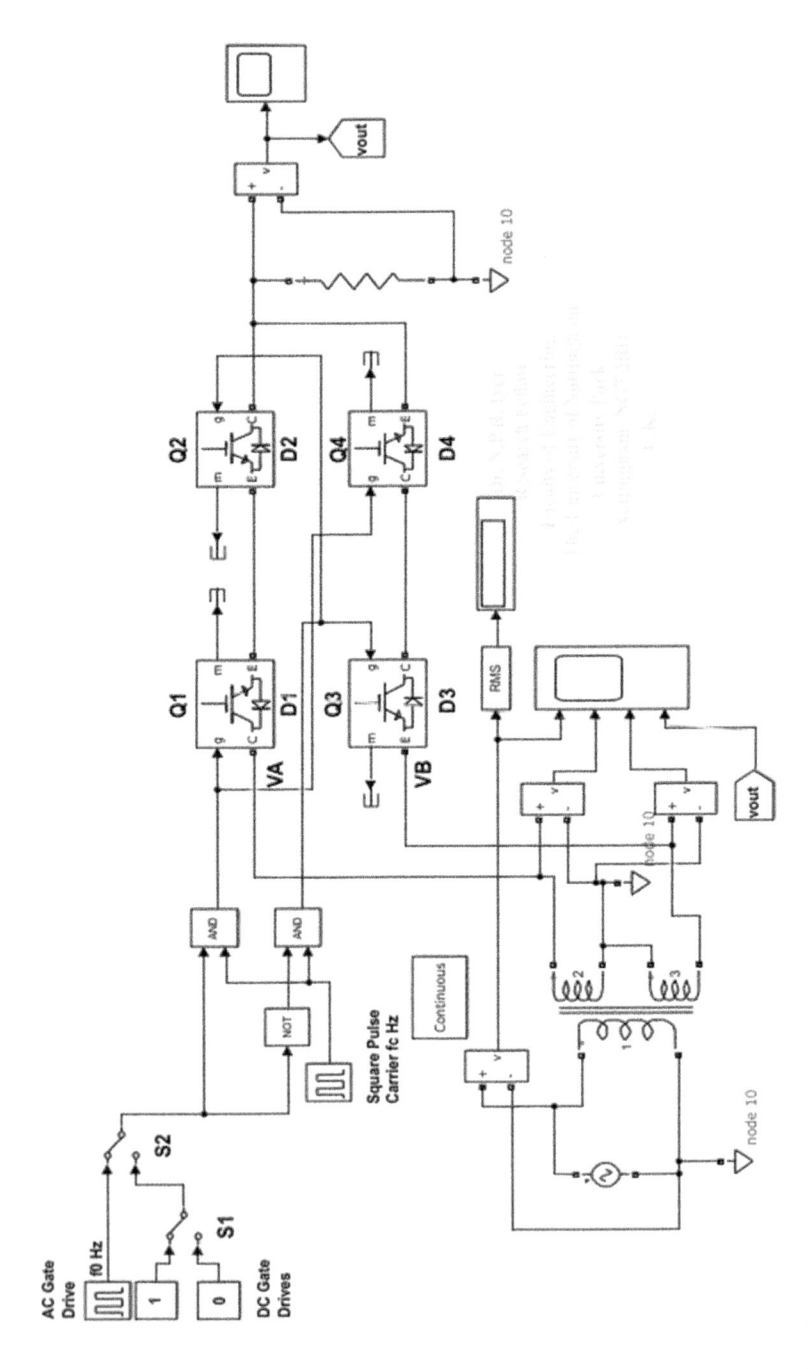

FIGURE 12.8
Single-phase sine-wave PWM AC to AC and AC to DC converter.

TABLE 12.1

Single-Phase Sine-Wave PWM AC to AC and AC to DC Converter – Model Parameters

Sl. No.	RMS Input Voltage (V)	Input Frequency f_i (Hz)	Output Frequency f_o (Hz)	Carrier Frequency f_c (Hz)	Carrier Duty Cycle D	Mode
1	220	50	20, 100	5,000	0.2, 0.8	AC
2	220	50	0	5,000	0.2, 0.8	DC

12.2.4 Simulation Results

The simulation of the above converter is carried out using SIMULINK [3]. The ode 15s (Stiff/NDF) solver is used. The simulation is carried out for the parameters shown in Table 12.1. Simulation results for the AC mode are shown in Figure 12.9a–d, and those for the DC mode are shown in Figure 12.10a–d [2]. Simulation results are also tabulated in Table 12.2.

12.3 Three-Phase Sine-Wave PWM AC to AC and AC to DC Converter

The block diagram of the three-phase sine-wave PWM AC to AC and AC to DC converter is shown in Figure 12.11. The block diagram for the single-phase AC input is presented in Section 12.2. The same block diagram applies for the three-phase AC input voltage except for the following modifications.

Here, three-phase sine-wave AC voltages V_A, V_B and V_C are the inputs and v_a, v_b and v_c are the three-phase output voltages. Each input phase has selector switches S1A, S2A, S1B, S2B, S1C and S2C for selection of the gate drive. Hence, by proper switch contact selection, all the three-phase can give three-phase AC output voltage, three DC output voltage or a combination of AC and DC output voltages. Only one variable duty-cycle square pulse carrier is used for all the three-phase. The clock pulse gate drive for input phase A, B and C all have a frequency f_0 Hz, 50% duty cycle and their phase delays are 0, $-120[1/(3f_0)]$ s] and $-240[2/(3f_0)]$ s] degrees, respectively. Three winding transformers, IGBT bidirectional switch pairs, output filter and load for each input phase are connected as shown in Figure 12.11.

12.3.1 Model Development

The principle of operation of this three-phase sine PWM AC to AC and AC to DC converter is the same as explained in Section 12.2.2. A SIMULINK [3] model of this three-phase converter is shown in Figure 12.12. The model is the same

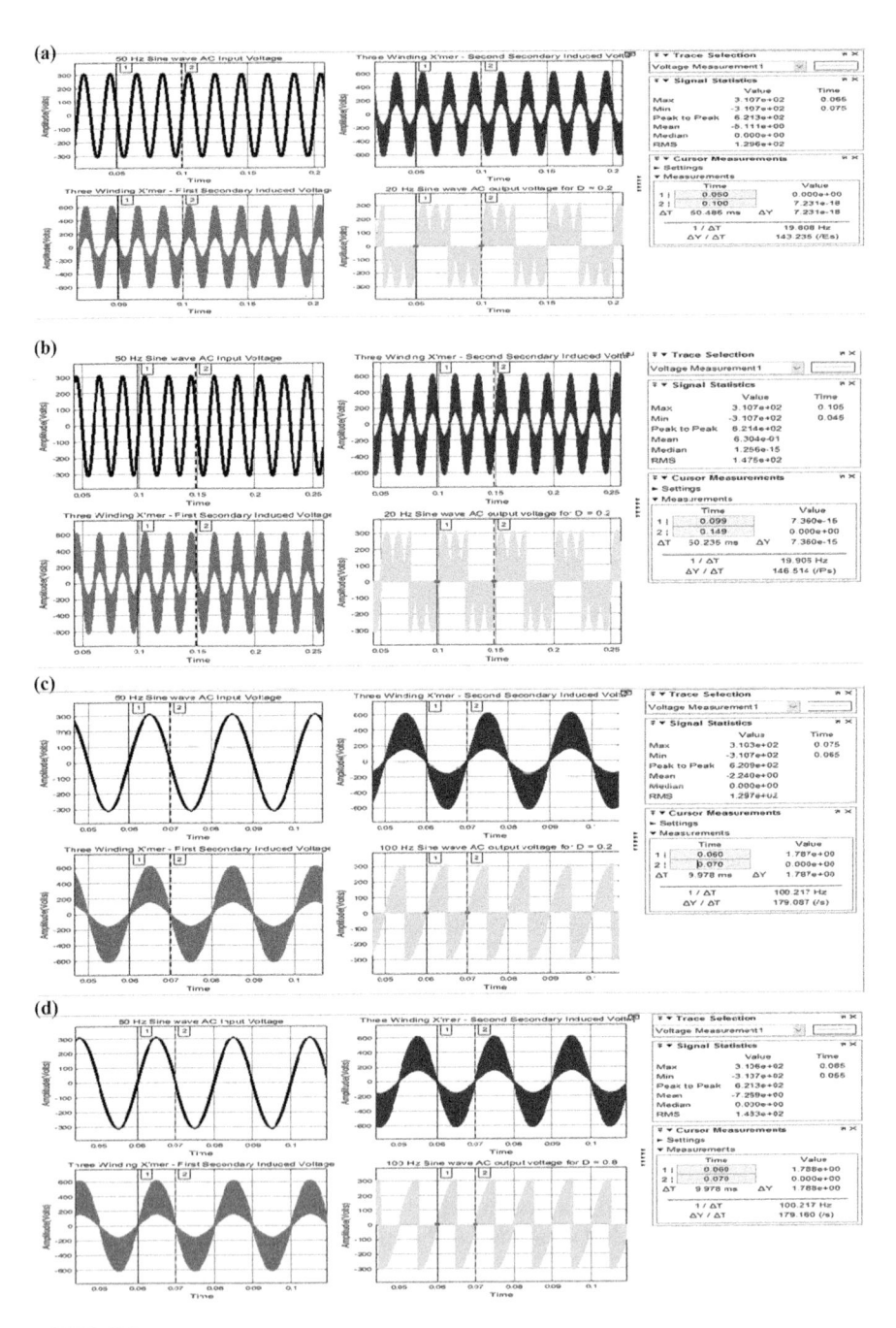

FIGURE 12.9

Single-phase sine-wave PWM AC to AC and AC to DC converter – simulation results for primary input, two secondary input and output voltages. (a) $f_0 = 20\,\text{Hz}$, $D = 0.2$, (b) $f_0 = 20\,\text{Hz}$, $D = 0.8$, (c) $f_0 = 100\,\text{Hz}$, $D = 0.2$ and (d) $f_0 = 100\,\text{Hz}$, $D = 0.8$.

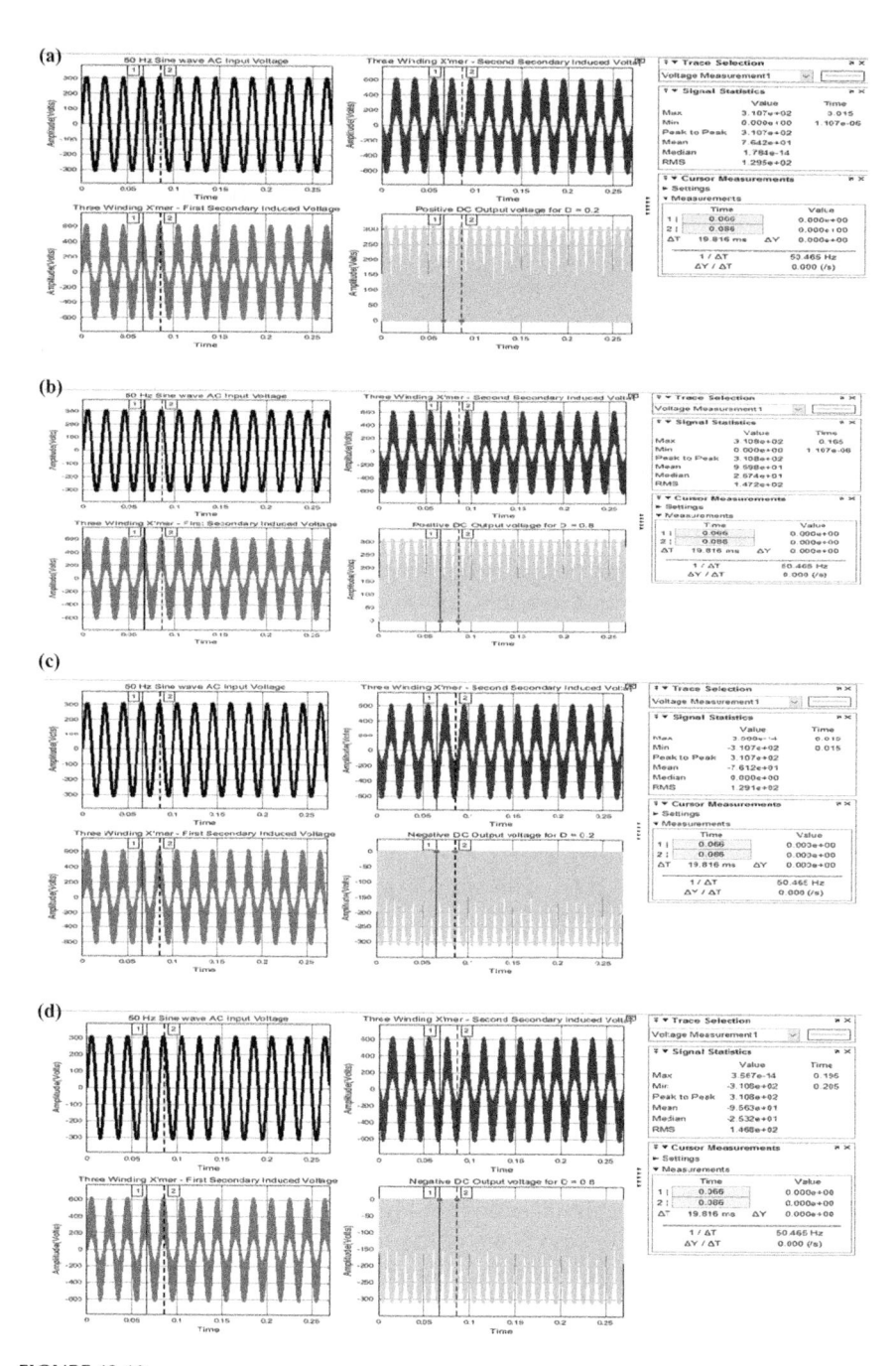

FIGURE 12.10
Single-phase sine-wave AC to AC and AC to DC converter – simulation results for primary input, two secondary input and positive and negative DC output voltages. (a) $D = 0.2$, (b) $D = 0.8$, (c) $D = 0.2$ and (d) $D = 0.8$.

TABLE 12.2

Single-Phase Sine-Wave PWM AC to AC and AC to DC Converter – Simulation Results

Sl. No.	RMS Input Voltage (V)	Input Frequency (Hz)	Output Frequency (Hz)	Carrier Duty Cycle D	Mode	RMS Output Voltage (V)	Mean Output Voltage (V)
1	220	50	20	0.2, 0.8	AC	129.4, 146.6	–
2	220	50	100	0.2, 0.8	AC	131.1, 147.4	–
3	220	50	0	0.2, 0.8	DC	–	+75.31, +95.08
4	220	50	0	0.2, 0.8	DC	–	–75.42, –95.85

as explained in Section 12.2.3. The gate pulse has a frequency f_0 Hz, 50% duty cycle and phase delay of 0, $T_0/3$ and $2T_0/3$ s for input phase A, B and C, respectively, where $T_0 = 1/f_0$ is the period of the output voltage. The square pulse carrier with frequency f_c Hz and variable duty cycle is the same for the three input phase. Here, three identical resistive loads are used.

12.3.2 Simulation Results

The simulation of the above converter is carried out using SIMULINK [3]. The ode 23tb (stiff/TR-BDF2) solver is used. The simulation is carried out for the parameters shown in Table 12.1. The simulation results for the AC mode are shown in Figure 12.13a–d and those for the DC mode are shown in Figure 12.14a–d [2]. The simulation results are also tabulated in Table 12.3.

12.4 RMS and Average Value of a Uniform PWM Sine-Wave AC Voltage

The formula for the RMS and average output voltage of a uniform square PWM sinusoidal voltage signal shown in Figure 12.15 is derived below.

Initially, assume a pure sine wave of the form $V_m*\sin(\omega t)$ with NO carrier pulse-width modulation. The RMS and average value of this pure sinusoidal signal can be derived as follows:

$$V_{rms}^2 = \frac{V_m^2}{\pi} * \int_0^\pi \sin^2(\omega t) d(\omega t) \qquad (12.1)$$

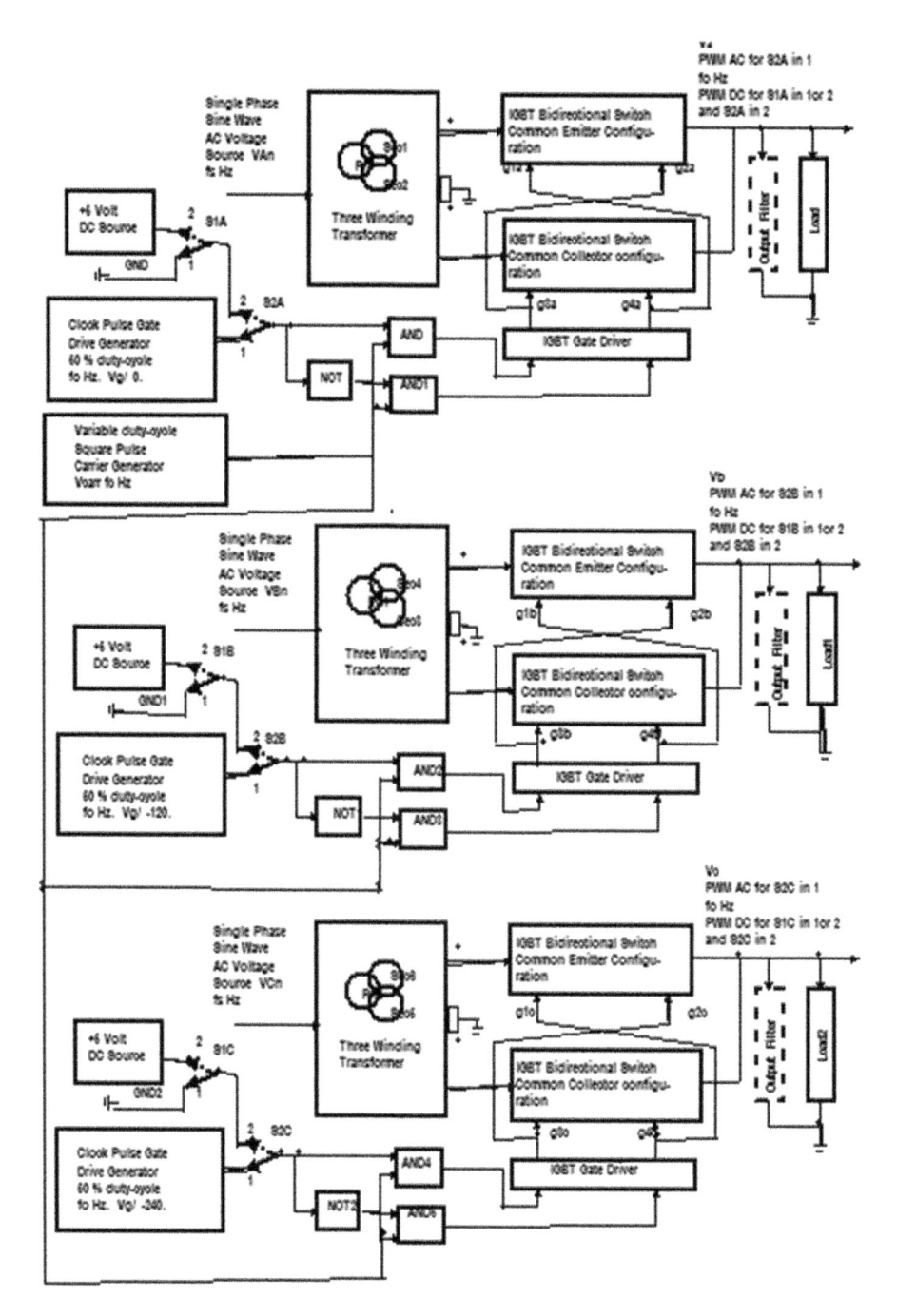

FIGURE 12.11
Block diagram of a three-phase sine-wave PWM AC to AC and AC to DC converter.

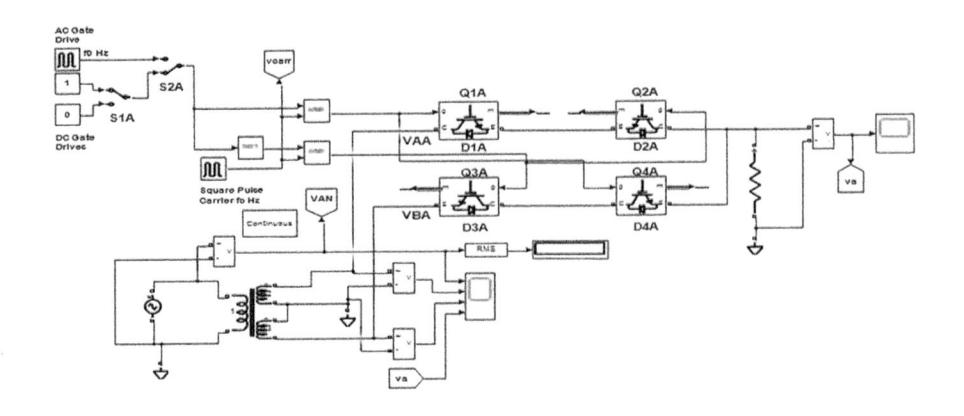

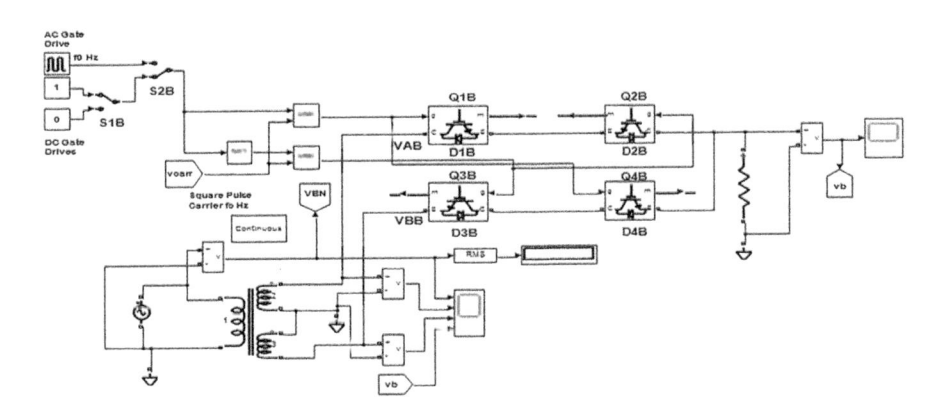

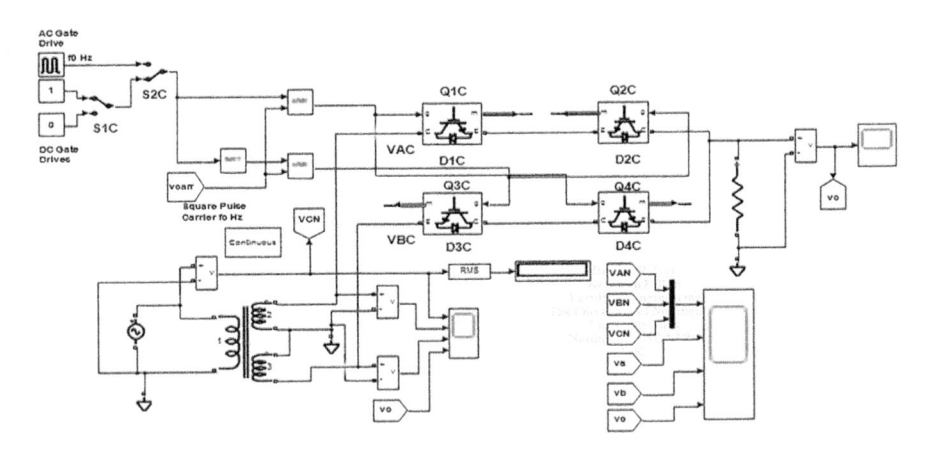

FIGURE 12.12
Three-phase sine-wave PWM AC to AC and AC to DC converter.

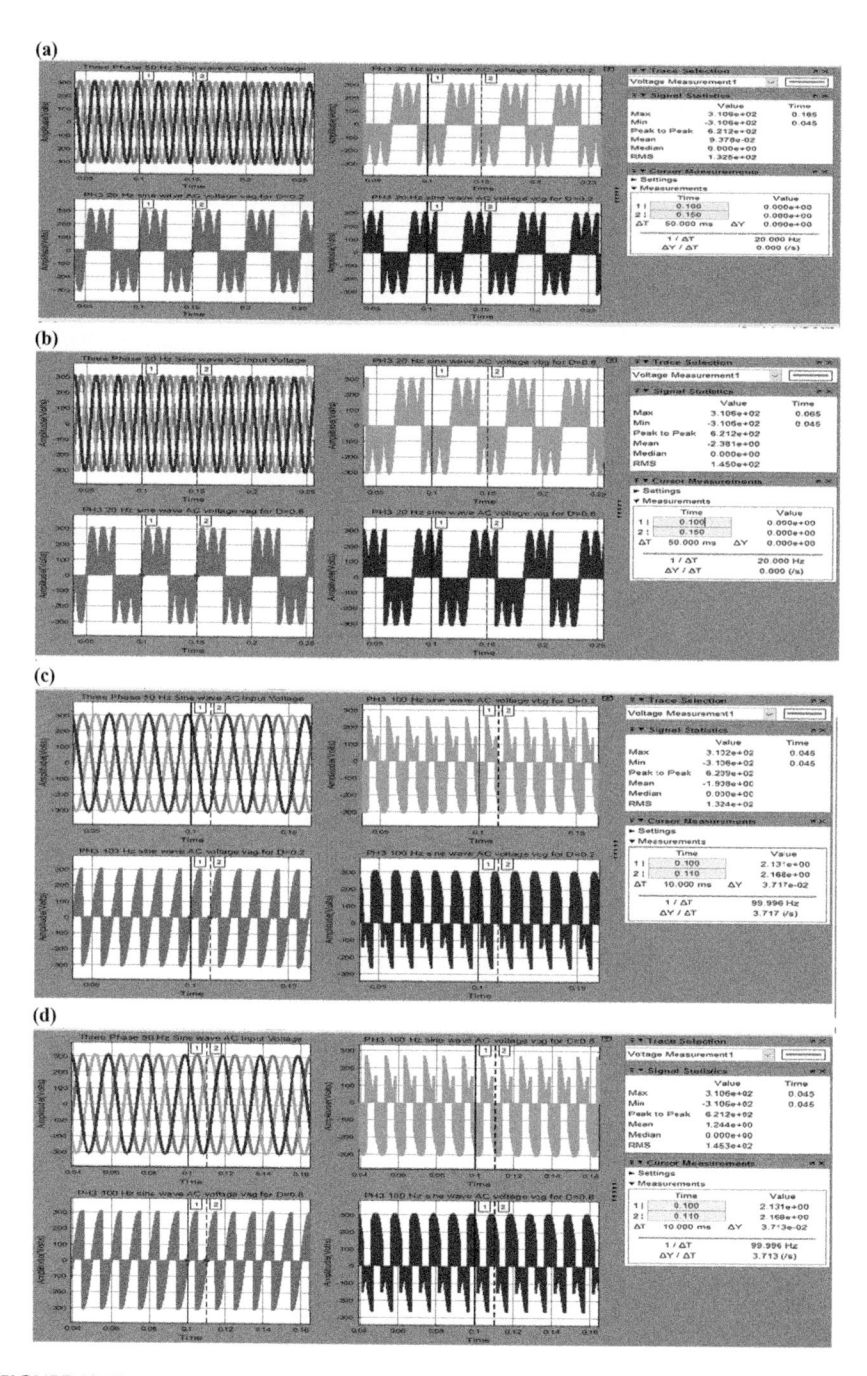

FIGURE 12.13

Three-phase sine-wave AC to AC and AC to DC converter – simulation results for three-phase input voltage and three-phase output voltages. (a) $f_0 = 20\,\text{Hz}$, $D = 0.2$, (b) $f_0 = 20\,\text{Hz}$, $D = 0.8$, (c) $f_0 = 100\,\text{Hz}$, $D = 0.2$ and (d) $f_0 = 100\,\text{Hz}$, $D = 0.8$.

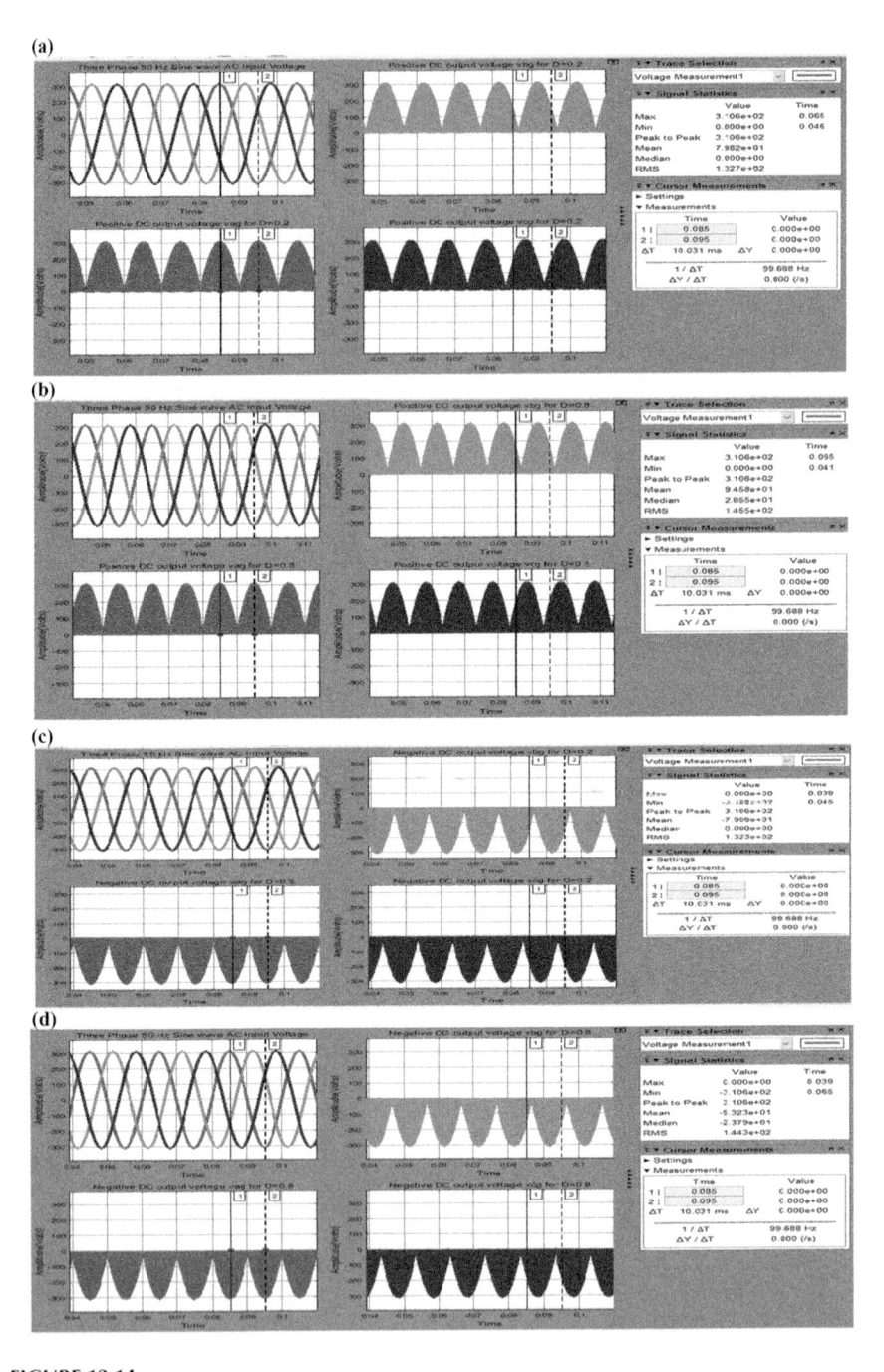

FIGURE 12.14
Three-phase sine-wave AC to AC and AC to DC converter – simulation results for three-phase input voltages and three positive and three negative DC output voltages. (a) $D = 0.2$, (b) $D = 0.8$, (c) $D = 0.2$ and (d) $D = 0.8$.

TABLE 12.3

Three-Phase Sine-Wave PWM AC to AC and AC to DC Converter – Simulation Results

Sl. No.	RMS Line-to-Neutral Input Voltage (V)	Input Frequency (Hz)	Output Frequency (Hz)	Carrier Duty Cycle D	Mode	RMS Line-to-Ground Output Voltage (V)	Mean DC Output Voltage (V)
1	220	50	20	0.2, 0.8	AC	132.4, 144.9	–
2	220	50	100	0.2, 0.8	AC	134.3, 146.6	–
3	220	50	0	0.2, 0.8	DC	–	+78.89, +94.9
4	220	50	0	0.2, 0.8	DC	–	−80.64, −94.74

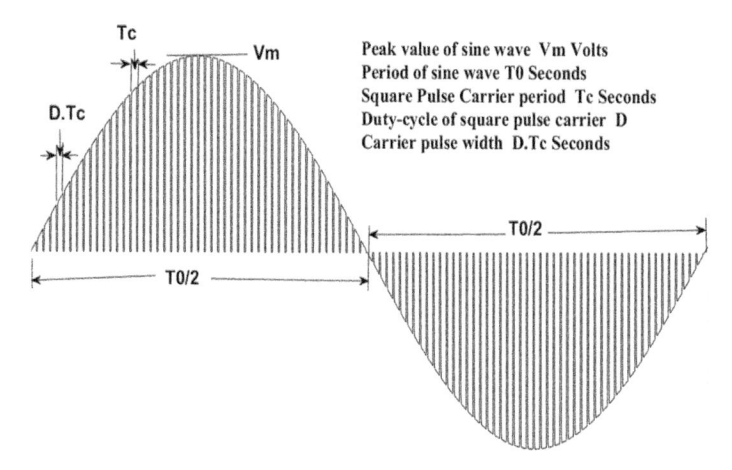

Peak value of sine wave Vm Volts
Period of sine wave T0 Seconds
Square Pulse Carrier period Tc Seconds
Duty-cycle of square pulse carrier D
Carrier pulse width D.Tc Seconds

FIGURE 12.15
Uniform pulse-width-modulated sinusoidal voltage.

$$V_{rms} = \left[\sqrt{\frac{V_m^2}{\pi} * \int_0^\pi \sin^2(\omega t)d(\omega t)} \right] \tag{12.2}$$

$$V_{avg} = \frac{V_m}{\pi} * \int_0^\pi \sin(\omega t)d(\omega t) \tag{12.3}$$

With a uniform square PWM sinusoidal voltage signal shown in Figure 12.15, equations 12.1 and 12.3 have to be multiplied by a constant k1 defined below:

$$k_1 = \frac{\text{Area under one square pulse PWM signal at any time } t_1 \text{ seconds}}{\text{Area under one carrier period at any time } t_1 \text{ seconds}}. \tag{12.4}$$

$$k_1 = \frac{[V_m * \sin(\omega t_1)] * D * T_C}{[V_m * \sin(\omega t_1)] * T_C} = D. \tag{12.5}$$

Multiplying equations 12.1 and 12.3 by equation 12.5 and simplifying, the following equations are obtained for the RMS and average value of a uniform square PWM sinusoidal voltage signal:

$$V_{rms}^2 = \frac{V_m^2 * D}{\pi} * \int_0^\pi \sin^2(\omega t)d(\omega t) \tag{12.6}$$

$$V_{rms} = \sqrt{\frac{V_m^2 * D}{\pi} * \left[\frac{\omega t}{2} - \frac{\sin(2\omega t)}{4}\right]_0^\pi} \tag{12.7}$$

$$V_{rms} = \frac{V_m * \sqrt{D}}{\sqrt{2}} = V_m * \sqrt{\frac{D}{2}} \tag{12.8}$$

$$V_{avg} = \frac{V_m * D}{\pi} * \int_0^\pi \sin(\omega t)d(\omega t) \tag{12.9}$$

$$V_{avg} = \frac{-V_m * D}{\pi} * \left[\cos(\omega t)\right]_0^\pi \tag{12.10}$$

$$V_{avg} = \frac{2V_m * D}{\pi} \tag{12.11}$$

Equations 12.8 and 12.11 give the RMS and average value of the uniform square PWM sine-wave voltage signal. In these equations, D represents the duty cycle of the carrier square pulse; V_m, V_{avg} and V_{RMS} represent the peak, average and RMS values of the sinusoidal voltage signal, respectively. The accuracy of equations 12.8 and 12.11 is higher as long as $T_c \ll T_0$, where T_c and T_0 are as defined in Figure 12.15.

12.5 Discussion of Results

This method is a patent submitted by the author [2]. Simulation results for single-phase sine PWM AC to AC and AC to DC converter for both AC and DC mode indicate controllable AC and DC output voltages [2]. AC and DC modes are selected using a selector switch for the gate drive [2]. Square PWM is a well-known technique for output voltage control. The converter can also be used with three-phase AC input voltage to get three-phase AC output voltage in the AC mode and up to three DC output voltages in the DC mode. A combination of AC and DC modes can be selected with three-phase configuration to run simultaneously AC and DC loads. This method is a direct AC to AC cum AC to DC conversion technique and requires no DC link capacitor storage element [2]. With three-phase AC input voltage and either with all three converter in DC mode or a combination of DC and AC modes, independent square pulse carrier can be used for each input phase to control the three loads independently which is a special feature.

12.6 Conclusions

A new method of direct sine-wave AC to AC and AC to DC conversion is presented. The simulation results indicate that such a method is a feasible proposition. This method invariably requires three winding transformers. In the AC and DC modes, magnitude of the AC and DC output voltage can be controlled by varying the duty cycle of the square pulse carrier. In the DC mode, the DC output voltage polarity can also be reversed by changing the gate drive pattern. Thus, in the DC mode this converter can be used for the speed control and brake by plugging of SEDC motor and PMDC motor. In the DC mode, this converter is suitable for HEV, ET and other control applications. In the AC mode, this converter can be used for the speed control of single-phase and three-phase induction motors.

References

1. N.P.R. Iyer: Modelling, simulation and real time implementation of a three phase AC to AC matrix converter, Ph.D. thesis, Ch. 13, Department of ECE, Curtin University, Perth, WA, Australia, February 2012.
2. N.P.R. Iyer: A single phase AC to PWM single phase AC and DC converter, (alternative title: "Swamy converter"), Patent submitted to Technology Transfer Office, The University of Nottingham, UK, 2013.
3. The Mathworks Inc.: www.mathworks.com, MATLAB/SIMULINK user's manual, MATLAB R2017b, 2017.

13

Cycloconverters, Indirect Matrix Converters and Solid-State Transformers

13.1 Introduction

In this chapter, the modelling of traditional thyristor cycloconverters, other forms of AC to AC converters such as indirect matrix converter (IMC) and solid-state transformers (SST), is presented. Cycloconverters are AC to AC converters which operate as frequency changers giving an AC output voltage at a different frequency from that of the input AC voltage source [1,2]. By proper switching arrangement, cycloconverters construct an AC voltage of lower frequency from successive segments of the input AC voltage source. IMCs convert the input AC voltage with a known frequency to DC and then convert the DC to AC voltage using semiconductor switches having the same or different output frequency [2]. SST is basically a high-frequency transformer (HFT) isolated AC to AC conversion technique. Models of the single-phase and three-phase SCR cycloconverters, conventional and multilevel IMC, SST using dual active bridge (DAB) and direct AC to AC converter topology are presented.

13.2 Single-Phase AC to Single-Phase AC Cycloconverters

The circuit configuration of the single-phase AC to single-phase AC bridge type cycloconverter is shown in Figure 13.1 [1,2]. Here gate pulses for all the SCRs in the P-converter and N-converter have a frequency f_0 Hz which is the desired output frequency. The gate pulse delay for the SCRs T1P and T2P is 0 rad and that for T3P and T4P is $\pi/2$ rad. Similarly, the gate pulse delay for SCRs T1N and T2N is π rad and that for T3N and T4N is $3\pi/2$ rad. The output frequency f_0 Hz is always less than input frequency f_s Hz, and hence it is known as single-phase four-pulse step-down cycloconverter.

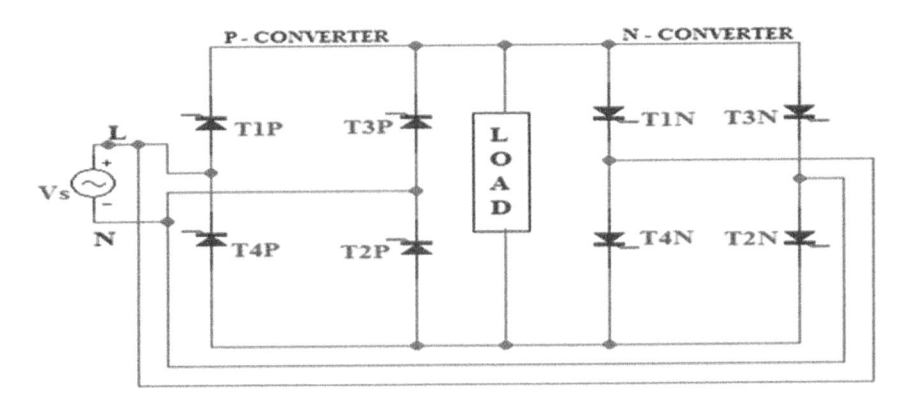

FIGURE 13.1
Single-phase SCR bridge cycloconverter.

Another topology is the single-phase midpoint SCR cycloconverter which is shown in Figure 13.2. Here, either a three winding transformer or two single-phase transformers are used. The two secondary midpoints are grounded. The two secondary output voltages are 180° out of phase.

The gate pulses for the four SCRs TH1P, TH2P, TH1N and TH2N have a frequency f_0 Hz which is the desired output frequency and fired with a phase delay of $3\pi/2$, $\pi/2$, 0 and π rad, respectively.

13.2.1 Simulation Results

The model of single-phase SCR bridge cycloconverter is developed using SIMULINK [3]. The model is shown in Figure 13.3. SCRs T1P to T4P form the P-converter and T1N to T4N forms the N-converter. The converter has already been explained in Section 13.2. Pulse generator blocks are used to generate gate pulse for individual SCRs with required amplitude, desired

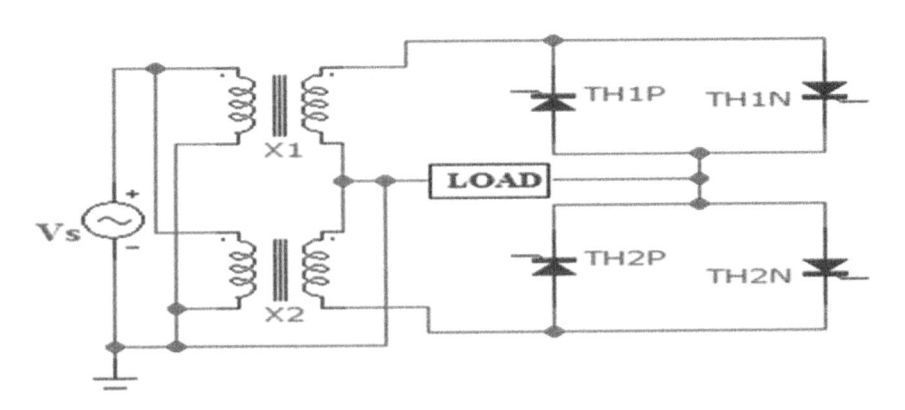

FIGURE 13.2
Single-phase midpoint SCR cycloconverter.

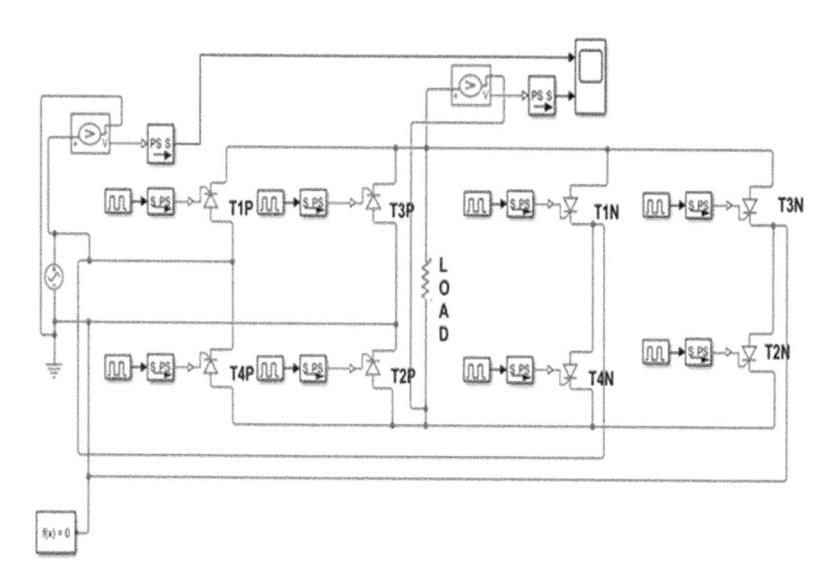

FIGURE 13.3
Single-phase SCR bridge cycloconverter.

output frequency, pulse width and the phase delay. The sine-wave AC source has a voltage of 100 V (RMS) and a frequency of 60 Hz. The desired output frequency is 20 Hz. The ode15s (stiff/NDF) solver is used. Simulation results are shown in Figure 13.4. The model of single-phase midpoint SCR

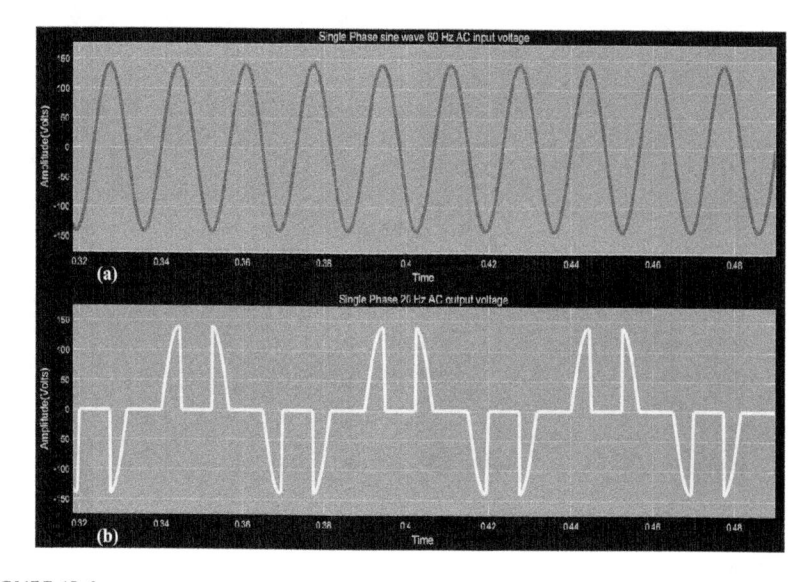

FIGURE 13.4
Single-phase SCR bridge cycloconverter – simulation results. Single-phase 60 Hz sine-wave AC input voltage (a) and single-phase 20 Hz AC output voltage (b).

cycloconverter is shown in Figure 13.5. This is explained in Section 13.2. The sine-wave AC input and output voltage parameters are the same as for SCR bridge cycloconverter given above. The ode23tb (stiff/TR-BDF2) solver is used. The simulation results are shown in Figure 13.6.

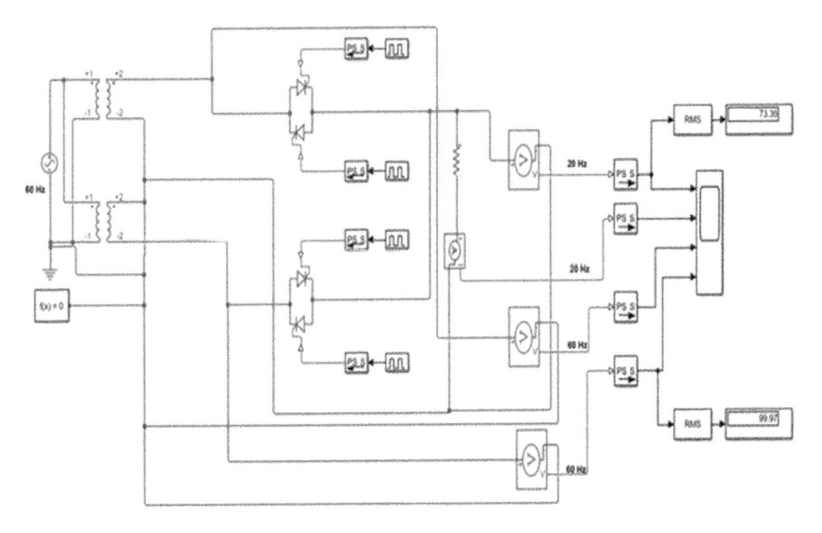

FIGURE 13.5
Single-phase midpoint SCR cycloconverter.

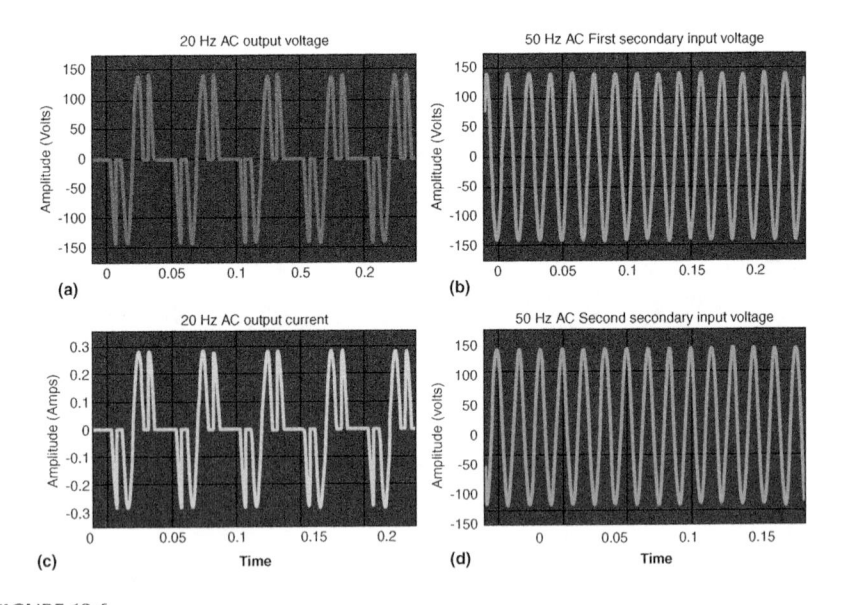

FIGURE 13.6
Single-phase midpoint SCR cycloconverter – simulation results: (a) 20 Hz AC output b voltage, (b) 50 Hz AC first primary input voltage, (c) 20 Hz AC load current and (d) 50 Hz AC second secondary input voltage.

13.3 Three-Phase Cycloconverters

The three-phase half wave three pulse SCR cycloconverter delivering power to a single-phase load is shown in Figure 13.7. The SCRs TRp, TYp, TBp form the P-converter and TRn, TYn and TBn form the N-converter. The gate pulses for all the SCRs have a frequency f_0 Hz which is the desired output frequency. Three-phase AC input voltage frequency is f_s Hz. The gate pulse for SCRs TRp, TYp and TBp are fired with a phase delay of $\pi/3$, π and $5\pi/3$ rad. Similarly, SCRs TRn, TYn and TBn are fired with a phase delay of $4\pi/3$, 0 and $2\pi/3$ rad, respectively.

The three-phase half wave three pulse SCR cycloconverter delivering power to a three-phase load is shown in Figure 13.8. Each phase has P- and N-converter group. The SCR gate pulse firing is the same as for single-phase load given above. In Figure 13.8, SCR TRp2 in output phase b is fired with a phase delay of π rad and TRp3 is fired with a delay of $5\pi/3$ rad. The gate pulse delay for the other SCRs in Figure 13.8 can be similarly calculated.

The three-phase four-pulse SCR bridge cycloconverter delivering power to a three-phase load is shown in Figure 13.9. Here gate pulses for all the SCRs in the P-converter and N-converter have a frequency f_0 Hz which is the desired output frequency. The gate pulse delay for the SCRs T1P1 and T2P1 are 0 rad and that for T3P1 and T4P1 are $\pi/2$ rad. Similarly, the gate pulse delay for SCRs T1N1 and T2N1 is π rad and that for T3N1 and T4N1 is $3\pi/2$ rad.

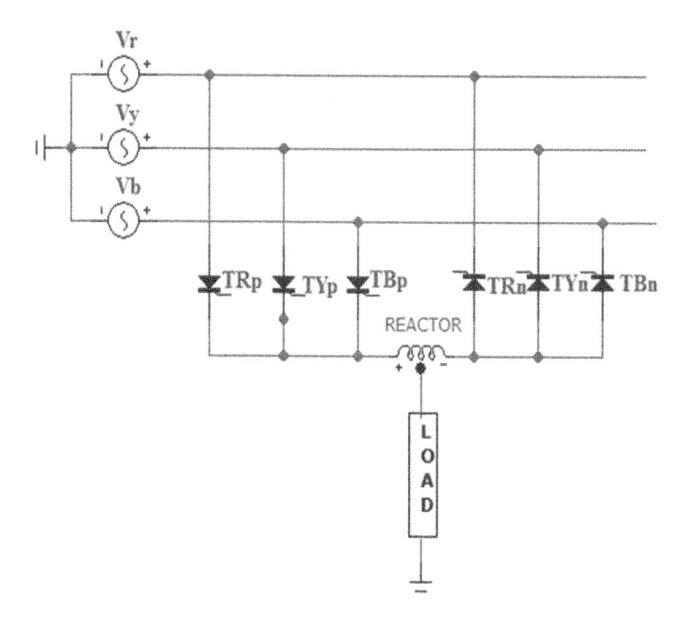

FIGURE 13.7
Three-phase half wave three pulse SCR cycloconverter.

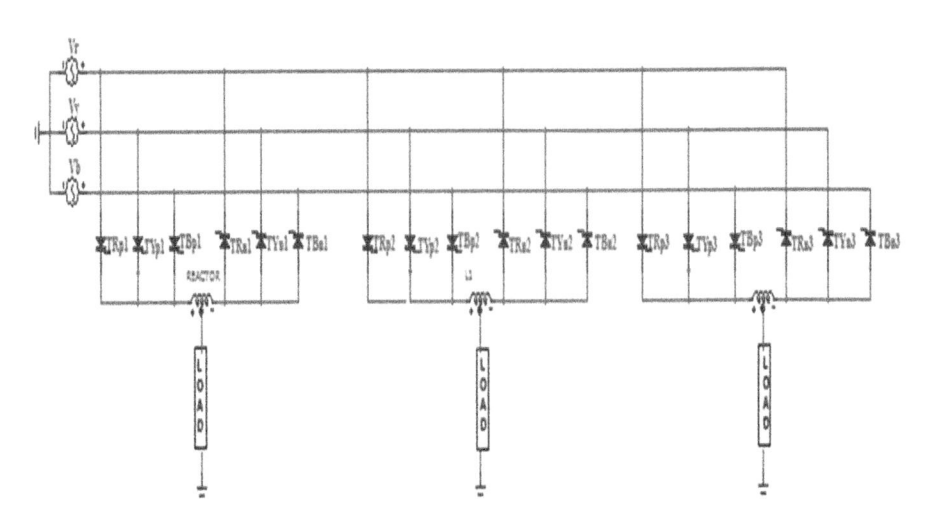

FIGURE 13.8
Three-phase half wave three pulse SCR cycloconverter delivering power to three-phase load.

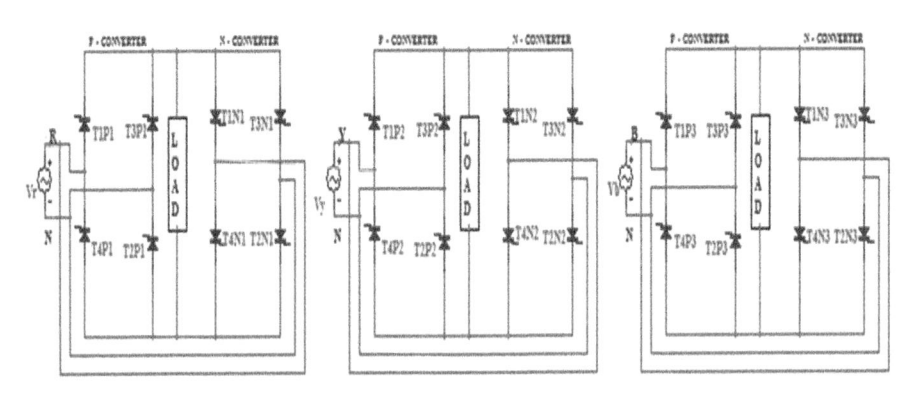

FIGURE 13.9
Three-phase SCR bridge cycloconverter delivering power to three-phase load.

Similarly, the phase delay for SCRs in the input phase Y and B can be calculated. The SCRs T1P2 and T2P2 are fired with a phase delay of $2\pi/3$ rad
and T1P3 and T2P3 with a phase delay of $4\pi/3$ rad. The gate pulse phase
delay for other SCRs can be similarly calculated. The output frequency f_0 Hz
is always less than input frequency f_s Hz, and hence it is known as three-
phase four-pulse step-down cycloconverter.

13.3.1 Simulation Results

The model of the three-phase half wave three pulse SCR cycloconverter
delivering power to a three-phase load is developed using SIMULINK [3].
The model is shown in Figure 13.10. The ode23t (mod.stiff/Trapezoidal)

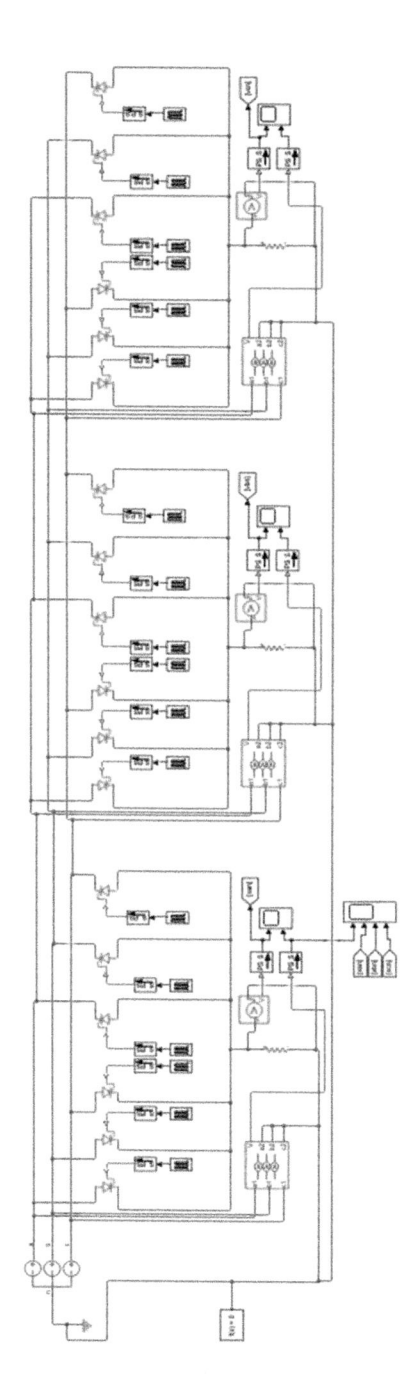

FIGURE 13.10
Three-phase half wave three pulse SCR cycloconverter delivering power to three-phase load.

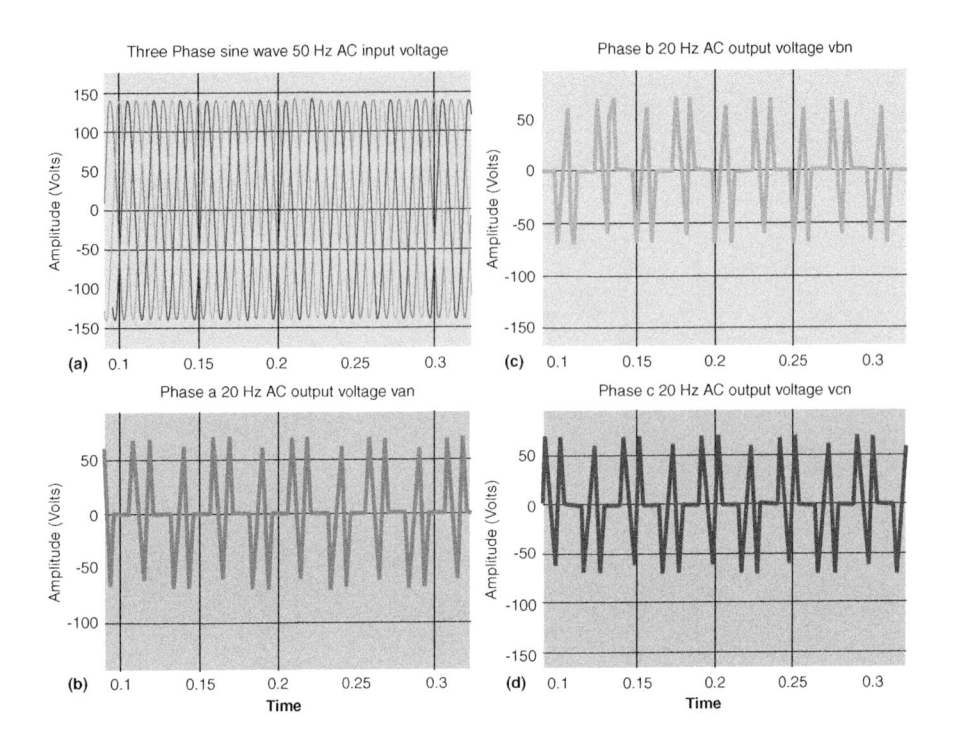

FIGURE 13.11
Three-phase half wave three pulse SCR cycloconverter simulation results. Three-phase 50 Hz sine-wave input voltage (a), phase a output voltage (b), phase b output voltage (c) and phase c output voltage (d).

solver is used. The parameters are the same as given in Section 13.2.1. The simulation results are shown in Figure 13.11.

The model of the three-phase four-pulse SCR bridge cycloconverter delivering power to a three-phase load is shown in Figure 13.12. This model is developed using SIMULINK [3]. The ode23t (mod.stiff/Trapezoidal) solver is used. The parameters are the same as given in Section 13.2.1. The simulation results are shown in Figure 13.13.

13.4 Discussion of Results

Models for single-phase AC to single-phase AC as well as three-phase AC to three-phase AC SCR cycloconverters are presented. Here, the frequency of output voltage is less than the input sine-wave AC source frequency. The three-phase AC output voltage magnitude is well balanced and displaced by 120° and 240° from the reference AC voltage waveform.

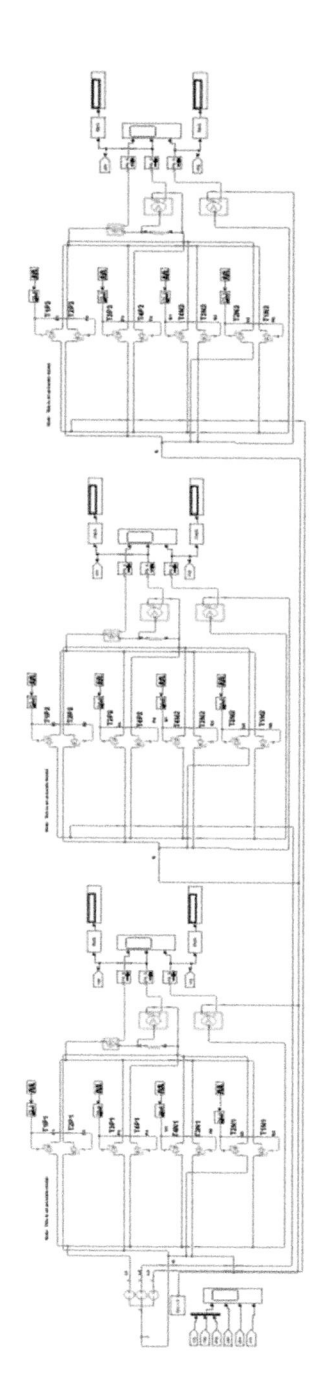

FIGURE 13.12
Three-phase four-pulse SCR cycloconverter delivering power to three-phase load.

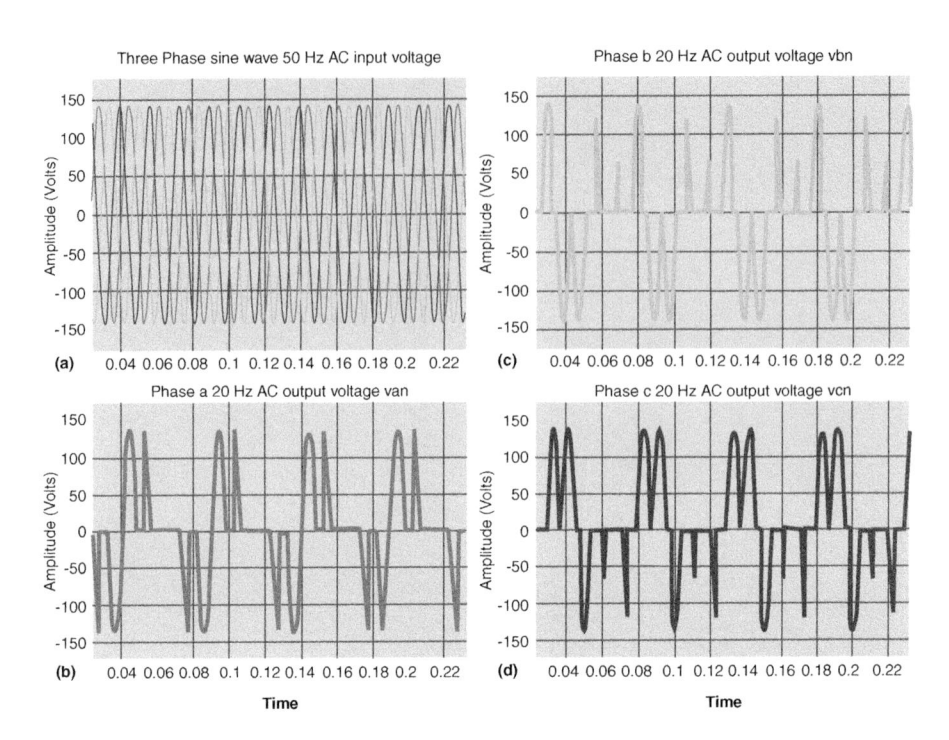

FIGURE 13.13

Three-phase four-pulse SCR bridge cycloconverter simulation results. Three-phase 50 Hz sine-wave AC input voltage (a), 20 Hz phase a output voltage (b), 20 Hz phase b output voltage (c) and 20 Hz phase c output voltage (d).

13.5 Three-Phase Conventional Indirect Matrix Converter

The block diagram of the simple form of IMC is shown in Figure 13.14. The three-phase sine-wave AC source has a frequency f_s Hz. The six semiconductor switches S11 to S16 in order are turned on and off using gate pulse having frequency f_s Hz, duty cycle 50% and phase delay of 0, π [1/(2f_s) s], 2π/3, 5π/3, 4π/3 and π/3 rad, respectively. The DC output is filtered and this forms the DC-link voltage for the three-phase output inverter. The output inverter has six semiconductor switches from S21 to S26. These are switched on with gate pulse having frequency f_0 Hz, duty-cycle 50% and phase delays as mentioned above for switches S11–S16. The inverter output is a three-phase square-wave AC voltage having frequency f_0 Hz.

13.5.1 Simulation Results

The model of the indirect MC is developed using SIMULINK [3]. This model is shown in Figure 13.15. The IGBT blocks form the semiconductor switches. Gate pulses for individual switches are generated using pulse generator

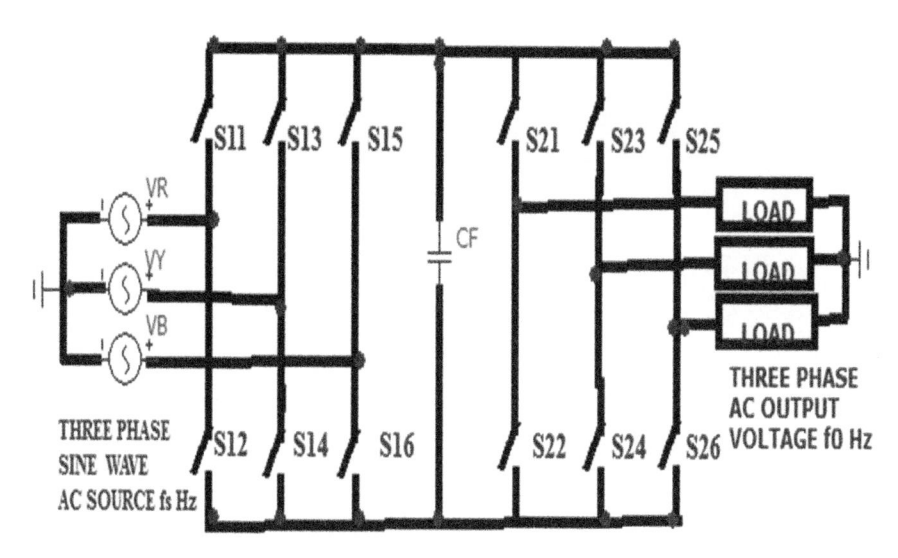

FIGURE 13.14
Three-phase indirect matrix converter block diagram.

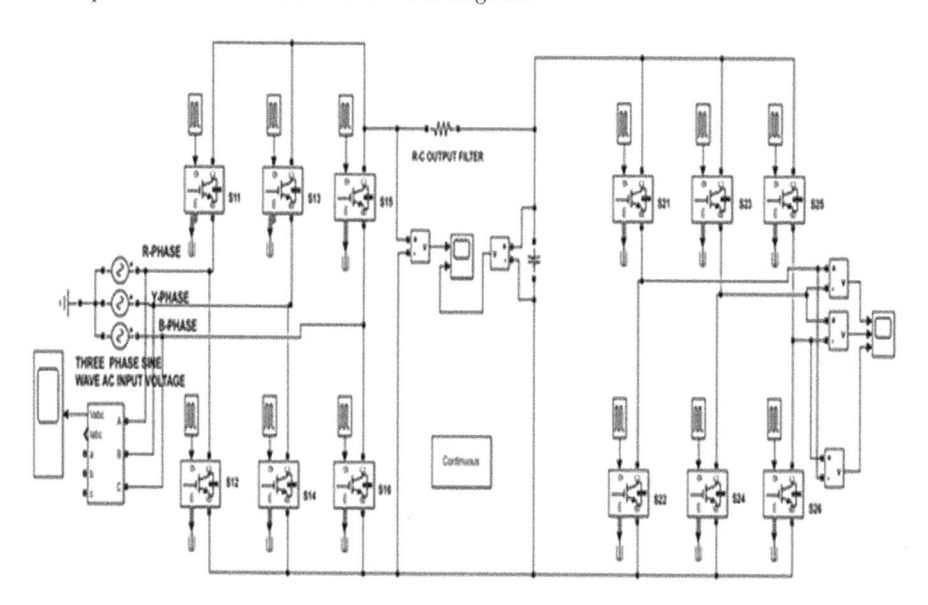

FIGURE 13.15
Model of three-phase indirect matrix converter.

block. The three-phase sine-wave AC voltage has RMS line-to-neutral voltage of 100 V and frequency 60 Hz. Two output frequencies 20 and 100 Hz are considered. Ode23tb (stiff/TR-BDF2) solver is used. The simulation results for 20 and 100 Hz output voltage are shown in Figures 13.16 and 13.17. Three-phase 60 Hz sine-wave input voltage is shown in Figure 13.18.

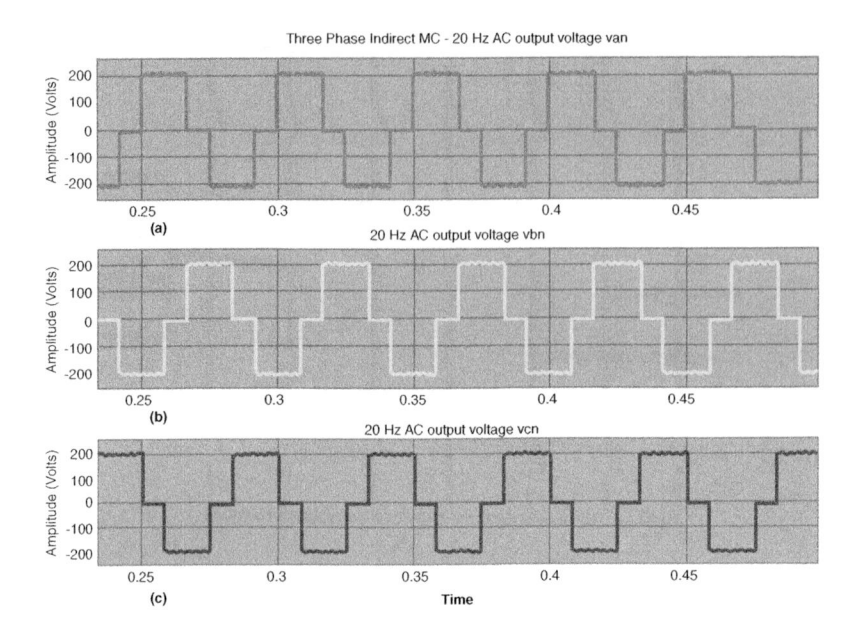

FIGURE 13.16
Three-phase indirect MC simulation results: (a) 20 Hz phase a output voltage $v_{an,}$ (b) 20 Hz phase b output voltage v_{bn} and (c) 20 Hz phase c output voltage v_{cn}.

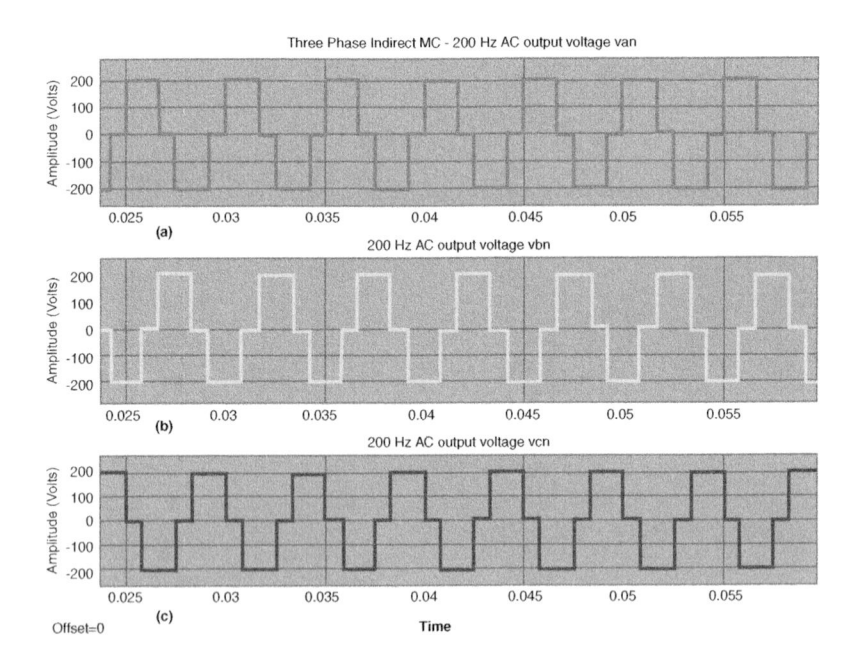

FIGURE 13.17
Three-phase indirect MC simulation results: (a) 200 Hz phase a output voltage $v_{an,}$ (b) 200 Hz phase b output voltage v_{bn} and (c) 200 Hz phase c output voltage v_{cn}.

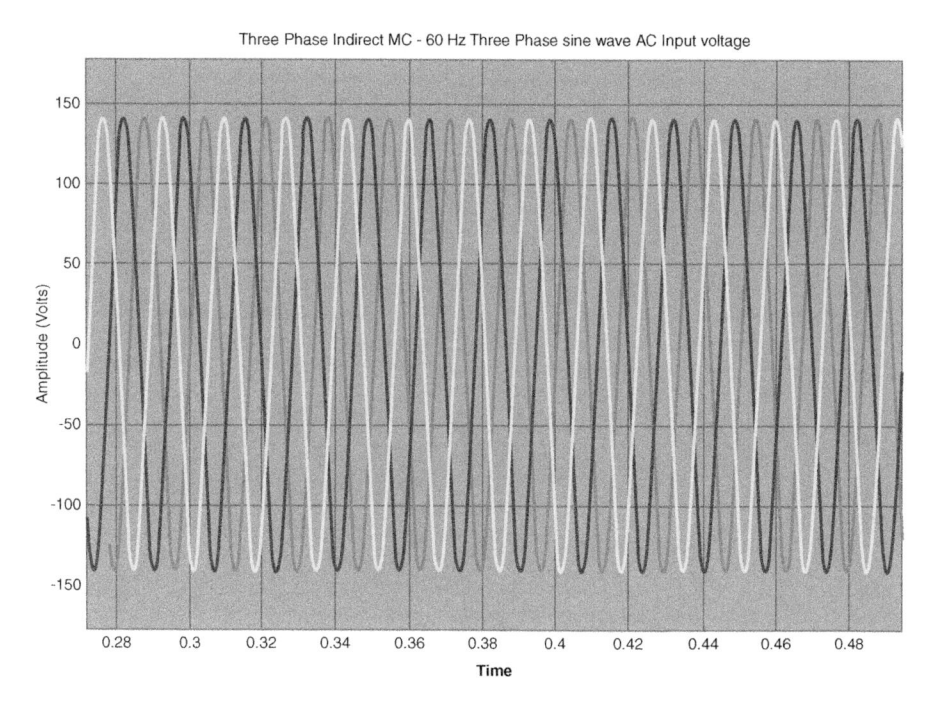

FIGURE 13.18
Three-phase indirect MC simulation results – 60 Hz three-phase sine-wave AC input voltage.

13.6 Three-Phase Multilevel Indirect Matrix Converter

The block diagram schematic of the three-phase diode-clamped multilevel IMC is shown in Figure 13.19. Here a three-phase diode-clamped three-level AC to DC rectifier is used on the input side and a similar three-phase inverter is used on the output side. Three-phase sine-wave AC voltage source having frequency f_s Hz is the input. The rectified output voltage V_{dc} appears across the DC-link capacitor CF. This DC output voltage is inverted using three-phase diode-clamped three-level inverter (DCTLI). The output voltage is a three-phase square-wave AC having frequency f_0 Hz. Output frequency f_0 can be greater than, equal to or less than f_s.

To understand the operation, consider the three-phase DCTLI connected to the DC-link voltage V_{dc} on the output side of Figure 13.19. The truth Table 13.1 applies to the four semiconductor switches connected to output phase a. The gate pulse frequency for all the switches in the output phase a, b and c is f_0 Hz. From Table 13.1, the duty cycle and phase delay in radians for the gate pulses S1a, S2a, S3a and S4a are 0.25, 0; 0.75, $3\pi/2$; 0.75, $\pi/2$; 0.25, π [$1/(2f_0)$ s], respectively. The switches in output phase b and c have a phase delay of $2\pi/3$ and $4\pi/3$ rad added with respect to the corresponding switch in phase a.

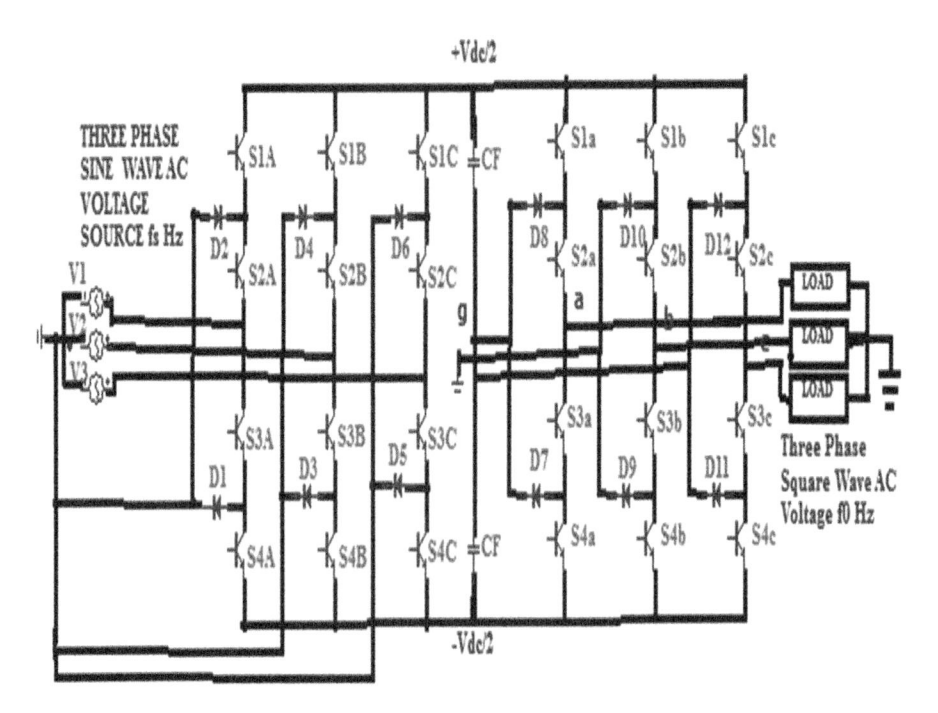

FIGURE 13.19
Block diagram of the three-phase diode-clamped multilevel indirect matrix converter.

TABLE 13.1

Three-Phase DCTLI Truth Table

Sl. No.	S1a	S2a	S3a	S4a	Output Voltage (V)	Remarks
1	1	1	0	0	$+V_{dc}/2$	Phase a start
2	0	1	1	0	0	–
3	0	0	1	1	$-V_{dc}/2$	–
4	0	1	1	0	0	–
5	1	1	0	0	$+V_{dc}/2$	Next cycle

The duty cycle and phase delay for the corresponding switches in phase A, B and C on the input rectifier side remain the same.

13.6.1 Simulation Results

The model of three-phase diode-clamped three-level indirect matrix converter (DCTLIMC) is developed using SIMULINK [3]. The ode 23tb (stiff/TR-BDF2) solver is used. This model is shown in Figure 13.20. The model is developed using IGBT and diode blocks in Power Systems block set. Pulse generator blocks form the gate drive. Pulse generator in the input rectifier side has a frequency

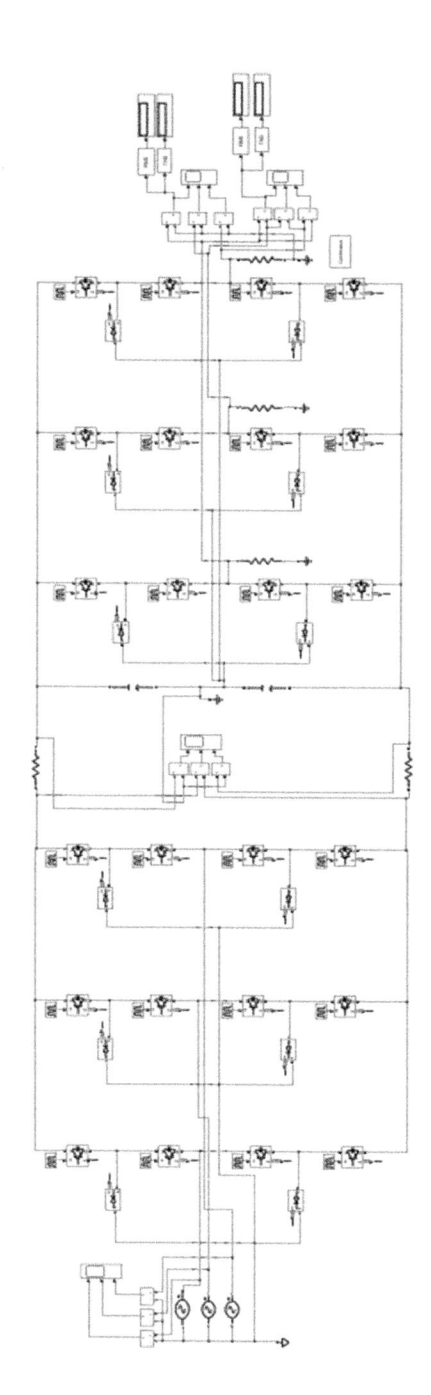

FIGURE 13.20
Three-phase diode-clamped three-level indirect matrix converter.

of 60 Hz corresponding to the frequency of the three-phase sine-wave AC input voltage. On the output inverter side, gate pulses frequency of 20 and 200 Hz are used. Gate pulse duty cycle and phase delay for individual switches are given in Section 13.6. An R–C filter is used at the output side of rectifier. The simulation results for 20 and 200 Hz output voltage are shown in Figures 13.21 and 13.22, respectively, and the three-phase input voltage is shown in Figure 13.23.

13.7 Discussion of Results

Models for three-phase conventional indirect and DCTLI-based multilevel indirect MC are presented. Simulation results are successful. Simulation results show that output AC voltage with frequency above and below the input frequency can be obtained. Three-phase DCTLI-based indirect MC line-to-ground and line-to-line output voltages have three and five levels, respectively.

13.8 Solid-State Transformer

Modernization of power distribution system is a key task in smart grid. Transformers based on power electronics also called SSTs play a key role in Future Renewable Electric Energy Delivery and Management (FREEDM) system [4]. SST is basically an HFT isolated AC to AC conversion technique. The HFT can be operated in the step-up or step-down mode depending on output voltage requirement. The block diagram schematic of SST is shown in Figure 13.24. The high-voltage (HV), low-frequency (LF) sine-wave AC voltage source is the input to a high-frequency (HF) AC to AC converter. The output voltage of the HF AC to AC converter is stepped down to a low value using HFT and then converted to LF, low-voltage (LV) AC using an LF AC to AC converter. The LF LV AC voltage is fed to the load. The LF at the input and output sides is the power system frequency. The HF output of AC to AC converter is usually of the order of several kilohertzs much higher than power system frequency. Compared to conventional 50/60 Hz power transformer, the HFT is compact in size, controls active/reactive power flow and protects power system from voltage sag and over current [4,5]. The AC to AC converter on the input and output sides of Figure 13.24 can be either direct converter or indirect converter first converting (a) AC to DC and then DC to AC in two stages and (b) AC to DC, DC to DC and then DC to AC in three stages. Accordingly, SST topology can be classified as (a) single stage with no DC link, i.e. by using HFT only, (b) two stages with LVDC link, (c) two stages with HVDC link and (d) three stages with HVDC and LVDC links which is also known as Dual Active Bridge (DAB) topology.

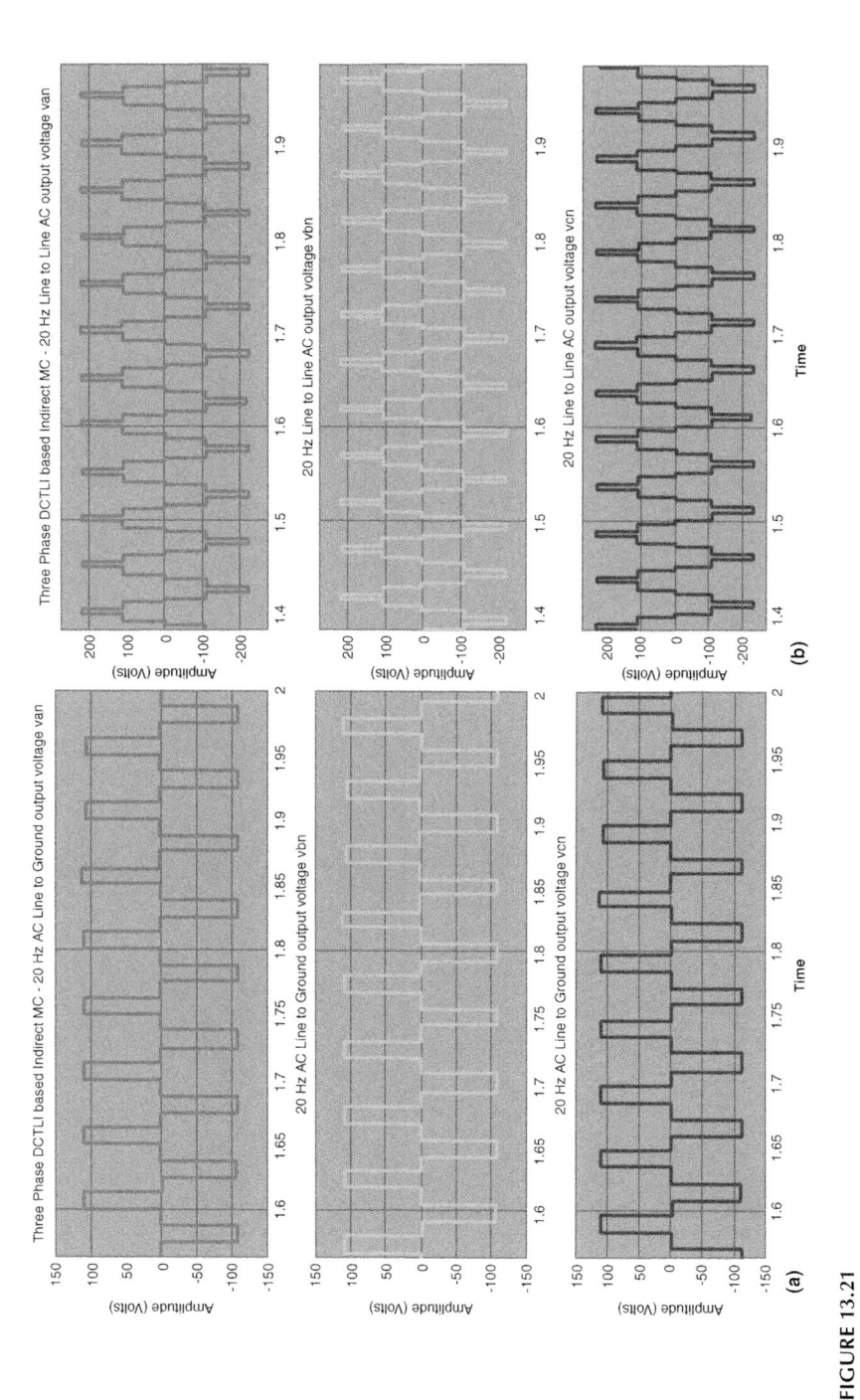

FIGURE 13.21

Three-phase DCTLI-based indirect MC simulation results: (a) 20 Hz three-phase line-to-ground output voltage and (b) 20 Hz three-phase line-to-line output voltage.

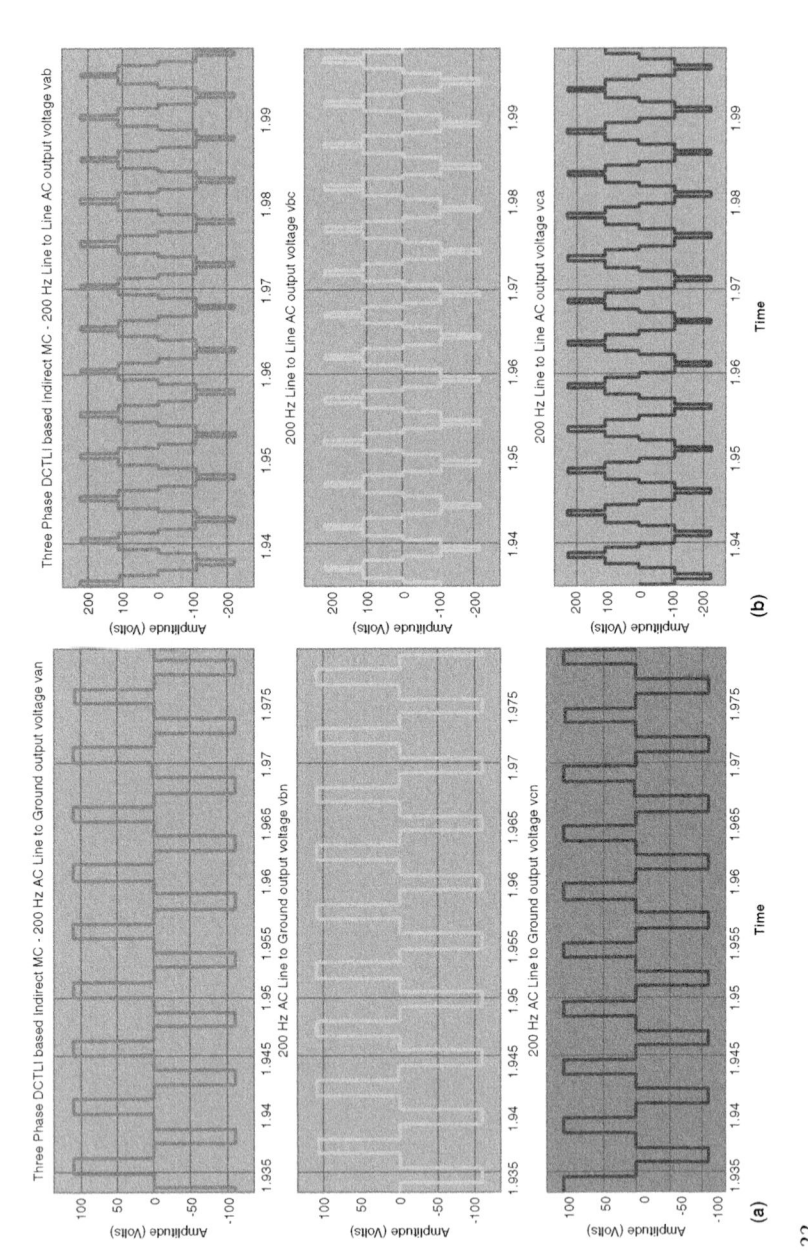

FIGURE 13.22

Three-phase DCTLI-based indirect MC simulation results. 200 Hz AC three-phase line-to-ground output voltage (a) and 200 Hz AC three-phase line-to-line output voltage (b).

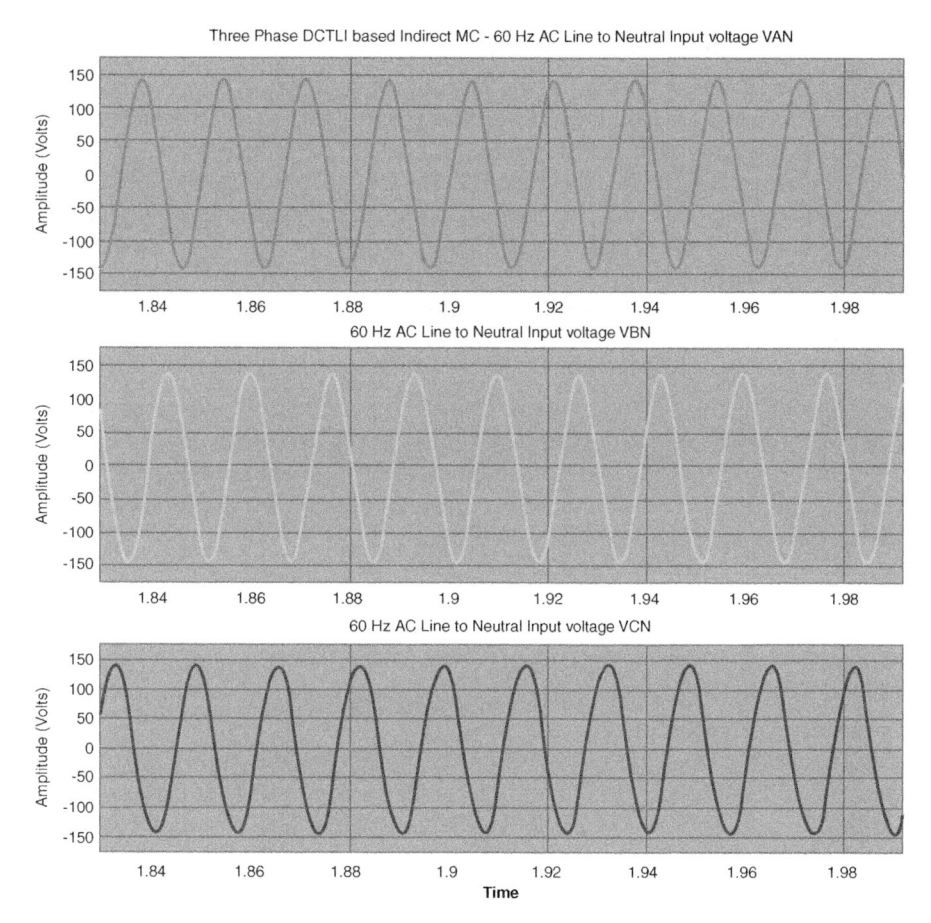

FIGURE 13.23
Three-phase DCTLI-based indirect MC simulation results – three-phase 60 Hz sine-wave AC input voltage.

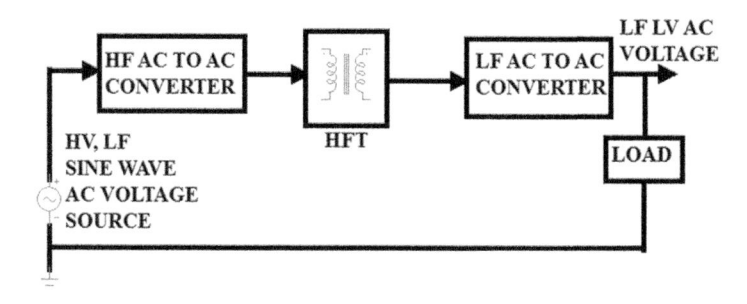

FIGURE 13.24
Block diagram of a solid-state transformer.

13.8.1 SST Using Dual Active Bridge Topology

The block diagram schematic of the DAB topology is shown in Figure 13.25. The input is an HV LF sine-wave AC voltage source, having frequency f Hz that of the power system. H-bridges S11, S12, S13 and S14 rectify this AC input voltage. Gate pulses for switches S11, S12, S13 and S14 have a period $1/_f$ s, duty-cycle 50%, phase delay of 0, $\pi[1/(2f)$s], π and 0 rad, respectively. The filtered DC output voltage is inverted to HF HVAC using H-bridges S21, S22, S23 and S24. The gate pulses for S21, S22, S23 and S24 have a frequency f_{sw} which is far higher than f, duty-cycle 50% and phase delay of 0, π, π and 0 rad, respectively. The HFT steps down this to HF LV AC. The HF LV AC is again rectified to LV DC using H-bridges S31, S32, S33 and S34 whose gate pulse have frequency f_{sw}, duty-cycle 50% and phase delay are 0, $\pi\left[1/(2f_{sw})s\right]$, π and 0 rad, respectively. This LVDC is then inverted using three-phase inverter. The inverter switching frequency is f Hz and the method of gate pulse generation for S41 to S46 is the same as explained in Section 13.5. The output of inverter is a three-phase LVAC having frequency f Hz that of power system.

13.8.2 Simulation Results

The model of the SST using DAB topology developed using SIMULINK [3] is shown in Figure 13.26. The model parameters are shown in Table 13.2. The fixed-step discrete solver is used.

Here, single-phase sine-wave AC source, IGBT switches and HFT are from Power Systems Specialized Technology block set. Pulse generator from Simulink Sources block set provides the gate pulse. The simulation results for the single-phase input voltage, HVDC link voltage, HF LVAC, LVDC link voltage and the two-phase AC output voltage are shown in Figure 13.27a–g, respectively. The three 50 Hz AC output voltages in Figure 13.27e–g are two-phase voltages. Output voltages v_{ab} and v_{ca} are in exact 180° phase opposition, whereas v_{ab} and v_{bc} are 90° phase apart. The gate pulse for S41 to S46 in order has a period of 1/50 s, 50% duty cycle with phase delays of 0, π [1/(2*50) s],

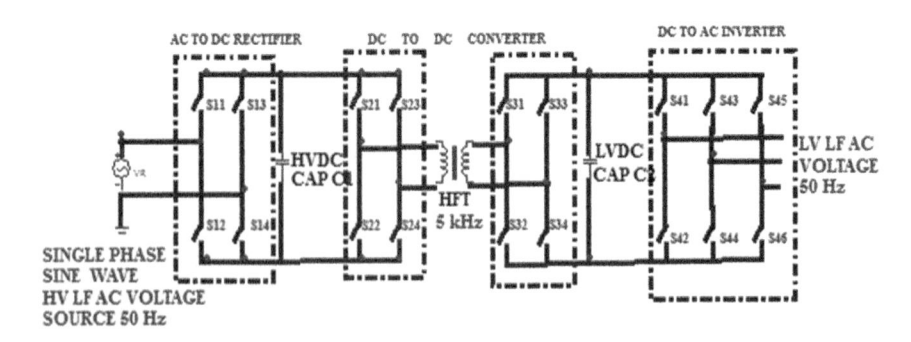

FIGURE 13.25
Dual active bridge topology.

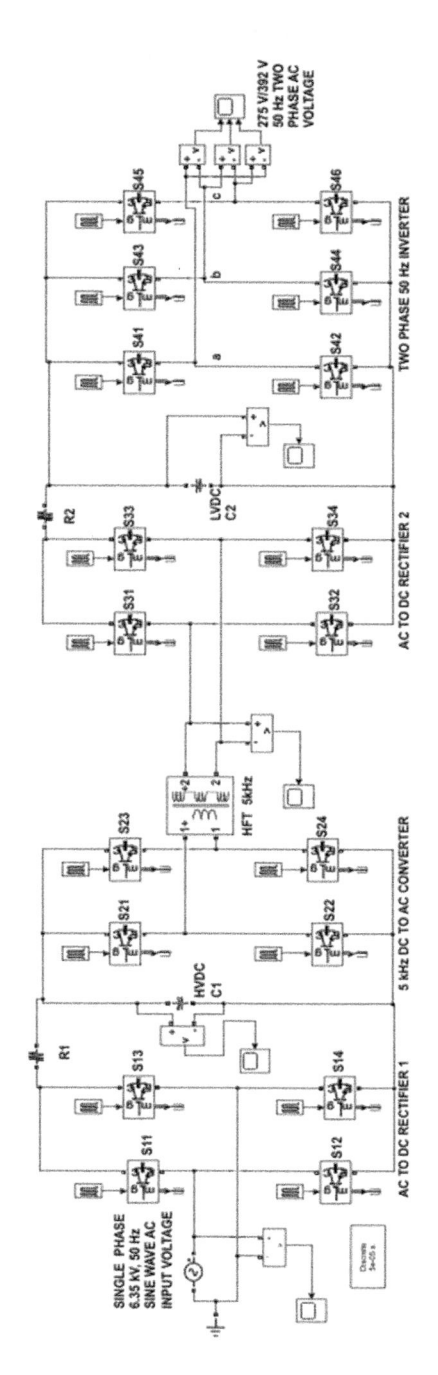

FIGURE 13.26
Model of solid-state transformer using dual active bridge topology.

TABLE 13.2

SST Using DAB Topology – Model Parameters

Sl. No.	Item Specification	Value	Units
1	Single-phase AC input voltage	6.351	kV (RMS)
2	Input/output frequency	50/50	Hz
3	DAB DC–DC converter switching frequency	5	kHz
4	HFT power transformer	75	kV A
5	HVDC link voltage	5.72	kV
6	LVDC link voltage	0.392	kV
7	Output voltage	0.275/0.392	kV (RMS)

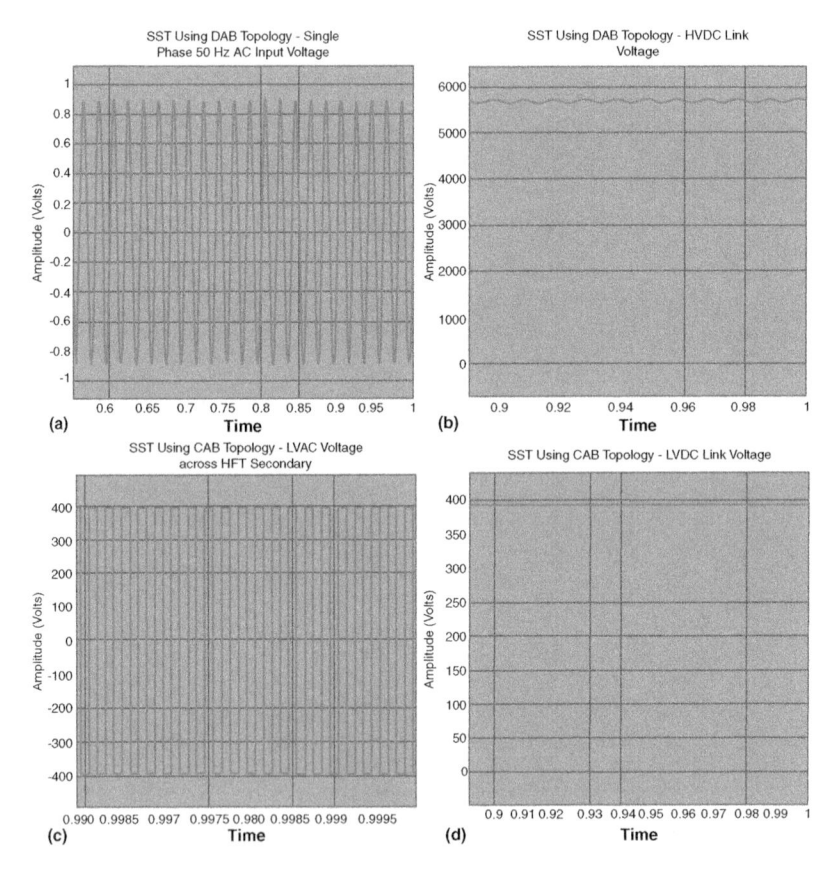

FIGURE 13.27

SST using DAB topology – simulation results: (a) HV 50 Hz AC sine-wave input voltage, (b) HVDC link voltage, (c) HF AC voltage across HFT secondary, (d) LVDC link voltage (e), (f) and (g) two-phase 50 Hz LVAC output voltages v_{ab}, v_{bc} and v_{ca}.

(Continued)

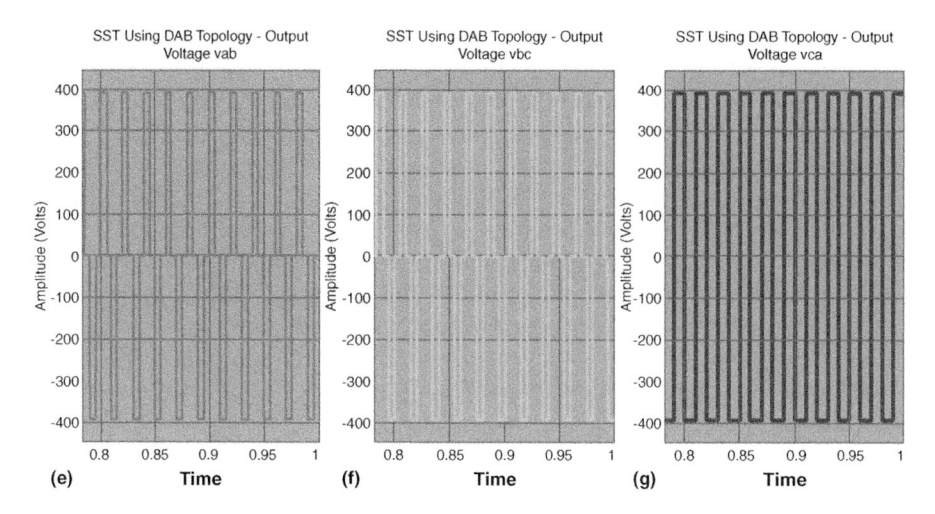

FIGURE 13.27 (CONTINUED)
SST using DAB topology – simulation results: (a) HV 50 Hz AC sine-wave input voltage, (b) HVDC link voltage, (c) HF AC voltage across HFT secondary, (d) LVDC link voltage (e), (f) and (g) two-phase 50 Hz LVAC output voltages v_{ab}, v_{bc} and v_{ca}.

$\pi/2$, $3\pi/2$, π and 0 rad. The method of gate pulse generation for H-bridge AC to DC and DC to AC converter is explained in Section 13.8.1.

13.8.3 SST Using Direct AC to AC Converter Topology

The AC to AC converter discarding square pulse carrier PWM of the combined PWM AC to AC and AC to DC converter presented in Section 12.2 can be used to develop SST [6,7]. The block diagram is shown in Figure 13.28. The principle of operation of this converter for the AC mode is given in Section 12.2.2. The mode of operation of this converter as SST is given below.

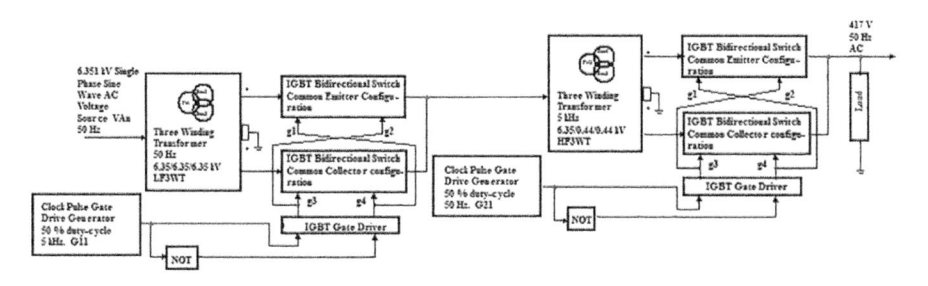

FIGURE 13.28
Block diagram of SST using direct AC to AC converter topology.

The parameters for the input voltage and output voltage magnitude and frequency are the same as given in Table 13.2. Here, two three winding transformers are used: one is HV/HV/HV LF type and the other is the HV/LV/LV HF type. These are, respectively, marked LF3WT and HF3WT. The HV LF (50 Hz) sine-wave AC voltage is applied to the primary of LF3WT. The two secondary voltages of LF3WT are given to a first pair of bidirectional IGBTs which are switched by square pulse G11 and inverted G11 having a switching frequency of 5 kHz. The outputs of first pair of bidirectional IGBTs are given to the primary of HF3WT. The two secondary outputs of HF3WT are given to a second pair of bidirectional IGBTs switched by square pulse G21 and inverted G21 having a switching frequency of 50 Hz. The output of second pair of bidirectional IGBTs is the LV LF (50 Hz) sine-wave AC voltage [7].

13.8.4 Simulation Results

The model of SST using direct AC to AC converter topology is developed using SIMULINK [3]. The model is shown in Figure 13.29. The IGBTs, three winding transformers and sine-wave AC source are from Power Systems Specialized Technology block set. The pulse generators for gate drives are from Simulink Sources block set. The model operation is explained in Section 13.8.3. The fixed-step ode3 (Bogacki–Shampine) solver is used. The parameters are shown in Table 13.2. The simulation results are shown in Figure 13.30a–c. The simulation results indicate that the single-phase sine-wave AC input voltage has an RMS value of 6.35 kV and frequency 50 Hz, output voltage v_{out1} of first pair of bidirectional IGBTs is 6.019 kV and frequency 5 kHz and the output voltage v_{out2} of second pair of bidirectional IGBTs is 416.2 V and frequency 50 Hz.

13.9 Discussion of Results

Two SST topologies, one using DAB and the other using direct AC to AC converter, are presented. Simulation results indicate that with a sine-wave AC input voltage, DAB topology gives square-wave output voltage [4,5] whereas the direct AC to AC topology gives a sine-wave AC output voltage [7]. In both cases, the output frequency is the same as the input frequency. DAB topology requires one two winding HFT, one HVDC and one LVDC link. Direct AC to AC converter topology requires two three winding transformers in which one is HFT, and there is no AC to DC conversion.

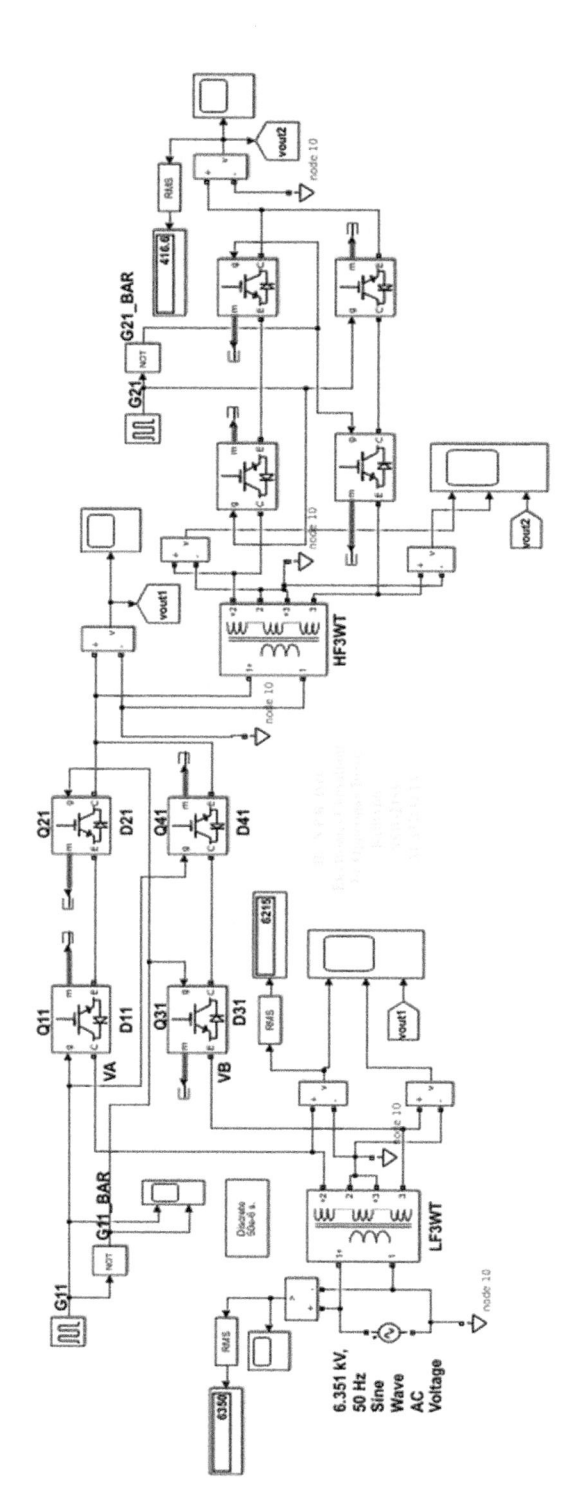

FIGURE 13.29
Model of a solid-state transformer using direct AC to AC converter topology.

13.10 Conclusions

The models for the three-phase SCR cycloconverters, IMCs and SSTs are presented. Model simulation results are successful. The SCR cycloconverter models are shown for an output frequency less than that of the input. The model performance for indirect MC shows that it is possible to obtain square-wave AC voltage with frequency less than, equal to or greater than that of the sine-wave AC input voltage. Similarly, models for SST using DAB

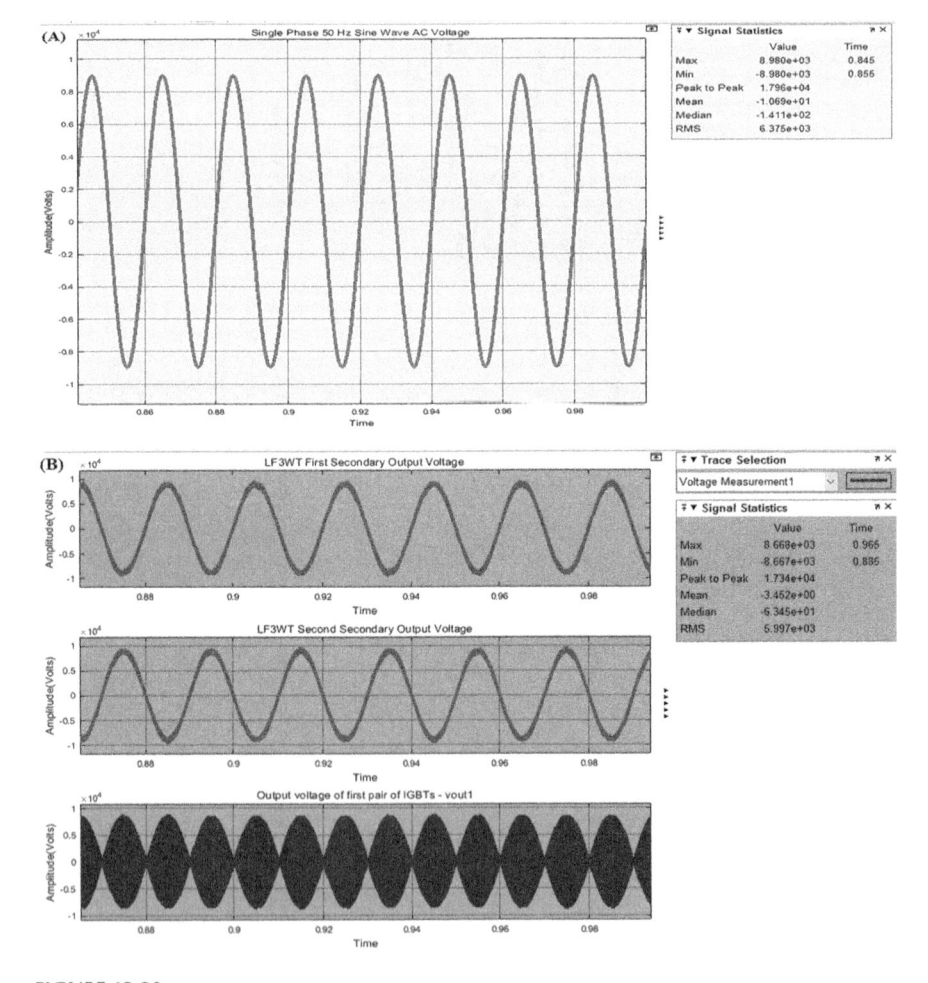

FIGURE 13.30

SST using direct AC to AC converter topology – simulation results: (a) single-phase HV LF sine-wave AC voltage, (b) LF3WT two secondary output voltage and v_{out1} and (c) HF3WT two secondary output voltage and v_{out2}.

(Continued)

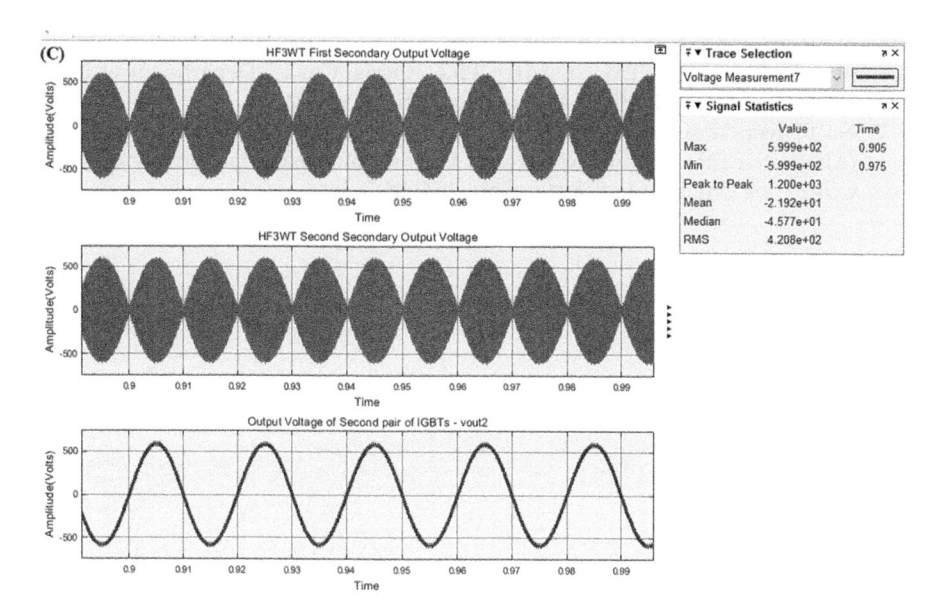

FIGURE 13.30 (CONTINUED)
SST using direct AC to AC converter topology – simulation results: (a) single-phase HV LF sine-wave AC voltage, (b) LF3WT two secondary output voltage and v_{out1} and (c) HF3WT two secondary output voltage and v_{out2}.

and direct AC to AC converter topology are presented with successful simulation results. DAB gives LV square-wave AC voltage, whereas direct AC to AC converter gives LV sine-wave AC output voltage with frequency same as that of the input frequency in both cases.

References

1. B.R. Pelly: *Thyristor Phase-Controlled Converters and Cycloconverters*, John Wiley & Sons, New York, 1971.
2. A.K. Chattopadhyay: AC-AC converters. In: *Power Electronics Handbook*, ed. M.H. Rashid, Chapter 18, pp. 497–506, Elsevier, Atlanta, GA, 2011.
3. The Mathworks Inc.: www.mathworks.com, MATLAB/SIMULINK users' manual, MATLAB R2017b, 2017.
4. Y. Du, S. Baek, S. Bhattacharya and A.Q. Huang: High-voltage high-frequency transformer design for a 7.2kV to 120V/240V 20kVA solid state transformer, *IEEE-IECON*, Glendale, AZ, November 2010, pp. 487–492.
5. H. Qin and J.W. Kimball: AC-AC dual active bridge converter for solid state transformer, *IEEE Energy Conversion Congress and Exposition (ECCE)*, San Jose, CA, November 2009, pp. 3039–3044.

6. N.P.R. Iyer: Modelling, simulation and real time implementation of a three phase AC to AC matrix converter, Ph.D. thesis, Ch. 13, Department of ECE, Curtin University, Perth, WA, Australia, February 2012.
7. N.P.R. Iyer: A single phase AC to PWM single phase AC and DC converter (Alternative title: "Swamy converter"), Patent submitted to Technology Transfer Office, The University of Nottingham, UK, 2013.

Appendix A: Matrix Converter Derivations

A.1 Introduction

In this appendix, selected derivations relating to matrix converter (MC) from selected chapters are presented.

A.2 Venturini Modulation Algorithm

Now consider the three-phase input and output voltages defined in equations 2.9 and 2.11 all made to phase lag by 90°, so that they are now three-phase sine-wave input and output voltages at the respective frequencies. The nine modulation functions are defined in equation 2.17, using which the modulation functions for the three bidirectional switches between input phase A and output phase a can be expressed as follows:

$$
\begin{aligned}
M_{Aa} &= \frac{1}{3} + \frac{2}{3V_{im}^2}\left[V_{im} * \sin(\omega_i t) * q * V_{im} * \sin(\omega_o t)\right] \\
&= \frac{1}{3} + \frac{2}{3}\left[q * \sin(\omega_i t) * \sin(\omega_o t)\right]
\end{aligned}
\tag{A.1}
$$

Similarly, modulation functions M_{Ba} and M_{Ca} can be expressed as follows:

$$
M_{Ba} = \frac{1}{3} + \frac{2}{3}\left[q * \sin\left(\omega_i t - \frac{2\pi}{3}\right) * \sin(\omega_o t)\right]
\tag{A.2}
$$

$$
M_{Ca} = \frac{1}{3} + \frac{2}{3}\left[q * \sin\left(\omega_i t - \frac{4\pi}{3}\right) * \sin(\omega_o t)\right]
\tag{A.3}
$$

Output voltage for phase a can be expressed as follows:

$$
v_a = q * V_{im} * \sin(\omega_o t) = M_{Aa}v_A + M_{Ba}v_B + M_{Ca}v_C
\tag{A.4}
$$

Using equations A.1–A.3 and the newly defined three-phase sine-wave input voltages in equation A.4, and letting $\omega_i t=\theta$, the RHS of equation A.4 reduces to the following:

$$v_a = \frac{V_{im}}{3}\underbrace{\left[\sin(\theta)+\sin\left(\theta-\frac{2\pi}{3}\right)+\sin\left(\theta-\frac{4\pi}{3}\right)\right]}_{\text{Term 1}}$$

$$+\frac{2}{3}V_{im}*q*\sin(\omega_o t)*\underbrace{\left[\sin^2(\theta)+\sin^2\left(\theta-\frac{2\pi}{3}\right)+\sin^2\left(\theta-\frac{4\pi}{3}\right)\right]}_{\text{Term 2}} \quad (A.5)$$

The term 1 in equation A.5 reduces to zero. Noting that $\sin^2(\theta)$ is $[0.5-0.5\cos(2\theta)]$, and using this in term 2 of equation A.5, we have the following:

$$v_a = \frac{2}{3}V_{im}*q*\sin(\omega_o t)*\left[\frac{3}{2}-0.5\left\{\underbrace{\cos(2\theta)-\cos\left(2\theta-\frac{4\pi}{3}\right)-\cos\left(2\theta-\frac{8\pi}{3}\right)}_{\text{Term 3}}\right\}\right]$$

$$= q*V_{im}*\sin(\omega_o t) \quad (A.6)$$

Three-phase AC voltage term 3 in equation A.6 reduces to zero. Equation A.6 well agrees with the LHS of equation A.4. This proves the Venturini algorithm with three-phase sine-wave input and output voltages and with unity input phase displacement factor.

A.3 Indirect Space Vector Modulation Algorithm

Tables 5.1–5.6 are derived using the phasor diagram of the output voltage and input current shown in Figures 5.3, 5.4, 5.7 and 5.8. For example, let the reference output voltage and input current vectors are in sectors II and IV, respectively, then $SV=II$ and $SI=IV$ and hence Figures A.1 and A.2 must be considered. The switching time order sequence is $T_{a\mu}$, $T_{\beta\mu}$, $T_{\beta\lambda}$, $T_{a\lambda}$ and T_o, respectively. Referring to Figures A.1 and A.2, during $T_{a\mu}$ time interval, corresponding to I_a-V_μ vector I_3-V_1, input phase B corresponds to positive terminal p of rectifier and input phase A corresponds to negative terminal n of rectifier. Three-phase output voltage vector V_1 is p, n, n and therefore output phase a corresponds to input phase B and the output phase b and c correspond to input phase A and A or a, b, c is connected to B, A, A, respectively. Similarly, during time interval $T_{\beta\mu}$, for $I_\beta - V_\mu$ vector corresponding to $I_4 - V_1$, input phase C and A correspond to p and n terminal of rectifier, V_1 corresponds to p, n, n and the output phase a, b and c are connected to C, A, A, respectively. During time intervals $T_{\beta\lambda}$ and $T_{\alpha\lambda}$, the vectors are $I_4 - V_2$ and $I_3 - V_2$, respectively. During time interval $T_{\beta\lambda}$, input phase C and A are connected to p and n terminal of rectifier, V_2 corresponds to p, p, n and the output phase a, b, c are connected to C, C, A, respectively. Similarly, during time interval $T_{\alpha\lambda}$, input phase B and A are connected to p and n terminal of rectifier, and as output voltage V_2 corresponds to p, p, n, output phase a, b, c are connected to B, B, A, respectively. During time interval T_o, as the current vector SI is in IVth quadrant, the output phase a, b, c are connected to A, A, A, respectively, as A is the common terminal to I_3 and I_4 input current vector. Tables 5.1–5.6 are filled in this way.

The matrix T_{phl} in equation 5.13 can be expressed as follows:

$$T_{phl} = m \begin{bmatrix} \cos(\omega_o t - \varphi_o + 30) \\ \cos(\omega_o t - \varphi_o + 30 - 120) \\ \cos(\omega_o t - \varphi_o + 30 + 120) \end{bmatrix} \tag{A.7}$$

$$* \begin{bmatrix} \cos(\omega_i t - \varphi_i) & \cos(\omega_i t - \varphi_i - 120) & \cos(\omega_i t - \varphi_i + 120) \end{bmatrix}$$

Let

$$(\omega_o t - \varphi_o) = \alpha \text{ and } (\omega_i t - \varphi_i) = \beta \tag{A.8}$$

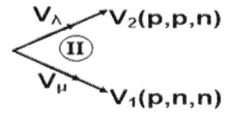

FIGURE A.1
Output voltage $Sv = \text{II}$.

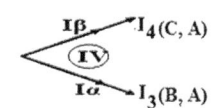

FIGURE A.2
Input current $S_i = \text{IV}$.

Simplifying equations A.7 and A.8, we have the following equation:

$$T_{phl} = \begin{bmatrix} m*\cos(\alpha+30)\cos(\beta) & m*\cos(\alpha+30)\cos(\beta-120) \\ m*\cos(\alpha-90)\cos(\beta) & m*\cos(\alpha-90)\cos(\beta-120) \\ m*\cos(\alpha+150)\cos(\beta) & m*\cos(\alpha+150)\cos(\beta-120) \end{bmatrix}$$

$$\begin{matrix} m*\cos(\alpha+30)\cos(\beta+120) \\ m*\cos(\alpha-90)\cos(\beta+120) \\ m*\cos(\alpha+150)\cos(\beta+120) \end{matrix} \Bigg] \tag{A.9}$$

Using equations 5.3, 5.11, 5.12 and A.9, we have the following:

$$v_{ab} = \sqrt{3}V_{om}*\cos(\alpha+30)$$

$$= V_{im}*m*\cos(\alpha+30)\cos(\beta+\varphi_i)\cos(\beta)$$

$$+ V_{im}*m*\cos(\alpha+30)\cos(\beta-120+\varphi_i)\cos(\beta-120)$$

$$+ V_{im}*m*\cos(\alpha+30)\cos(\beta+120+\varphi_i)\cos(\beta+120) \tag{A.10}$$

i.e.

$$\sqrt{3}V_{om}*\cos(\alpha+30) = V_{im}*m*\cos(\alpha+30)$$

$$*\Big[\cos(\varphi_i)*\{\cos^2(\beta)+\cos^2(\beta-120)+\cos^2(\beta+120)\}$$

$$-\sin(\varphi_i)*\{\sin(\beta)*\cos(\beta)+\sin(\beta-120)$$

$$*\cos(\beta-120)+\sin(\beta+120)*\cos(\beta+120)\}\Big] \tag{A.11}$$

Simplifying equation A.11, we have the following:

$$\sqrt{3}V_{om}*\cos(\alpha+30) = \frac{3}{2}V_{im}*m*\cos(\alpha+30)\cos(\varphi_i) \tag{A.12}$$

$$V_{om} = \frac{\sqrt{3}}{2}V_{im}*m*\cos(\varphi_i) \tag{A.13}$$

In equation 5.15, let

$$(\omega_o t - \varphi_o - \varphi_L) = \gamma \tag{A.14}$$

$$i_{oL} = \frac{I_{om}}{\sqrt{3}} * \begin{bmatrix} \cos(\gamma + 30) \\ \cos(\gamma - 90) \\ \cos(\gamma + 150) \end{bmatrix} \tag{A.15}$$

$$\left[T_{phl}\right]^T = \begin{bmatrix} m*\cos(\alpha + 30)\cos(\beta) & m*\cos(\alpha - 90)\cos(\beta) \\ m*\cos(\alpha + 30)\cos(\beta - 120) & m*\cos(\alpha - 90)\cos(\beta - 120) \\ m*\cos(\alpha + 30)\cos(\beta + 120) & m*\cos(\alpha - 90)\cos(\beta + 120) \end{bmatrix}$$

$$\begin{bmatrix} m*\cos(\alpha + 150)\cos(\beta) \\ m*\cos(\alpha + 150)\cos(\beta - 120) \\ m*\cos(\alpha + 150)\cos(\beta + 120) \end{bmatrix} \tag{A.16}$$

Using equations 5.4 and A.15, we have the following:

$$\begin{bmatrix} i_a \\ i_b \\ i_c \end{bmatrix} = \left[T_{phl}\right]^T * \frac{I_{om}}{\sqrt{3}} * \begin{bmatrix} \cos(\gamma + 30) \\ \cos(\gamma - 90) \\ \cos(\gamma + 150) \end{bmatrix} \tag{A.17}$$

Simplifying for i_a, we have the following:

$$i_a = m*\cos(\beta) * \frac{I_{om}}{\sqrt{3}} * \Big[\{\cos(\alpha + 30)\cos(\gamma + 30)\} + \{\cos(\alpha - 90)\cos(\gamma - 90)\}$$

$$+ \{\cos(\alpha + 150)\cos(\gamma + 150)\}\Big] \tag{A.18}$$

i.e.

$$i_a = m*\cos(\beta) * \frac{I_{om}}{\sqrt{3}} * \Big[\frac{1}{2}\cos(\alpha + \gamma + 60) + \frac{1}{2}\cos(\alpha - \gamma) + \frac{1}{2}\cos(\alpha + \gamma - 180)$$

$$+ \frac{1}{2}\cos(\alpha - \gamma) + \frac{1}{2}\cos(\alpha + \gamma + 300) + \frac{1}{2}\cos(\alpha - \gamma)\Big] \tag{A.19}$$

Simplifying,

$$i_a = \frac{I_{om}}{\sqrt{3}} * \frac{3}{2} m*\cos(\beta)\cos(\alpha - \gamma) \tag{A.20}$$

Substituting α, β and γ and further simplifying, we have the following:

$$I_{im} = \frac{\sqrt{3}}{2} I_{om} * m * \cos(\varphi_L) \qquad (A.21)$$

From equation 5.13, $\left[T_{VSR}(\omega_i) \right]^T$ is given as follows:

$$\left[T_{VSR}(\omega_i) \right]^T = \left[\begin{array}{ccc} \cos(\omega_i t - \varphi_i) & \cos(\omega_i t - \varphi_i - 120) & \cos(\omega_i t - \varphi_i + 120) \end{array} \right] \qquad (A.22)$$

Multiplying equation A.21 with equation 5.11, we have the following:

$$\left[T_{VSR}(\omega_i) \right]^T * v_{iph} = V_{im} * \left[\cos(\omega_i t)\cos(\omega_i t - \varphi_i) + \cos(\omega_i t - 120)\cos(\omega_i t - \varphi_i - 120) \right.$$
$$\left. + \cos(\omega_i t + 120)\cos(\omega_i t - \varphi_i + 120) \right]$$
$$= \frac{V_{im}}{2} * \left[\cos(2\omega_i t - \varphi_i) + \cos(2\omega_i t - 240 - \varphi_i) \right. \qquad (A.23)$$
$$\left. + \cos(2\omega_i t + 240 - \varphi_i) + 3\cos(\varphi_i) \right]$$

$$= \frac{3V_{im}}{2} * \cos(\varphi_i) \qquad (A.24)$$

Equation 5.41 can be rewritten as follows:

$$\begin{bmatrix} v_{ab} \\ v_{bc} \\ v_{ca} \end{bmatrix} = \begin{bmatrix} (d_\mu + d_\lambda)(d_\alpha + d_\beta) & (d_\mu + d_\lambda)(-d_\alpha) & (d_\mu + d_\lambda)(-d_\beta) \\ -d_\mu(d_\alpha + d_\beta) & d_\mu d_\alpha & d_\mu d_\beta \\ -d_\lambda(d_\alpha + d_\beta) & d_\lambda d_\alpha & d_\lambda d_\beta \end{bmatrix} * \begin{bmatrix} v_{AO} \\ v_{BO} \\ v_{CO} \end{bmatrix} \qquad (A.25)$$

Simplifying, we have the following:

$$v_{ab} = (d_\mu + d_\lambda) * \left[d_\alpha v_{AB} + d_\beta v_{AC} \right] \qquad (A.26)$$

$$v_{bc} = (-d_\mu) * \left[d_\alpha v_{AB} + d_\beta v_{AC} \right] \qquad (A.27)$$

$$v_{ca} = (-d_\lambda) * \left[d_\alpha v_{AB} + d_\beta v_{AC} \right] \qquad (A.28)$$

$$\begin{bmatrix} v_{ab} \\ v_{bc} \\ v_{ca} \end{bmatrix} = \begin{bmatrix} \left(d_\mu d_\alpha + d_\lambda d_\alpha\right) & \left(d_\mu d_\beta + d_\lambda d_\beta\right) \\ -d_\mu d_\alpha & -d_\mu d_\beta \\ -d_\lambda d_\alpha & -d_\lambda d_\beta \end{bmatrix} * \begin{bmatrix} v_{AB} \\ v_{AC} \end{bmatrix} \tag{A.29}$$

Let

$$d_\mu d_\alpha = d_{\alpha\mu}$$
$$d_\lambda d_\alpha = d_{\alpha\lambda}$$
$$d_\mu d_\beta = d_{\beta\mu} \tag{A.30}$$
$$d_\lambda d_\beta = d_{\beta\lambda}$$

Using equation A30 in A29, we have the following:

$$\begin{bmatrix} v_{ab} \\ v_{bc} \\ v_{ca} \end{bmatrix} = \begin{bmatrix} \left(d_{\alpha\mu} + d_{\alpha\lambda}\right) & \left(d_{\beta\mu} + d_{\beta\lambda}\right) \\ -d_{\alpha\mu} & -d_{\beta\mu} \\ -d_{\alpha\lambda} & -d_{\beta\lambda} \end{bmatrix} * \begin{bmatrix} v_{AB} \\ v_{AC} \end{bmatrix} \tag{A.31}$$

Equation A.31 corresponds to equation 5.42.

A.4 Three-Phase Voltage-Fed Z-Source Direct Matrix Converter Algorithm

Equation 11.10 in Section 11.2 is derived below:
Let $\omega t = \theta$, $(\omega t + \varphi_C) = \beta$ and $[D(1-D)] = K$
Squaring the RHS of equation 11.9 and summing up, we have the following:

$$(1-D)^2 * v_i^2 * \left[\sin^2(\theta) + \sin^2\left(\theta - \frac{2\pi}{3}\right) + \sin^2\left(\theta - \frac{4\pi}{3}\right) \right] \tag{A.32}$$

As derived in equation A.6, the RHS of equation 11.9 simplifies to the following:

$$(1-D)^2 * v_i^2 * \frac{3}{2} \tag{A.33}$$

Squaring the LHS of equation 11.9, we have the following:

$$D^2 * v_C^2 * \left[\sin^2(\beta) + \sin^2\left(\beta - \frac{2\pi}{3}\right) + \sin^2\left(\beta - \frac{4\pi}{3}\right) \right] = D^2 * v_C^2 * \frac{3}{2} \quad \text{(A.34)}$$

$$(1-D)^2 * v_C^2 * \left[\sin^2(\beta) + \sin^2\left(\beta - \frac{2\pi}{3}\right) + \sin^2\left(\beta - \frac{4\pi}{3}\right) \right] = (1-D)^2 * v_C^2 * \frac{3}{2}$$

$$\text{(A.35)}$$

$$2K * v_C^2 * \left[\sin(\beta)*\sin\left(\beta + \frac{2\pi}{3}\right) + \sin(\beta)*\sin\left(\beta - \frac{2\pi}{3}\right) + \sin\left(\beta - \frac{2\pi}{3}\right)*\sin\left(\beta + \frac{2\pi}{3}\right) \right]$$

$$\text{(A.36)}$$

Noting that the relation $\sin(A + B)$ is $\sin(A)*\cos(B) + \cos(A)*\sin(B)$ and $\sin(A - B)$ is $\sin(A)*\cos(B) - \cos(A)*\sin(B)$, we have the following from equation A.36:

$$2K * v_C^2 * \left[\frac{-3}{4} * \{\sin^2(\beta) + \cos^2(\beta)\} \right] = \frac{-3K * v_C^2}{2} = \frac{-3D(1-D) * v_C^2}{2} \quad \text{(A.37)}$$

Sum of the square of LHS term in equation 11.9 can be obtained by adding equations A.34, A.35 and A.37. This simplifies to the following:

$$v_C^2 * \left[\frac{3D^2}{2} + \frac{3(1-D)^2}{2} - \frac{3D(1-D)}{2} \right] \quad \text{(A.38)}$$

Equating A.33 and A.38, we have the following:

$$v_C^2 \left[3D^2 - 3D + 1 \right] = V_i^2 (1-D)^2$$

$$\frac{V_C}{V_i} = \frac{(1-D)}{\sqrt{3D^2 - 3D + 1}} \quad \text{(A.39)}$$

Index